Forests and Food

Addressing Hunger and Nutrition Across Sustainable Landscapes

*Edited by Bhaskar Vira,
Christoph Wildburger and Stephanie Mansourian*

Contents

Preface: Connecting the Dots

With the establishment of the Global Forest Expert Panels (GFEP) initiative in the year 2007, the Collaborative Partnership on Forests (CPF) created an international mechanism which effectively links scientific knowledge with political decision-making on forests. The GFEP responds directly to key forest-related policy questions by consolidating available scientific knowledge and expertise on these questions at a global level. It provides decision-makers with the most relevant, objective and accurate information, and thus makes an essential contribution to international forest governance.

This book is based on the report entitled "Forests, Trees and Landscapes for Food Security and Nutrition" which presented the results of the fourth global scientific assessment undertaken so far in the framework of GFEP. Previous assessments addressed the adaptation of forests and people to climate change; international forest governance; and the relationship between biodiversity, carbon, forests and people. All assessment reports were prepared by internationally recognised scientists from a variety of biophysical and social science disciplines. They have all been presented to decision-makers across relevant international policy fora. In this way, GFEP supports a more coherent policy dialogue about the role of forests in addressing broader environmental, social and economic challenges.

The current volume reflects the importance of policy coherence and integration more than any previous GFEP assessment. It comes at a time when the United Nations General Assembly has adopted a set of Sustainable Development Goals (SDGs) which build upon the Millennium Development Goals (MDGs) and converge with the post-2015 development agenda. In this context, the eradication of hunger, realisation of food security and the improvement of nutrition are of particular relevance. By 2050, the international community will face the challenge of providing 9 billion people with food, shelter and energy. Despite impressive productivity increases, there is growing evidence that conventional agricultural strategies will fall short of eliminating global hunger and malnutrition. The assessment in hand provides comprehensive scientific evidence on how forests, trees and landscapes can be – and must be – an integral part of the solution to this global problem. In other words, we must connect the dots and see the bigger picture.

The review of the International Arrangement on Forests by the member states of the United Nations Forum on Forests provides a unique opportunity to integrate forests into the SDGs in a holistic manner and to promote synergies in the implementation of the post-2015 development agenda across multiple levels of governance. It is my hope that those with a responsibility for forests, food security and nutrition at all levels will find this book a useful source of information and inspiration.

Alexander Buck
IUFRO Executive Director

Acknowledgements

This publication is the product of the collaborative work of scientific experts in the framework of the Global Forest Expert Panel on Forests and Food Security, who served in different capacities as panel members and authors. We express our sincere gratitude to all of them:

Bina Agarwal, Sarah Ayeri Ogalleh, Frédéric Baudron, Sammy Carsan, Paolo Cerutti, Josephine Chambers, Ian K. Dawson, Neil M. Dawson, Beatrice Darko Obiri, Elizabeth Deakin, Ann Degrande, Jason Donovan, Jennie Dey de Pryck, Samson Foli, Lisa Fuchs, Amos Gyau, Gordon Hickey, Amy Ickowitz, Miyuki Iiyama, Ramni Jamnadass, Katy Jeary, Gudrun Keding, Katja Kehlenbeck, Daniela Kleinschmit, Christophe Kouame, Godwin Kowero, Patti Kristjanson, Adrian Martin, Stepha McMullin, Henry Neufeldt, Mary Njenga, Vincent O. Oeba, Daniel Ofori, Hemant R. Ohja, Pablo Pacheco, Christine Padoch, John A. Parrotta, Bronwen Powell, Nitin D. Rai, Patrick Ranjatson, James Reed, Mirjam Ros-Tonen, Chris Sandbrook, Jolien Schure, Anca Serban, Bimbika Sijapati Basnett, Carsten Smith-Hall, Barbara Stadlmayr, Terry Sunderland, Celine Termote, Tran Nam Tu, Patrick Van Damme, Nathalie van Vliet, Barbara Vinceti and Solomon Zena Walelign.

Without their voluntary efforts and commitment the preparation of this publication would not have been possible.

We acknowledge and sincerely thank the reviewers of the full report and the various chapters whose comments have greatly improved the quality of this publication: Eduardo Brondizio, Carol Colfer, Martina Kress, Eric Lambin, Kae Mihara, Sarah Milne, Ellen Muehlhoff, Ben Phalan, Dominique Reeb, Patricia Shanley and Ingrid Visseren-Hamakers. And to Ella Walsh for her help in embedding links to digital resources.

We also gratefully acknowledge the generous financial and in-kind support provided by the Ministry for Foreign Affairs of Finland, the United States Forest Service, and the Austrian Federal Ministry of Agriculture, Forestry, Environment and Water Management. Our special thanks go to the IUFRO Secretariat for providing indispensable administrative and technical support to the work of the Panel. Furthermore, we would like to thank the member organisations of the

Collaborative Partnership on Forests for providing overall guidance to the Panel. We are particularly grateful also to the Food and Agriculture Organization of the United Nations (FAO, Rome, Italy), to the Center for International Forestry Research (CIFOR, Bogor, Indonesia), the University of Cambridge (UK) and to the World Agroforestry Centre (ICRAF, Delhi, India) for hosting expert meetings.

Bhaskar Vira
GFEP Panel Chair

Christoph Wildburger
GFEP Coordinator

Stephanie Mansourian
Content Editor

Acronyms, Units and Symbols

Acronyms

AFTP	Agroforestry Tree Product
AIPP	Asia Indigenous Peoples' Pact
BNDES	Brazilian Development Bank
CBD	Convention on Biological Diversity
CCBA	Climate, Community and Biodiversity Alliance
CFS	UN Committee on World Food Security
CFUG	Community Forest User Group
CGF	Consumer Goods Forum
CGIAR	Consultative Group on International Agricultural Research
CIE	Center for Independent Evaluations
CIFOR	Center for International Forestry Research
CINE	Centre for Indigenous Peoples' Nutrition and Environment
CITES	Convention on International Trade in Endangered Species of Wild Fauna and Flora
CPF	Collaborative Partnership on Forests
CSA	Climate Smart Agriculture
CTA	Technical Centre for Agricultural and Rural Cooperation
DRC	Democratic Republic of the Congo
ECOSOC	Economic and Social Council of the United Nations
EIA	Environmental Impact Assessment
EMBRAPA	Empresa Brasileira de Pesquisa Agropecuária (Brazilian Agricultural Research Corporation)
EP	Equator Principles
EU	European Union
FAO	Food and Agriculture Organization of the United Nations
FAOSTAT	FAO statistics
FDI	Foreign Direct Investment

FLEGT	Forest Law Enforcement, Governance and Trade
FMNR	Farmer-managed Natural Regeneration
FRA	Forest Resources Assessment
FSC	Forest Stewardship Council
GACSA	Global Alliance for Climate-Smart Agriculture
GAPKI	Indonesian Palm Oil Association
GAR	Golden Agri-Resources
GBM	Greenbelt Movement
GDP	Gross Domestic Product
GEF	Global Environment Facility
GFEP	Global Forest Expert Panels
GFRA	Global Forest Resources Assessment
GI	Government of Indonesia
GN	Government of Nepal
GSCP	Global Social Compliance Programme
HCS	High Carbon Stock
IAASTD	International Assessment of Agricultural Knowledge, Science and Technology for Development
ICCO	International Cacao Organization
ICO	International Coffee Organization
ICRAF	World Agroforestry Centre
IEA	International Energy Agency
IFAD	International Fund for Agricultural Development
IFC	International Finance Corporation
IFF	International Forum on Forests
IFOAM	International Federation of Organic Agriculture Movements
IFPRI	International Food Policy Research Institute
ILO	International Labour Organization
IP	Indigenous People
IPBES	Intergovernmental Platform on Biodiversity and Ecosystem Services
IPCC	Intergovernmental Panel on Climate Change
ISPO	Indonesian Sustainable Palm Oil
ITTO	International Tropical Timber Organization
IUFRO	International Union of Forest Research Organizations
IWGIA	International Work Group for Indigenous Affairs
IWMI	International Water Management Institute
KADIN	Indonesian Chamber of Commerce

LSP	Livelihood Support Programme
MA	Millennium Ecosystem Assessment (also MEA)
MDG	Millennium Development Goal
MEA	Millennium Ecosystem Assessment (also MA)
NGO	Non-governmental Organisation
NTFP	Non-timber Forest Product
OECD	Organisation for Economic Co-operation and Development
PEN	Poverty Environment Network
PES	Payment for Ecosystem Services
PFM	Participatory Forest Management
PPP	Purchasing Power Parity
RAI	Responsible Agricultural Investment
RECOFTC	The Center for People and Forests
REDD	Reducing Emissions from Deforestation and Forest Degradation
REDD+	Reducing Emissions from Deforestation and Forest Degradation and the role of conservation, sustainable management of forests and enhancement of forest carbon stocks in developing countries
RRI	Rights and Resources Initiative
RSPO	Roundtable on Sustainable Palm Oil
RTRS	Round Table on Responsible Soy
SARS	Severe Acute Respiratory Syndrome
SCI	Sustainable Cocoa Initiative
SFM	Sustainable Forest Management
SMAP	Soil Moisture Active Passive Observatory
SMS	Short Message Service
SNV	Netherlands Development Organization
SSU	Shamba Shape-up
UK	United Kingdom of Great Britain and Northern Ireland
UNCCD	United Nations Convention to Combat Desertification
UNDP	United Nations Development Programme
UNEP-WCMC	United Nations Environment Programme – World Conservation Monitoring Centre
UNESCO	United Nations Educational, Scientific and Cultural Organization
UNFCCC	United Nations Framework Convention on Climate Change
UNFF	United Nations Forum on Forests
UNICEF	United Nations Children's Fund (formerly United Nations International Children's Emergency Fund)
US	United States of America

USAID	United States Agency for International Development
USD	United States Dollars
USDA	United States Department of Agriculture
VFC	Village Forest Committee
WFP	World Food Programme
WHO	World Health Organization
WOCAT	World Overview of Conservation Approaches and Technologies
WTO	World Trade Organization
WWF	Worldwide Fund for Nature

Units and Symbols

The International System of Units (SI) is used in the publication.

ha = hectare (100 ha = 1 km^2)

yr = year

W = Watts

C = carbon

CO2 = carbon dioxide

SO2 = sulphur dioxide

NOX = nitrogen oxides (refers to NO and NO2)

HNO3 = nitric acid

NH3 = ammonia

Ca = Calcium

Cu = Copper

Fe = Iron

K = Potassium

Mn = Manganese

Mg = Magnesium

Na = Sodium

P = Phosphorus

Zn = Zinc

1. Forests, Trees and Landscapes for Food Security and Nutrition

Coordinating lead author: *Bhaskar Vira*

Lead authors: *Bina Agarwal, Ramni Jamnadass, Daniela Kleinschmit, Stepha McMullin, Stephanie Mansourian, Henry Neufeldt, John A. Parrotta, Terry Sunderland and Christoph Wildburger*

1.1 Problem Statement: Can Forests and Tree-based Systems Contribute to Food Security and Nutrition?

As population estimates for 2050 reach over 9 billion, issues of *food security* and *nutrition* have been dominating academic and policy debates, especially in relation to the global development agenda beyond 2015.[1] A total of 805 million people are undernourished worldwide, even though the trend appears to be slowly reversing (FAO et al., 2014) and *malnutrition* – defined as either under-5 stunting, anaemia among women of reproductive age or adult obesity – affects nearly every country on the planet (IFPRI, 2014). Despite impressive productivity increases, there is growing evidence that conventional agricultural strategies fall short of eliminating global hunger, result in unbalanced diets that lack nutritional diversity, enhance exposure of the most vulnerable groups to volatile food prices, and fail to recognise the long-term ecological consequences of intensified agricultural systems (FAO, 2013; FAO et al., 2013). In parallel, there is considerable evidence that suggests that *forests and tree-based systems* can play an important role in complementing agricultural production in providing better and more nutritionally-balanced diets (Vinceti et al., 2013); woodfuel for cooking; greater control over food consumption choices, particularly during lean seasons and periods of vulnerability (especially for marginalised groups); and deliver a broad set of *ecosystem services* which enhance and support crop production (FAO

1 All terms that are defined in the glossary (Appendix 1), appear for the first time in italics in a chapter.

 http://dx.doi.org/10.11647/OBP.0085.01

2011a; Foli et al., 2014). Already, while precise figures are difficult to come by, it has been estimated that approximately 1.2-1.5 billion people (just under 20 percent of the global population) are forest dependent (Chao, 2012, cited by FAO, 2014a; Agrawal et al., 2013). These estimates include about 60 million indigenous people who are almost wholly dependent on *forests* (World Bank, 2002).

Despite these figures, much of these forests remain under government control (even if the trend suggests a slight increase in community control of forests; see Figure 1.1). Ultimately, who controls forests has important implications for the role of forests in food security and nutrition.

The loss and *degradation* of forests exacerbate the problem of *food insecurity* both directly and indirectly: directly, by affecting the availability of fruits and other forest- and tree-based food products, and indirectly by modifying ecological factors relevant for crop and livestock and thereby affecting the availability of food (van Noordwijk et al., 2014). As of 1990, an estimated nearly 2 billion ha of the world's land surface could be classified as degraded, the legacy of extended periods of mismanagement in some long-settled areas (Oldeman et al., 1991). Models of current global trends in land (soil) degradation indicate that between 1981 and 2003, approximately 24 percent of the global land area (in which 1.5 billion people live) could be classified as degrading (Bai et al., 2008). Evidence suggests that cropland and forests are disproportionately represented in these areas undergoing degradation, with consequent implications for net primary productivity, and associated impacts on populations that depend on these *landscapes* for food and nutrient provisioning.

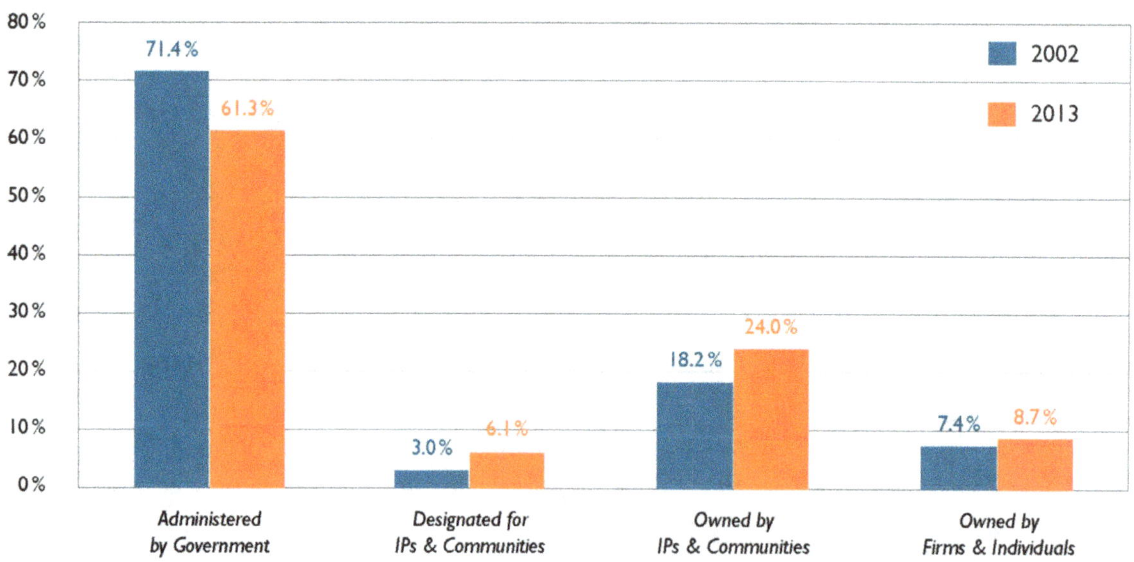

Fig. 1.1 Changes in statutory forest land tenure in low and middle income countries, 2002-2013, by percent. ***Source:*** **RRI (2014)**

While there is growing recognition that forests and tree-based systems complement farmland agriculture in providing food security and nutrition, responsibility for managing these diverse elements of the productive landscape is typically fragmented across different government departments and administrative jurisdictions in most countries. The complex, overlapping and interconnecting processes which link tree products and services to food security and nutrition are currently not adequately represented in forestry, agriculture, food or nutrition-related strategies at global and national levels, though their importance is often well known at more local scales by consumers, forest producers and farmers.

While the evidence base for the role of forests and tree-based systems for food security and nutrition is growing (see for example, Johnston et al., 2013; Ickowitz et al., 2014) there remain many gaps in our understanding of this relationship and its potential contribution to reducing global hunger and malnutrition. There is a need to explore the forest-food nexus in much more detail, particularly in relation to the integrated management of multi-functional landscapes, and the multi-scalar and cross-sectoral *governance* approaches that are required for the equitable delivery of these benefits.

1.2 Prevailing Paradigms about Forests, Agriculture, Food Security and Nutrition

In 2012, at the UN Conference on Sustainable Development: Rio+20, the UN Secretary General proposed an ambitious goal to eliminate global hunger by 2025 – the so-called 'Zero Hunger Challenge'. Zero Hunger was adopted as one of seventeen Sustainable Development Goals at the United Nations Sustainable Development Summit in September 2015, setting the global Agenda for Sustainable Development until 2030. Fulfilling these goals requires not just providing universal and year-round access to food for the world's growing population, but doing so in a nutritionally-balanced way, while enhancing *livelihood* security for smallholders, reducing waste from consumption and production systems and also ensuring that these systems are sustainable. Evolving strategies to respond to these challenges primarily focus on achieving '*sustainable intensification*', by improving the productivity of agricultural systems, without causing ecological harm or compromising *biodiversity* and ecosystem services (FAO, 2011b; Garnett et al., 2013). Plant biologists, crop scientists and agronomists are working hard to find solutions both on-farm and in the laboratory, which may be able to achieve this desired increase in productivity without the sorts of ecological side-effects that were associated with the Green Revolution of the 1960s and 1970s (Struik and Kuyper, 2014).

There are reasons to be cautious about these production-centric approaches to the food security dilemma. As Amartya Sen demonstrated through his seminal work on famine, what keeps people hungry is not just the lack of food, but the lack of access to that food and control over its production (Sen, 1983). Enhancing global production of food through productivity increases will therefore not guarantee that those who are hungry will have the means to increase their intake of food. The resource poor, in particular, may not have the means by which to purchase the increased output of food that these new technologies promise, and may continue to rely on more locally-appropriate and accessible means of fulfilling their nutritional needs (Pinstrup-Andersen, 2009). What is needed is recognition of the ways in which people command access to food, how this varies by season, and how the inter-personal dynamics and biases (especially due to gender) of intra-household food allocation result in differential nutritional outcomes within families. Enhanced *food sovereignty* (encompassing food security, the right to food and healthy diets, as well as the right to control over one's own food system (Patel, 2009; Edelman et al., 2014)) can help ensure that local people have control over their own diets and are engaged in efforts to improve the nutritional quality of their diets.

Production is also constrained by the lack of equitable access to land, technology and capital, which typically remain unavailable to the large majority of smallholder farmers (there are an estimated over 500 million family farms worldwide) (FAO, 2014a; Pretty et al., 2011; Vanlauwe et al., 2014). In these contexts, food from forests and tree-based systems is likely to continue to form an essential part of household strategies to eliminate hunger and achieve nutritionally balanced diets. Unfortunately, there is little current appreciation of the diverse ways in which these tree-based landscapes can supplement agricultural production systems in achieving global food security amongst the international and national decision-making communities. Many forms of *forest management* for food (whether strictly traditional or contemporary) including the creation of multi-storied agroforests, the planting of diverse forest gardens or, as discussed at greater length in this book, the management of *swidden*-fallows for food, have remained, with few exceptions, either invisible to researchers and planners or condemned by governments and conservationists. Even the many contributions that woodlands make to agricultural production outside of forests have been largely overlooked.

Paradigms for forest and tree management have also evolved considerably in the last fifty years, away from a state-controlled, production-centric approach to more collaborative systems which prioritise the needs of local people, and also value the roles of forests in providing critical ecosystem services, especially habitats for biodiversity (including *agrobiodiversity*), pollination, soil protection, water and climate regulation (Mace, 2014). Decentralised management systems now better reflect local demands, especially for woodfuel, fodder and small timber (Larson et al., 2010). More recently, new management regimes which take account of the key roles that forests and trees play in biodiversity conservation, the regulation of carbon fluxes, and the hydrological cycle have meant that these landscapes are being managed for a much more diverse (often non-local) set of purposes (Ribot et al., 2006). What has been relatively neglected, however, in these reconfigurations of forests and tree-based landscapes so far is an explicit recognition of the continued role that they play in food security and nutrition, especially in providing resilient and accessible production and consumption systems in general, and particularly for some of the most vulnerable groups. For many of these groups, linking the health of forests and landscapes to food sovereignty also provides a potential mechanism and argument to enhance greater autonomy over local food and agricultural systems, as well as their wider landscapes and bio-cultural environments. In many ways, this is a missed opportunity for stakeholders and decision-makers, as a greater emphasis on these roles could allow forestry debates to engage more actively with wider concerns about poverty alleviation and sustainable human well-being, which are at the centre of global, national and local agendas.

1.3 Policy Context and Scope

The contribution of forests to sustainable land use approaches which balance livelihood security and nutritional needs of people with other management goals is of high significance for the implementation of existing international commitments, including Agenda 21 and the three Rio Conventions (UNFCCC, CBD, UNCCD) adopted by the 1992 Earth Summit; the Global Objectives on Forests; the Millennium Development Goals; the UN Declaration on the Rights of Indigenous Peoples; as well as the ILO Indigenous and Tribal Peoples Convention (1989) No. 169. In the context of the United Nations 2030 Agenda for Sustainable Development, which seeks to establish a more integrated approach to poverty reduction under the framework of the Sustainable Development Goals, the contribution of forests to food security and nutrition, and the impact of food production on forests and landscapes are of particular relevance, bringing together goals on poverty, hunger and well-being with those concerned with sustainable use and management of terrestrial ecosystems and forests, combating desertification and addressing land degradation and biodiversity loss.

Against this backdrop, the Collaborative Partnership on Forests (CPF) tasked the Global Forest Expert Panel (GFEP) on Forests and Food Security to carry out a comprehensive global assessment of available scientific information on the relationship between forests and trees on the one hand, and food security and nutrition on the other, and to prepare a report to inform relevant international policy processes and the discussions on the post-2015 development agenda. The report was targeted particularly at decision-makers – policymakers, investors and donors – in order to provide a strong scientific basis for interventions and projects related to forests, *agroforestry* and landscapes aimed at addressing food security and nutrition.

The work of the GFEP on Forests and Food Security focused on three key objectives:

- To clarify the different dimensions and the role that forests and tree-based systems play in food security and nutrition;

- To analyse the social, economic and environmental synergies and trade-offs between forests and food security and nutrition, and related management interventions; and
- To assess relevant frameworks and responses, as an input to research, international policy processes, and evolving development agendas in different regions of the world.

This book is based on the GFEP report (Vira *et al.* 2015), published in May 2015, and released at the United Nations Forum on Forests in New York. Since the release, a number of key international meetings have further highlighted the importance of this work, and its potential to inform the agenda for future action. From the summer of 2015, the UN Committee on World Food Security (CFS) has convened a High Level Panel of Experts (HLPE) to produce a report on Sustainable Forestry for Food Security in Nutrition, to contribute to CFS debates in October 2017. In September 2015, the World Forestry Congress set out its vision for forests and forestry in 2050, and the Durban Declaration calls for forests to be fundamental for food security and improved livelihoods, while also recommending the need to to integrate forests and trees with other land uses, especially agriculture. The Sustainable Development Summit at the United Nations later that month reaffirmed the commitment to ending hunger and poverty, and to the sustainable management of land and forests.

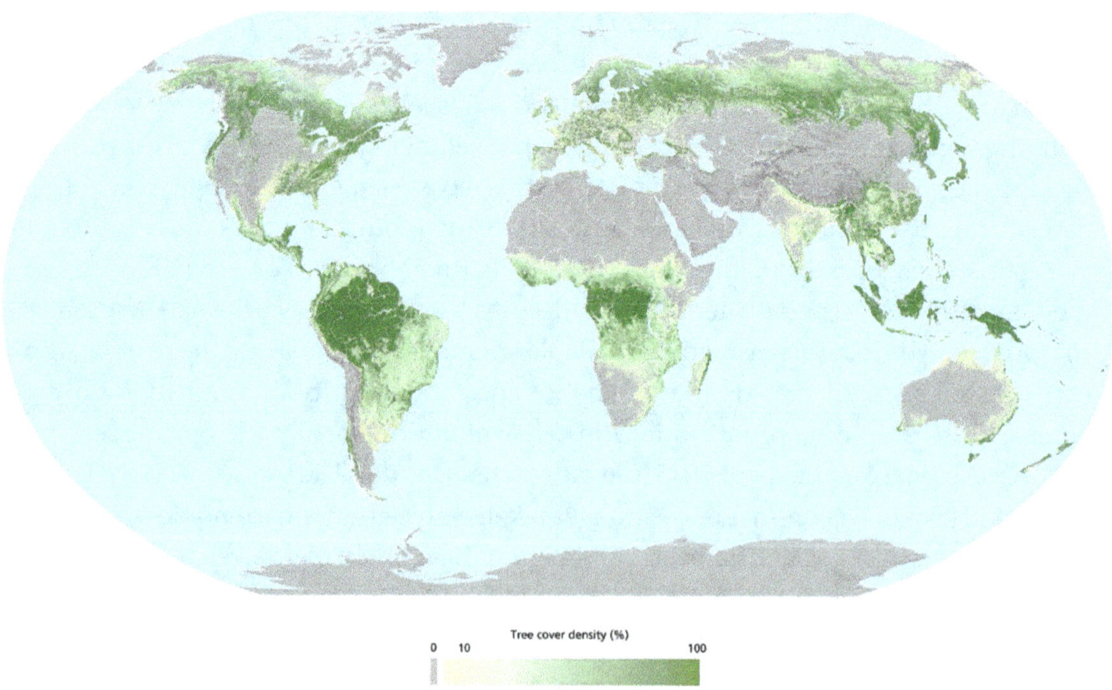

Fig 1.2a FAO world's forest map 2010. *Source:* FAO, 2010

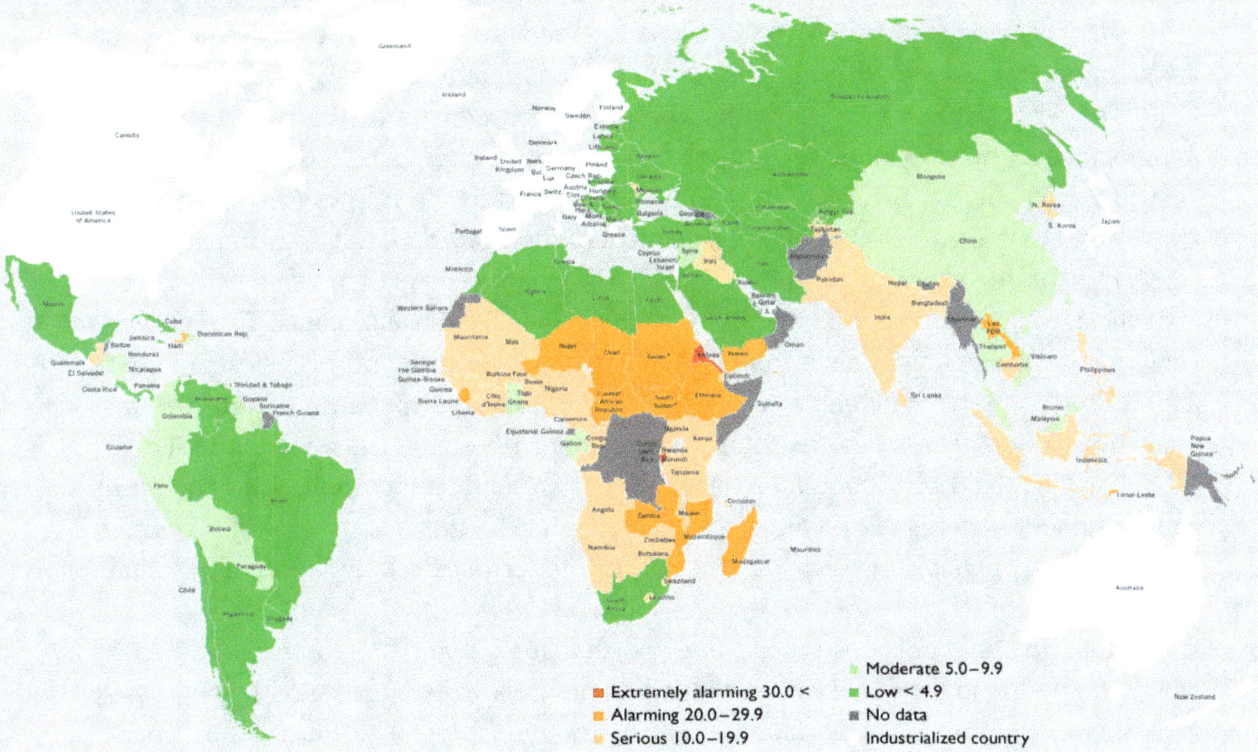

Fig. 1.2b Global Hunger Index 2014. *Source:* Von Grebmer et al., 2014

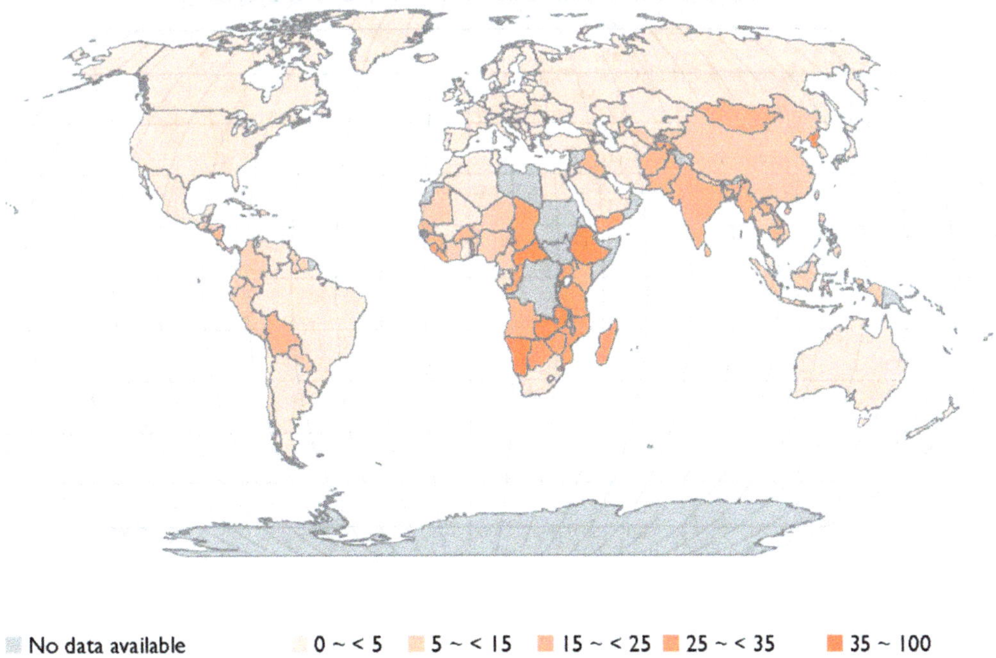

Fig. 1.2c Prevalence of people undernourished (percent, 2012-14). *Source:* FAO, 2014b

This book documents evidence of the relationships between forests and tree-based systems and food security and nutrition from different agro-ecological zones in all continents. However, a particular concern is those parts of the world that are characterised by deep-rooted hunger and malnutrition, where food security is a particular challenge, primarily in poorer nations and in the tropics (see Figure 1.2). Our discussion includes not only management of forests, woodlands, agroforests, and *tree crops* for direct food provisioning, but also the management of forested landscapes for the conditions they create that in turn affect all agricultural systems. The systems included in our analysis range from management of forests to optimise yields of wild foods and fodder, to *shifting cultivation*, through the broad spectrum of agroforestry practices, to single-species tree crop management (these systems are discussed in detail in Chapter 3 of this book). We consider the variability and applicability of these management systems within and across geographical regions, agro-ecological zones and biomes, highlighting the traditional and modern science and technology that underpin them.

Although this book documents the role that forests and tree-based landscapes play in relation to food security and nutrition at a relatively aggregated level, it also highlights the important variations in these relationships. This includes regional variability, depending on agro-ecological conditions and their relative suitability for different forms of wild and cultivated harvests; seasonal variability, indicating the role that forest- and tree-based diets might play at particularly lean periods of the agricultural cycle; and socio-economic variability, which especially emphasises the roles that land and tree *tenure* and governance, human capital, financial capital, and gender play in mediating the ways in which people have access to, and consume, food from forests and tree-based landscapes.

Throughout the book, there is specific attention to a number of important cross-cutting issues. Prominent amongst these is the role of gender specifically, and inequality more generally. Women and female children's roles in contributing to household *food systems* – both directly and indirectly – are substantial and often greater than men's, since they are the primary collectors of food, fodder and fuel from forests. In framing our discussion around the UN Secretary General's Zero Hunger Challenge, it is important to recognise the salience and importance of forest- and tree-based diets for these most vulnerable groups, even when the aggregate contribution to global food production from such landscapes might not be quite as significant. In addition, given the increasing feminisation of rural livelihoods and especially agriculture, as well as women's continued role in food provisioning for families, the book highlights the need to reach women as producers (by enhancing access to land, technologies, information etc.) and consumers who shape important behavioural choices in relation to food security and nutrition.

1.4 Structure of the Narrative

This book consists of six further chapters. Figure 1.3 provides a conceptual overview of the structure, and the broad linkages between the material presented in the different substantive chapters.

In Chapter 2, the available evidence on the direct and indirect roles that forests and tree-based landscapes play in providing food security and nutrition is presented, and critically assessed. Chapter 3 focuses on the forest-agriculture continuum, and the role of different landscape configurations in food production, the ways in which a mosaic of forest, agroforest and crop production systems combine and interact, and the importance of the social, cultural and economic contexts in which these systems exist, focusing on three factors that affect the socio-economic organisation of forest and tree-based systems, namely: land and tree tenure and governance, human capital, and financial capital. Chapter 4 steps back from this landscape scale and examines the broader drivers – environmental, social, economic and political – that are impacting the forest-food security "nexus", and highlights the importance of these in framing available options for responding to hunger and malnutrition. Chapter 5 starts to discuss response options, at landscape scale, highlighting in particular the need for multifunctional landscapes to be governed for their ability to provide food security, natural resource conservation and sustainable livelihoods. In Chapter 6, these response options are examined in relation to the broader drivers of change, focusing in particular on the role of markets and incentives, different forms of governance and the public policy challenges associated with recognising and enhancing the role of forest-tree landscapes in food security and nutrition. Chapter 7 concludes with some key messages for a range of decision-makers in local and national governments, the inter-governmental community, as well as the business sector and civil society.

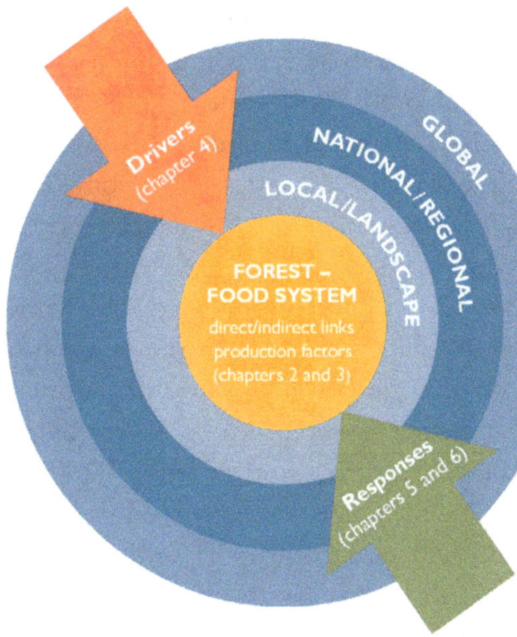

Fig. 1.3 Narrative structure of the book

1.5 Forests and Tree-based Landscapes for Food Security and Nutrition: A Brief Preview

1.5.1 Direct and Indirect Contributions of Forests and Tree-based Systems to Food Security and Nutrition

As this book explores in some detail, forests and tree-based systems provide a steady supply of wild and cultivated fruit, vegetables, seeds, nuts, oils, roots, fungi, herbs and animal protein, which complement more conventional staple diets derived from agricultural production systems (and, in some cases, provide dependable staple sources for food security and nutrition). Evidence reviewed in the book (especially in Chapters 2 and 3) suggests that some 50 percent of the fruit consumed globally comes from trees (much of this collected by women and children) and recent studies show that access to forests and tree-based systems is associated with increased vitamin intake from fruit and vegetable consumption. What this growing evidence suggests is that, while forests are not a solution for global hunger in themselves, in many circumstances they play a vital supplementary role, especially during periods of unpredictability (such as long dry spells). In some regions, food from forests plays a central role in providing calorific staples (such as açai palm fruit in the Amazon; Brondizio, 2008). It is also increasingly recognised that food from forests provides micronutrients and contributes to *dietary diversity*, thereby supporting a shift away from calorific intake as the primary metric for food security, towards a broader understanding of nutritionally-balanced diets (FAO, 2013).

Forests provide not only food items, they are also critically important for providing fuel for cooking. In developing countries, 2.4 billion households still use conventional biofuels (firewood, charcoal, crop residues and cattle dung) for cooking and heating. This includes 90 percent of rural households in large parts of sub-Saharan Africa and 70-80 percent in China (Modi et al., 2005). The most important biofuel used as rural domestic fuel is firewood, and the numbers dependent on it and other traditional biofuels are expected to increase over time (IEA, 2004). Firewood shortages can have negative nutritional effects, since efforts to economise on firewood can induce shifts to less nutritious foods which need less fuel to cook, or cause poor families to eat raw or partially cooked food which could be toxic, or to eat leftovers which could rot if left unrefrigerated, or even to miss meals altogether (Agarwal, 1986).

Apart from these direct roles, forests support the diversification of livelihoods through income earning opportunities that contribute to household food security (see Figure 1.4). Their role in providing ecosystem services which underpin the agricultural production system – through soil formation, nutrient cycling and provision of green manure, water provisioning, pollination and micro-climate regulation – further enhances synergies between the forest-tree landscape and the wider food production system (MA, 2005).

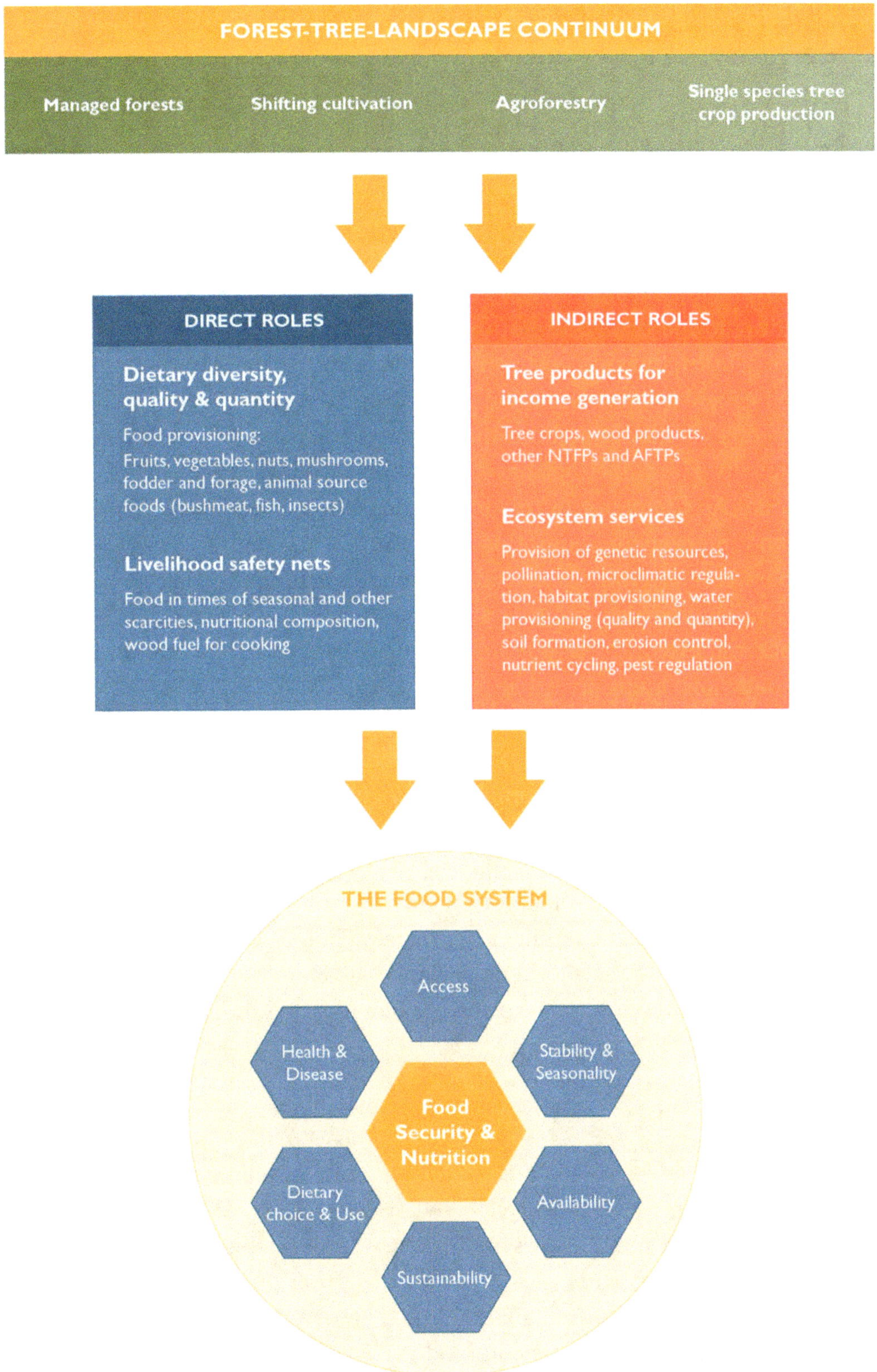

Fig. 1.4 The direct and indirect roles of forests and tree-based systems for food security and nutrition

1.5.2 Drivers Affecting the Relationship between Forest-tree Landscapes and Food

Demographic change and mobility

In 2013, the world population totalled 7.2 billion and it is projected to exceed 9 billion by 2050, with most of the increase being in developing regions, especially Africa (Roberts, 2011). Consequently the demand for food, feed and fibre will increase, while per capita land availability will decline. A continued focus, therefore, on understanding and responding to the drivers of population growth is likely to remain an essential component of efforts towards ensuring food security in the twenty-first century. In addition to the increase in absolute numbers, however, changes in the structure and location of people –with populations moving between rural and urban areas, as well as transnationally – are likely to have an important influence on the demand and supply of food. As Chapter 4 of this book discusses in some detail, the sheer scale of internal and international migration is unprecedented, and what is known about these numbers is likely to be a considerable underestimate due to undocumented movements. While international migration has become one of the defining features of globalisation, the world's population is also increasingly becoming urban, with more than half now living in urban areas (UN, 2014). Small cities and towns in Asia, Africa and Latin America that lie in or near tropical forest areas are likely to experience the greatest magnitude of urbanisation. Migration and urbanisation lead to profound changes in socio-economic systems, including the growing feminisation of rural landscapes in many of these regions. Urban migration is also resulting in major transformations in rural production-based economies, and associated loss of knowledge about forest foods and management. From a food security perspective, these trends have important implications for availability of, access to and relative dependence on forest products for food and income. However, research on the nexus between migration, urbanisation and forests remains very limited, let alone from a food security and nutrition perspective.

Shifts to market-driven economies

The last three decades have seen a considerable shift in public policy, encouraging the growth of markets and the private sector. The management of agrarian and forested landscapes for smallholders and their food needs is becoming less appealing to states in comparison to their desire to attract agro-industrial investors for large scale production systems, or for managing these landscapes in response to emergent global markets for carbon, biofuels and biodiversity (Fairhead et al., 2012). Pressures for the expansion of commodity exports are also adding to the degradation and loss of forest lands (Nevins and Peluso, 2008). As Chapter 4 points out, the resultant focus on enhancing production efficiency, specialisation and trade in agricultural commodities exposes vulnerable groups to the volatility of international commodity prices, and reduces their ability to access more localised food sources, over which they often have greater control. The food price spikes in 2008-09 demonstrated how the impacts of this volatility are felt, especially in those parts of the world that are least able to withstand such shocks, and contribute to undermining access to food for the poorest groups (Akter and Basher, 2014; Berazneva and Lee, 2013). As climate uncertainty adds to the potential volatility of global agricultural and commodity markets, developing more resilient production systems across the agricultural-forestry landscape is essential for ensuring food security and nutrition to the most vulnerable populations.

Consumer preferences and values

As discussed in Chapter 4, with increasing incomes, households' demand for food increases less than proportionally, and there is generally a dietary shift with decreasing importance of starchy staples (e.g. rice, wheat) and increased consumption of meat, fish, fruits and vegetables. Many forest foods are likely, in economic terms, to be seen as "inferior" goods (demand decreases with rising incomes and increases with declining incomes) and rising incomes would thus mean less forest food production, extraction and reliance. Delang (2006) notes, however, that forest food gathering is important in many rural communities with low economic growth, and likely to remain so, especially as per capita incomes rise relatively slowly in some parts of the world. Rising income and desire for meat consumption may also impact the demand for animal proteins, including bushmeat, with subsequent impacts on forests. Chapter 4 also suggests that forest food consumption is increasing in some high income countries, e.g. in northern Europe, apparently in response to perceptions that food should be locally grown, organic and aesthetic, indicating that we need to understand the dynamics of forest food consumption better.

As Chapter 2 of this book discusses, household decision-making (mostly by women) regarding food use and practice is influenced by levels of knowledge on nutrition (FAO, 1997; Jamnadass et al., 2011). Translating the harvest and cultivation of tree foods and other forest foods into improved dietary intakes therefore involves making nutrition education and behavioural-change communication to women a high priority. But, as Chapter 2 emphasises, the education of men should also not be neglected, since they often have most control over household incomes, and need to be aware of the importance of diverse cropping systems and the spending of income on healthy foods.

Environmental transformation and degradation

The effect of human activities on *ecosystems* has been profound, particularly during the past century. Many critical thresholds of the earth's biophysical systems have already been crossed as a result of human activities (Rockström et al., 2009; Steffen et al., 2015). Though the consequences are complex, there is considerable evidence that ongoing and future *climate change* will have drastic impacts, especially in the poorest regions of the world. As Chapter 4 elaborates, people living directly off the production from the earth's ecosystems are particularly affected by these changes. Forests are affected by increasing temperatures, variable precipitation, *fragmentation*, *deforestation*, loss of biological diversity and spread of *invasive species*. These factors affect not only the extent of forest but also the structure and species composition within forests (and therefore, forest products) thus impacting on the availability of food and nutrition. Environmentally-induced changes affecting forest cover imply both direct and indirect consequences for food security and nutrition: direct consequences result from changes in the availability and quality of food and nutrition, while indirect consequences result from changes in income and livelihoods related to forest products.

1.5.3 Trade-offs, Conflicts and Synergies in Land Use, and Responses

Chapter 5 of this book discusses possible responses across the landscape, that attempt to reconcile competing demands for agriculture, forestry and other uses. There is no single configuration of land uses in any landscape that can provide all the different outcomes that people might find desirable. For example, the "best" landscape configuration for biodiversity conservation might include large areas of forest strictly protected from human use, but this might support the livelihood needs of a very small human population or even displace previously resident people (and the resultant conflict may undermine conservation impacts in the long run). In contrast, the "best" landscape for cereal production might contain very little forest at all. Other desirable outcomes, like malaria mitigation or food security may be best provided by more diverse landscapes. With increasing pressure on biodiversity and ecosystem services across many landscapes from the growing footprint of human activities, choices have to be made about what is desirable and how landscapes should be managed. There may be difficult decisions about the relative merits of enhancing short term outputs through intensification of increasingly overworked landscapes versus maintaining their long term ecological productivity. In a context where views on these options are often deeply entrenched and conflicts of interest are difficult to reconcile, consensus on what constitutes success may be difficult to achieve.

While agriculture and forestry have been seen in conflict, especially at the forest frontier where conversion for crop production is a primary driver of deforestation, there is a risk that this framing over-simplifies a more complex set of relationships between different components of the landscape mosaic. Thinking more carefully about productive landscapes and the potential synergy between their different elements, ranging from intensified agriculture to relatively intact forests, with a complex mix of inter-cropped and multi-use systems in between, allows a more creative reconceptualization of the potential and possibilities of multi-functional production systems which are able to serve competing needs.

In a world characterised by increasing resource and land scarcity, these conflicts are likely to arise not just between the most desirable use of the agrarian-forest landscape, but also about how best to accommodate increased demands for land to allow for the expansion of urban settlements, industrial development and resource extraction. Dilemmas arise in relation to difficult choices about the most optimal configuration of land use in this mosaic, but also about who gets to decide when such choices need to be made, and whose interests are represented in the decision-making process. Trade-offs arise not just between alternative land use options, but also amongst different resource users and stakeholders in a landscape, and their associated preferences. Political economy issues have often meant that a theoretically optimal landscape is unrealistic or unachievable on the ground.

Chapters 5 and 6 of this book emphasise the significant shifts in governance that are required to manage these trade-offs and difficult choices, and to promote pathways to more integrated multi-functional agricultural-forest landscapes for food security and nutrition. As Chapter 6 elaborates, many of these responses lie outside the land sectors altogether. The growing demand for food, fibres, energy and other products from the land often result in market pressures for exploitation that can lead to forest destruction if they are not managed through appropriate governance systems and institutions. Perverse incentives, such as subsidies that have been set up to address the demand for cheap food without considering environmental externalities, may aggravate these pressures. Issues of presence and representation require the adoption of more open, participatory and deliberative forms of multi-stakeholder governance, which enhance linkages between food security and forests. Power needs to be exercised in ways that are seen to be legitimate and accountable, and transformative change requires innovative multi-level linkages, and creative cross-sectoral partnerships. There is also a need for market and natural resource governance-related responses focusing on global processes that support sustainable supply, and innovative corporate and multi-actor initiatives that support inclusive value chains of forest and tree products. These need to be coupled with social and cultural response options to enhance food security where the focus is on cultural norms and values including gender, and social mobilisation such as advocacy.

Governing such complex production systems at multiple scales requires the reimagining of institutional mechanisms. In particular, many governments have compartmentalized approaches to agriculture, food security and nutrition, and forestry, and little synergy between administrative departments that are tasked with these responsibilities. Recognizing the contribution of forests to food security and nutrition necessitates more integrated approaches to governance, as well as more inclusive mechanisms that acknowledge the rights (and responsibilities) of local stakeholders, as well as those of non-state actors and the private sector, at different stages of increasingly interconnected global value chains. Harnessing the potential of forest and tree-based commodities also requires vigilance and regulation, to ensure that high value products are managed and harvested sustainably, and not subject to the pressures of short-term commercialization for immediate returns.

1.6 Evidence and Knowledge Gaps

The diversity of the Earth's forest ecosystems and the human cultures associated with them has produced a vast array of food systems connected to forests and trees. These food systems are based on the traditional wisdom, knowledge, practices and technologies of societies. They are dynamic, developed and enriched through experimentation and adaptation to changing environmental conditions and societal needs, often over countless generations. Despite the huge potential of forest and tree foods to contribute to diets, knowledge on many forest foods, especially wild foods, is rapidly being lost due to social change and modernisation. Lack of knowledge in the community might be exacerbated by the effects of migration and movement, with considerable research demonstrating that information on forest-based foods is higher amongst long-term residents than migrants. Much of this knowledge is also associated with wisdom particularly held by the elderly and by women, with implications for its preservation and propagation within families and communities. Equally, many of these traditional forms of knowledge are non-formalised and have not been written down, which makes access to this information challenging. There are, of course, oral knowledge transmission traditions in many cultures (such as storytelling, folklore, music and informal learning within families) and there is a growing sensitivity in the research community to try and find ways of recording these non-formal forms of knowledge.

For the purposes of this book, however, this form of knowledge production and generation makes collation of evidence significantly more challenging. In reviewing the evidence, the authors have relied primarily on available literature, which has undergone processes of peer review and verification. Apart from work that is published in journals, they have used sources from a variety of organisations that have a repository of relevant information, and are reliable sources of data. Grey literature, where available, has been used and is indicated. What is largely missing are the voices of the poor, which are typically under-represented in these more formalised sources of knowledge. Despite our best efforts, for many of the analyses undertaken in the assessment, there are considerable limitations on the availability of useful information from the literature and other relevant sources. Recognising these constraints, the assessment tries to point out where the current knowledge base is strong, where it is currently weak or lacking, and the degree of consistency in the literature (and among experts) regarding research findings (and other knowledge sources), all of which influence the degree of certainty regarding conclusions that may be drawn from the available evidence.

The message of this book is nuanced. As the detailed chapters demonstrate, there is variability in the ways in which forests and tree-based landscapes interface with human food and nutritional systems. In particular places, and for particular groups of people (and individuals), these landscapes provide goods, services and livelihood options that can be critical for avoiding the worst forms of hunger, malnutrition and destitution. As the 2030 Agenda for Sustainable Development increasingly forces a recognition of the importance of nutrition-sensitive approaches to eliminating hunger, and to the wider role of natural ecosystems in supporting human well-being and development, these links between different forms of production across diverse landscapes will allow a much greater recognition of the role of forests and trees in global (and local) food security and nutrition. Reimagining the role of forests and tree-based systems as a critical element of productive and sustainable landscapes offers the possibility of more holistic and integrated approaches to the global development agenda.

References

Agarwal, B., 1986. *Cold Hearths and Barren Slopes: The Woodfuel Crisis in the Third World*. London: Zed Books.

Agrawal, A., Cashore, B., Hardin, R., Shepherd, G., Benson, C. and Miller, D., 2013. *Economic Contributions of Forests*. Background Paper to UNFF Tenth Session, Istanbul, 8-19 April 2013. http://www.un.org/esa/forests/pdf/session_documents/unff10/EcoContrForests.pdf

Akter, S., and Basher, S.A., 2014. The impacts of food price and income shocks on household food security and economic well-being: Evidence from rural Bangladesh. *Global Environmental Change* 25: 150-162. http://dx.doi.org/10.1016/j.gloenvcha.2014.02.003

Bai, Z.G., Dent, D.L., Olsson, L. and Schaepman, M.E., 2008. Proxy global assessment of land degradation. *Soil Use and Management* 24(3): 223-234. http://dx.doi.org/10.1111/j.1475-2743.2008.00169.x

Berazneva, J. and Lee, D.R., 2013. Explaining the African food riots of 2007-2008: An empirical analysis. *Food Policy* 39: 28-39. http://dx.doi.org/10.1016/j.foodpol.2012.12.007

Brondizio, E.S., 2008. *Amazonian Caboclo and the Acai Palm: Forest Farmers in the Global Market*. New York: New York Botanical Gardens Press.

Delang, C.O., 2006. The role of wild food plants in poverty alleviation and biodiversity conservation in tropical countries. *Progress in Development Studies* 6(4): 275-286. http://dx.doi.org/10.1191/1464993406ps143oa

Edelman, M., Weis, T., Bavsikar, A., Borras Jr, S. M., Holt-Gimenez, E., Kandiyoti, D. and Wolford, W., 2014. Introduction: Critical Perspectives on Food Sovereignty. *Journal of Peasant Studies* 41(6): 911-931. http://dx.doi.org/10.1080/03066150.2014.963568

Fairhead, J., Leach, M. and Scoones, I., 2012. Green grabbing: A new appropriation of nature? *Journal of Peasant Studies* 39(2): 237-261. http://dx.doi.org/10.1080/03066150.2012.671770

FAO, 1997. *Agriculture Food and Nutrition for Africa: A Resource Book for Teachers of Agriculture*. Food and Nutrition Division. Rome: FAO. http://www.fao.org/docrep/w0078e/w0078e00.HTM

FAO, 2010. *Global Forest Resources Assessment*. Rome: FAO. http://www.fao.org/forestry/fra/80298/en/

FAO, 2011a. *Forests for Improved Food Security and Nutrition Report*. Rome: FAO. http://www.fao.org/docrep/014/i2011e/i2011e00.pdf

FAO, 2011b. *Save and Grow: A Policy Makers Guide to the Sustainable Intensification of Crop Production*. Rome: FAO. http://www.fao.org/docrep/014/i2215e/i2215e.pdf

FAO, 2013. *The State of Food and Agriculture. Better Food Systems for Better Nutrition*. Rome: FAO. http://www.fao.org/docrep/018/i3300e/i3300e00.htm

FAO, 2014a. *Towards Stronger Family Farms*. Rome: FAO. http://www.fao.org/3/a-i4171e.pdf

FAO, 2014b. *Food and Nutrition in Numbers*. Rome: FAO. http://www.fao.org/3/a-i4175e.pdf

FAO, IFAD and WFP, 2013. *The State of Food Insecurity in the World* 2013. *The Multiple Dimensions of Food Security*. Rome: FAO. http://www.fao.org/docrep/018/i3434e/i3434e.pdf

FAO, IFAD and WFP, 2014. *The State of Food Insecurity in the World* 2014. *Strengthening the Enabling Environment for Food Security and Nutrition*. Rome: FAO. http://www.fao.org/3/a-i4030e.pdf

Foli, S., Reed, J., Clendenning, J., Petrokofsky, G., Padoch, C. and Sunderland, T., 2014. To what extent does the presence of forests and trees contribute to food production in humid and dry forest landscapes?: A systematic review protocol. *Environmental Evidence* 3: 15. http://dx.doi.org/10.1186/2047-2382-3-15

Garnett, T., Appleby, M., Balmford, A., Bateman, I., Benton, T., Bloomer, P. Burlingame, B., Dawkins, M., Dolan, L., Fraser, D., Herroro, M., Hoffman, I. Smith, P., Thornton, P., Toulmin, C., Vermeulen, S. and Godfray, C., 2013. Sustainable intensification in agriculture: Premises and policies. *Science* 341: 33-34. http://dx.doi.org/10.1126/science.1234485

Ickowitz, A., Powell, B., Salim, A. and Sunderland, T., 2014. Dietary quality and tree cover in Africa. *Global Environmental Change* 24: 287-294. http://dx.doi.org/10.1016/j.gloenvcha.2013.12.001

IEA, 2004. Energy and development. In: *World Energy Outlook*. Paris: International Energy Agency, OECD. http://dx.doi.org/10.1787/weo-2004-en

IFPRI, 2014. *Global Nutrition Report. Actions and Accountability to Accelerate the World's Progress on Nutrition*. Washington DC: IFPRI. http://dx.doi.org/10.2499/9780896295643

Jamnadass, R.H., Dawson, I.K., Franzel, S., Leakey, R.R.B., Mithöfer, D., Akinnifesi, F.K. and Tchoundjeu, Z., 2011. Improving livelihoods and nutrition in sub-Saharan Africa through the promotion of indigenous and exotic fruit production in smallholders' agroforestry systems: A review. *International Forestry Review* 13: 338-354. http://dx.doi.org/10.1505/146554811798293836

Johnston, K.B., Jacob, A. and Brown, M.E., 2013. Forest cover associated with improved child health and nutrition: Evidence from the Malawi Demographic and Health Survey and satellite data. *Global Health: Science and Practice* 1(2): 237-248. http://dx.doi.org/10.9745/ghsp-d-13-00055

Larson, A.M., Barry, D., Dahal, G.R. and Colfer, C.P. (eds.), 2010. *Forests for People: Community Rights and Forest Tenure Reform*. London: Earthscan. http://dx.doi.org/10.4324/9781849774765

MA (Millennium Ecosystem Assessment), 2005. *Ecosystems and Human Well-being: Synthesis*. Washington DC: Island Press. http://www.millenniumassessment.org/documents/document.356.aspx.pdf

Mace, G., 2014. Whose Conservation? *Science* 345(6204): 1558-1560. http://dx.doi.org/10.1126/science.1254704

Modi, V., McDade, S., Lallement, D. and Saghir, J., 2005. *Energy Services for the Millennium Development Goals*. Washington, DC: UNDP, World Bank, ESMAP. http://www.unmillenniumproject.org/documents/MP_Energy_Low_Res.pdf

Nevins, J. and Peluso, N.L., 2008. Introduction: Commoditization in Southeast Asia. In: *Taking Southeast Asia to Market*, edited by J. Nevins and N.L. Peluso. Ithaca: Cornell University Press, 1-24.

Oldeman, L.R., Hakkeling, R.T.A. and Sombroek, W.G., 1991. *World Map of the Status of Human-induced Soil Degradation* (2nd edn). Wageningen: ISRIC.

Patel, R., 2009. Grassroots voices: Food sovereignty. *Journal of Peasant Studies* 36(3): 663-706. http://dx.doi.org/10.1080/03066150903143079

Pinstrup-Andersen, P., 2009. Food Security: Definition and measurement. *Food Security* 1(1): 5-7. http://dx.doi.org/10.1007/s12571-008-0002-y

Pretty, J., Toulmin, C. and Williams, S., 2011. Sustainable intensification in African agriculture. *International Journal of Agricultural Sustainability* 9(1): 5-24. http://dx.doi.org/10.3763/ijas.2010.0583

Ribot, J.C., Agrawal, A. and Larson, A.M., 2006. Recentralizing while decentralizing: How national governments reappropriate forest resources. *World Development* 34(11): 1864-1886. http://dx.doi.org/10.1016/j.worlddev.2005.11.020

Roberts, L., 2011. 9 Billion? *Science* 333(6042): 540-543. http://dx.doi.org/10.1126/science.333.6042.540

Rockström, J., Steffen, W., Noone, K., Persson, Å., Chapin, F.S.III, Lambin, E.F., Lenton, T.M., Scheffer, M., Folke, C., Schellnhuber, H.J., Nykvist, B., de Wit, C.A., Hughes, T., van der Leeuw, S., Rodhe, H., Sörlin, S., Snyder, P.K., Costanza, R., Svedin, U., Falkenmark, M., Karlberg, L., Corell, R.W., Fabry, V.J., Hansen, J., Walker, B., Liverman, D., Richardson, K., Crutzen, P. and Foley, J.A., 2009. A safe operating space for humanity. *Nature* 461: 472-475. http://dx.doi.org/10.1038/461472a

RRI, 2014. *What Future for Reform? Progress and Slowdown in Forest Tenure Reform Since* 2002. Washington DC: Rights and Resource Initiative. http://www.rightsandresources.org/

Sen, A., 1983. *Poverty and Famines: An Essay on Entitlement and Deprivation*. Oxford: Oxford University Press.

Steffen, W., Richardson, K., Rockström, J., Cornell, S.E., Fetzer, I., Bennett, E.M., Biggs, R., Carpenter, S.R., de Vries, W., de Wit, C.A., Folke, C., Gerten, D., Heinke, J., Mace, G.M., Persson, L.M., Ramanathan, V., Reyers, B. and Sörlin, S., 2015. Planetary boundaries: Guiding human development on a changing planet. *Science*. Published online 15 January 2015. http://dx.doi.org/10.1126/science.1259855

Struik, P.C. and Kuyper, T.W. (eds.), 2014. Sustainable intensification to feed the world: Concepts, technologies and trade-offs. *Current Opinion in Environmental Sustainability*. Special Issue (Volume 8). http://dx.doi.org/10.1016/j.cosust.2014.10.008

UN, 2014. *World Urbanization Prospects: The* 2014 *Revision - Highlights*. New York: Department of Economic and Social Affairs, Population Division, United Nations. http://esa.un.org/unpd/wup/Highlights/WUP2014-Highlights.pdf

Vanlauwe, B., Coyne, D., Gockowski, J., Hauser, S., Huising, J., Masso, C., Nziguheba, G. Schut, M. and Van Asten, P., 2014. Sustainable intensification and the African smallholder farmer. *Current Opinion in Environmental Sustainability* 8: 15-22. http://dx.doi.org/10.1016/j.cosust.2014.06.001

Van Noordwijk, M., Bizard, V., Wangpakapattanawong, P., Tata, H.L., Villamor, G.B. and Leimona, B., 2014. Tree cover transitions and food security in Southeast Asia. *Global Food Security* 3(3-4): 200-208. http://dx.doi.org/10.1016/j.gfs.2014.10.005

Von Grebmer, K., Saltzman, A., Birol, E., Wiesmann, D., Prasai, N., Yin, S., Yohannes, Y., Menon, P., Thompson, J. and Sonntag, A., 2014. *2014 Gobal Hunger Index: The Challenge of Hidden Hunger*. Bonn, Washington, DC, and Dublin: Welthungerhilfe, International Food Policy Research Institute, and Concern Worldwide. http://www.ifpri.org/publication/2014-global-hunger-index

Vinceti, B., Ickowitz, A., Powell, P., Kehlenbeck, K., Termote, C., Cogill, B. and Hunter, D., 2013. The contributions of forest foods to sustainable diets. *Unasylva* 64(241): 54-64. http://dx.doi.org/10.3390/su5114797

Vira, B., Wildburger, C., Mansourian, S. (eds.) (2015) *Forests, Trees and Landscapes for Food Security and Nutrition: A Global Assessment Report*. Vienna: IUFRO World Series Volume 33.

World Bank, 2002. *A revised forest strategy for the World Bank Group*. Washington DC: World Bank.

2. Understanding the Roles of Forests and Tree-based Systems in Food Provision

Coordinating lead authors: *Ramni Jamnadass and Stepha McMullin*
Lead authors: *Miyuki Iiyama and Ian K. Dawson*
Contributing authors: *Bronwen Powell, Celine Termote, Amy Ickowitz, Katja Kehlenbeck, Barbara Vinceti, Nathalie van Vliet, Gudrun Keding, Barbara Stadlmayr, Patrick Van Damme, Sammy Carsan, Terry Sunderland, Mary Njenga, Amos Gyau, Paolo Cerutti, Jolien Schure, Christophe Kouame, Beatrice Darko Obiri, Daniel Ofori, Bina Agarwal, Henry Neufeldt, Ann Degrande and Anca Serban*

Forests and other tree-based systems such as agroforestry contribute to food and nutritional security in myriad ways. Directly, trees provide a variety of healthy foods including fruits, leafy vegetables, nuts, seeds and edible oils that can diversify diets and address seasonal food and nutritional gaps. Forests are also sources of a wider range of edible plants and fungi, as well as bushmeat, fish and insects. Tree-based systems also support the provision of fodder for meat and dairy animals, of "green fertiliser" to support crop production and of woodfuel, crucial in many communities for cooking food. Indirectly, forests and tree-based systems are a source of income to support communities to purchase foods and they also provide environmental services that support crop production. There are, however, complexities in quantifying the relative benefits and costs of tree-based systems in food provision. These complexities mean that the roles of tree-based systems are often not well understood. A greater understanding focuses on systematic methods for characterising effects across different landscapes and on key indicators, such as dietary diversity measures. This chapter provides a number of case studies to highlight the relevance of forests and tree-based systems for food security and nutrition, and indicates where there is a need to further quantify the roles of these systems, allowing proper integration of their contribution into national and international developmental policies.

 http://dx.doi.org/10.11647/OBP.0085.02

2.1 Introduction

The role played by *forests*[1] and trees in the lives of many people appears obvious through the many uses made of tree products, including foods, medicines, fodder, fibres and fuels, and for construction, fencing and furniture (FAO, 2010). Indeed, forests and other tree-based production systems such as agroforests have been estimated to contribute to the *livelihoods* of more than 1.6 billion people worldwide (World Bank, 2008), but just how they contribute – and the varying levels of dependency of different communities on tree products and services and how these change over time – has often not been well defined (Byron and Arnold, 1997). Complications arise for reasons that include the vast diversity and ubiquity of products and services these systems can supply, complexities of *tenure*, land-use-change dynamics, and the different routes by which products reach subsistence users and other consumers (FAO, 2010). At least until recently, this has been compounded by the inadequate attention that has been given to the characterisation of these systems, and the benefits and costs that are associated with them among different portions of the community (Dawson et al., 2014b; Turner et al., 2012).

Complexities in quantification and a general lack of proper appreciation of relative benefits help explain why the positive roles and limitations of tree-based production systems in supporting local peoples' livelihoods have frequently been neglected by policymakers, and why rural development interventions concerned with managing *forests and tree-based systems* have sometimes been poorly targeted (Belcher et al., 2005; Belcher and Schreckenberg, 2007; World Bank, 2008). The vast diversity of forest products available includes not only those derived from trees, but a wide range of (often) "less visible" products from other plants, fungi, animals and insects. "Natural" forests, agroforests and other tree-based production systems not only provide such direct products, but contribute indirectly tosupport people's livelihoods through the provision of a wide range of *ecosystem services* (FAO, 2010 and Figure 2.1).

In this chapter, we are concerned with describing the direct and indirect roles of forests and tree-based production systems (such as those based on commodity tree crops) in supporting the food and nutritional security of human communities. Our emphasis is on the tropics, where this role is often the greatest and where development interventions have been widely targeted in this regard (FAO, 2010). With the world food price "spikes" of the last decade, the political unrest and suffering caused by the lack of an adequate diet for many people, and the recognition of the threats of anthropogenic *climate change* and other global challenges to agricultural production, the importance of both food and nutritional security, and the roles of forests and farms in securing them, have come to the forefront politically (FAO, 2013c; Box 2.1). As a result, a greater understanding of how forests and tree-based production systems support *food security* and *nutrition*, both directly and indirectly is needed (Jamnadass et al., 2013; Padoch and Sunderland, 2013; Powell et al., 2013; Vinceti et al., 2013).

---

1 All terms that are defined in the glossary (Appendix 1), appear for the first time in italics in a chapter.

In the following sections of this chapter, we first introduce key concepts related to food security and nutrition. Both the direct and indirect roles of forests and tree-based production systems in food provision (depicted in Figure 2.1), including threats to these roles, and gender aspects that determine value and usage, are then discussed. Although our emphasis is primarily on tree products and services because of their high importance and to illustrate the concepts involved, we also consider other, mostly forest, products. In the concluding section, we provide indications where further work is required to optimise the use of forests and tree-based production systems to support food and nutritional security.

2.2 Food Security and Nutrition

Food security exists when communities "have physical and economic access to sufficient safe and nutritious food to meet their dietary needs and food preferences for a healthy and active life" (Pinstrup-Andersen, 2009). Well-nourished individuals are healthier, can work harder and have greater physical reserves, with households that are food- and nutrition-secure being better able to withstand and recover from external shocks. Despite advances in agricultural production globally, approximately one billion people are still chronically hungry, two billion people regularly experience periods of *food insecurity* and just over a third of humans are affected by micronutrient deficiencies (FAO et al., 2012; UN-SCN, 2010; Webb Girard et al., 2012). Most of the countries with "alarming" Global Hunger Index scores are in sub-Saharan Africa and this region therefore is a particular target for intervention (von Grebmer et al., 2014).

While rates of hunger (insufficient access to energy) have been falling in many parts of the world, there has been little change in the rates of micronutrient deficiencies (FAO et al., 2013). In particular, deficiencies of iron, vitamin A, iodine and zinc, are associated with poor growth and cognitive development in children, and increased mortality and morbidity in both adults and children (Black et al., 2013). Micronutrient deficiencies are often referred to as *"hidden hunger"*, as they can occur within the context of adequate energy intake, and can be overlooked using traditional measures of food security (FAO et al., 2012). *Malnutrition*, including under-nutrition, micronutrient deficiency and over-nutrition (obesity and over-weight, with the concomitant cardiovascular and chronic respiratory diseases, and diabetes) are key developmental challenges. Rates of obesity are increasing in virtually all regions of the world, affecting 1.4 billion adults globally (FAO et al., 2012) and obesity can no longer be viewed only as a disease of affluence. The burden of double (over- and under-) nutrition on the well-being of people in low-income nations is immense. As such, there have been calls for greater attention to "nutrition-sensitive" agriculture and *food systems* (Herforth and Dufour, 2013).

There has been growing recognition in the nutrition community that dietary behaviour is shaped by a broad range of psychological, cultural, economic and environmental factors (Fischler, 1988; Khare, 1980; Kuhnlein and Receveur, 1996; Sobal et al., 2014). This complexity indicates that to address food and nutritional security a multi-dimensional response is required (Bryce et al., 2008). Such a response must consider the production of sufficient food as well as its availability, affordability and utilisation, and the *resilience* of its production, among other factors (Ecker et al., 2011; FAO 2009). Nutrition-sensitive approaches across disciplines, including health, education, agriculture and the environment, are needed (Bhutta et al., 2013; Pinstrup-Andersen, 2013; Ruel and Alderman, 2013).

On the production side, nutritionists agree on the importance of bio-fortification of staple crops through breeding, as well as on the need for greater use of a more biodiverse range of nutritionally-higher-quality plants for more varied diets (i.e., not just enough food, but the right food), rather than just relying on a few "Green Revolution" staples (Keatinge et al., 2010). This diversity of plants can include locally-available and often little-researched species, including forest or once-forest taxa (Burlingame and Dernini, 2012; Frison, et al., 2011; Jamnadass et al., 2011; see Box 2.1.).

Many nutritionists now accept evidence of changes in intake of certain nutritious foods and a more diverse diet (*dietary diversity* being defined as the number of different foods or food groups consumed over a given reference period (Ruel, 2003)) as enough to determine impacts on nutrition and health, since the links between dietary diversity and energy and micronutrient adequacy, and child growth, are now well established (Arimond et al., 2010; Johns and Eyzaguirre, 2006; Kennedy et al., 2007; Kennedy et al., 2011; Ogle et al., 2001). Dietary diversity of individuals or households is thus recommended as a reliable indicator to assess if nutrition is adequate, and it is a useful measure of impact following project interventions.

Box 2.1 Fruit and vegetable consumption in sub-Saharan Africa

A good example where changes to a healthier and more diverse diet would be beneficial is illustrated by figures on fruit and vegetable consumption in sub-Saharan Africa, where consumption is on average low with mean daily intake, respectively, of between 36 g and 123 g in surveyed East African countries; 70 g and 130 g in Southern Africa; and 90 g and 110 g in West and Central Africa (Lock et al., 2005; Ruel et al., 2005). These figures add up to considerably less than the international recommendation of 400 g in total per day to reduce micronutrient deficiencies and chronic disease (Boeing et al., 2012; FAO, 2012; WHO, 2004; see also Siegel et al., 2014). In response, initiatives are underway to bring "wild" foods in Africa into cultivation (e.g., see Jamnadass et al., 2011 for the case of fruit trees) and such approaches are receiving increased attention globally (CGIAR, 2014). This is exemplified by a recent *State of Food and Agriculture* report by the Food and Agriculture Organization of t he United Nations (FAO), titled *Food Systems for Better Nutrition*, which states that "greater efforts must be directed towards interventions that diversify smallholder production such as integrated farming systems, including fisheries and forestry" (FAO, 2013c). Similarly, the World Health Organization (WHO) has recently agreed on criteria for a healthy diet that include: balanced energy intake and expenditure; the consumption of fruits, vegetables, legumes, nuts and whole grains; and the low intake of free sugars, fats and salt (WHO, 2014).

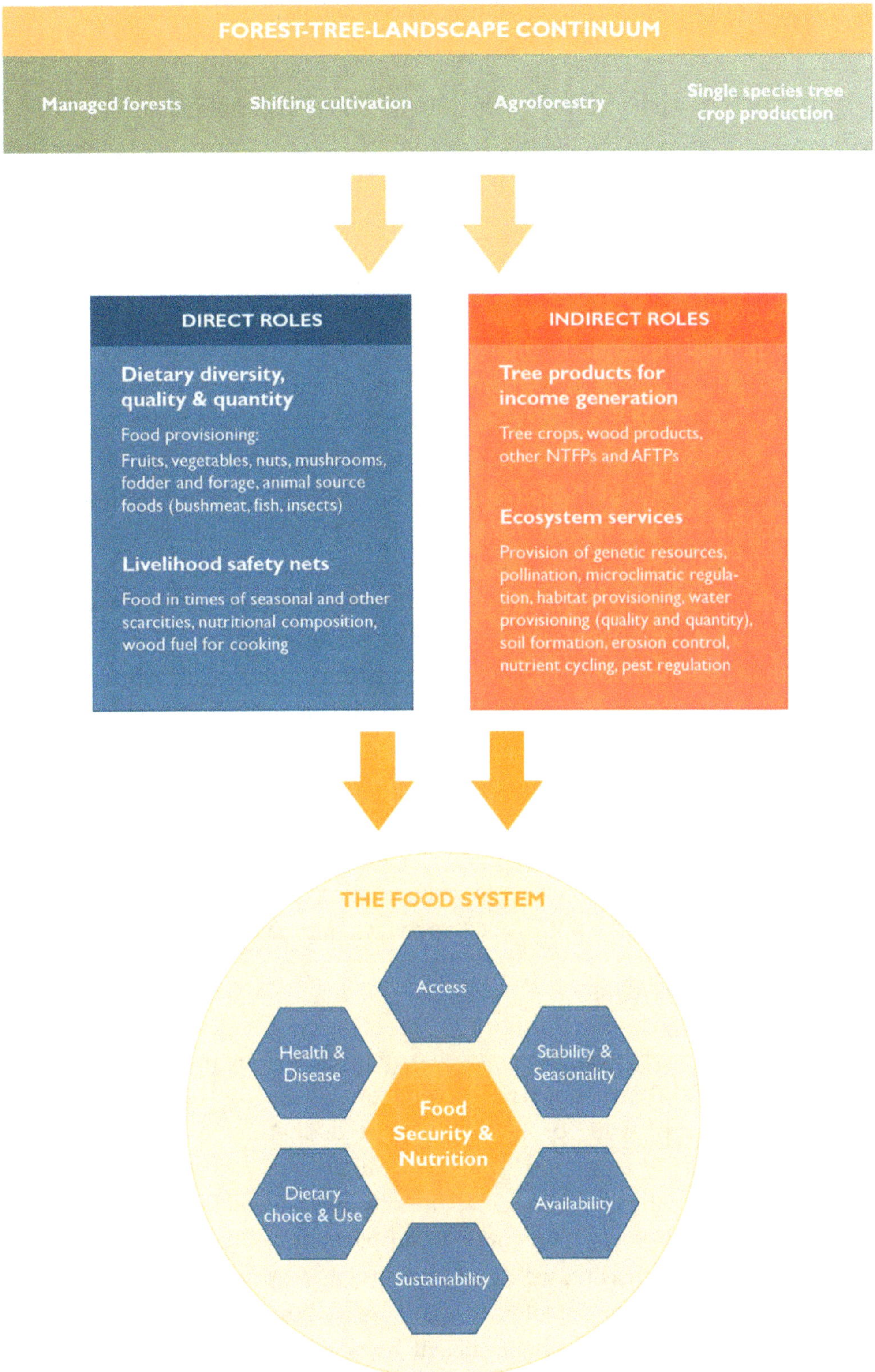

Fig. 2.1 A framework depicting the direct and indirect roles of forests and tree-based production systems in food provision. Components indicated in this framework are addressed in this chapter

2.3 The Direct Roles of Forests and Tree-based Systems

2.3.1 Foods Provided by Forests and Tree-based Systems

Access to forests and tree-based systems has been associated with increased fruit and vegetable consumption and increased dietary diversity. Powell et al. (2011), for example, found that in the East Usambara Mountains of Tanzania, children and mothers in households who ate more foods from forests, and who had more tree cover close to their homes, had more diverse diets. In another African example, Johnson et al. (2013) found that children in Malawi who lived in communities that experienced *deforestation* had less diverse diets than children in communities where there was no deforestation. Using data from 21 countries across Africa, Ickowitz et al. (2014) found a statistically significant positive association between the dietary diversity of children under five and tree cover in their communities. While the communities globally that depend completely on forest foods for their diets are relatively modest in number and size (Colfer, 2008), the above African examples illustrate that forest foods often play an important role as nutritious supplements in otherwise monotonous diets (Grivetti and Ogle, 2000). Since the productivity of trees is often more resilient to adverse weather conditions than that of annual crops, forest foods often provide a "safety net" during periods of other food shortages caused by crop failure, as well as making important contributions during seasonal crop production gaps (Blackie et al., 2014; Keller et al., 2006; Shackleton and Shackleton, 2004). Since different tree foods in the *landscape* have different fruiting phenologies (as well as different timings for the production of other edible products), particular nutrients such as vitamins can often be made available year-round (Figure 2.2), by switching from harvesting one species (or even variety) to another over the seasons (the "portfolio" approach; Jamnadass et al., 2011).

Human foods from trees

Globally, it is estimated that 50 percent of all fruit consumed by humans originate from trees (Powell et al., 2013), most of which come from cultivated sources. Many of these planted trees still have "wild" or "semi-wild" stands in "native" forest that are also harvested and which form important genetic resources for the improvement of planted stock (Dawson et al., 2014b). Although apparently wild, some forest fruit tree species have undergone a degree of domestication to support more efficient production (see for example Box 2.2), by increasing yields and quality, and by "clumping" trees together in forests to increase their density at particular sites and thus ease their harvesting. The classic case is in the Amazon, where ancient harvesting, managed regeneration and cultivation have led to genetic changes and high density aggregations, for example close to ancient anthropogenic "dark earth" soils (Clement and Junqueira, 2010) of several food tree species such as peach palm (*Bactris gasipaes*) and Brazil nut (*Bertholletia excelsa*) (Clement, 1989; Clement, 1999; Shepard and Ramirez, 2011).

Fig. 2.2 Fruit tree portfolio for year-round vitamin C and A supply

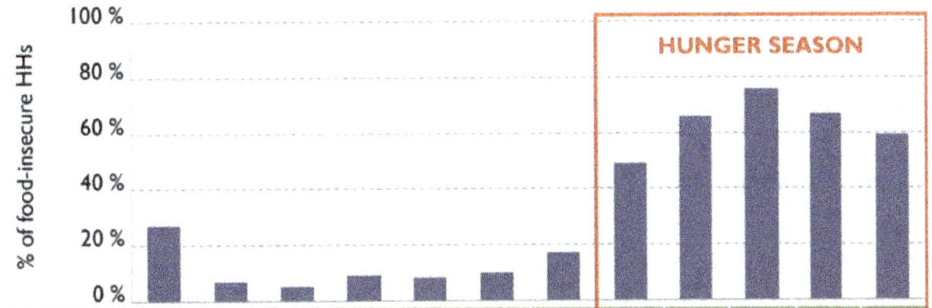

English name	Species name	Jan	Feb	Mar	Apr	May	Jun	Jul	Aug	Sep	Oct	Nov	Dec	Vit C	Vit A
Pawpaw	Carica papaya													+	+++
Mango	Mangifera indica													+	+++
Banana	Musa x paradisiaca														
Loquat	Eriobotrya japonica														+++
Mulberry	Morus alba													•	
Tamarind	*Tamarindus indica*														
Waterberry	Syzygium spp.														+++
Custard apple	Annona reticulata													•	
Guava	Psidium guajava													+++	+
White sapote	Casimiroa edulis													•	
Wild medlar	*Vangueria madagascariensis*														
Lemon	Citrus limon													+	
Orange	Citrus sinensis													+	
Chocolate berry	*Vitex payos*														+++
Avocado	Persea americana														
Passionfruit	Passiflora edulis														+
Jacket plum	*Pappea capensis*														
Desert date	*Balanites aegyptiaca*													•	
Bush plum	*Carissa edulis*														
Available vitamin C and A-rich fruit species		2	4	6	4	4	5	4	2	3	1	2	2		

Harvest time of vitamin C- and provitamin A-rich fruits (species given in red type)
Harvest time of vitamin C- and provitamin A-poor fruits (species given in black type)

Vitamin content levels:
+++ = very high + = intermediate • = moderate

Food security levels of smallholders' households and the harvest periods for the most important exotic and indigenous (in italics) fruits, for 300 households in Machakos County, Eastern Kenya. Fruit harvest periods are according to household respondents and the given ratings of vitamin C and provitamin A (a precursor of vitamin A) content are according to chemical analysis (several sources, including Tanzania Food Composition Tables and the USDA National Nutrient Database) Source: Katja Kehlenbeck (personal, previously unpublished observations).

Box 2.2 The case of allanblackia: integrating markets and cultivation to support the sustainable development of a new tree commodity crop

The seed of allanblackia (*Allanblackia spp.*), found wild in the humid forests of Central, East and West Africa, yields edible oil with a significant potential in the global food market, especially as a "hardstock" for the production of healthy spreads that are low in trans-fats. The tree is being brought into cultivation by improving seed handling and developing vegetative propagation methods, and through the selection of markedly superior genotypes. Tens of thousands of seedlings and clones have so far been distributed to smallholders.

The development of an allanblackia market has potential to improve smallholders' livelihoods and support global health. A private–public partnership known as Novella Africa is developing a sustainable allanblackia oil business that could be worth USD hundreds of millions annually for local farmers. The partnership allows different stakeholders with different interests and organisational capacities to work together.

A supply chain for seed has been established based on harvesting by local communities in natural forests and from trees remaining in farmland after forest clearance. The integration of allanblackia into small-scale cocoa farms is being promoted in West Africa to support more biodiverse and resilient agricultural landscapes. As allanblackia trees grow, cocoa trees provide the shade they need; when they are grown, they in turn will act as shade for cocoa. Cocoa and allanblackia provide harvests at different times of the year and – when the allanblackia trees have matured – will spread farmers' incomes.

Adapted from Jamnadass et al., 2010, 2014.

Traditional *agroforestry* systems often harbour high *biodiversity* and can deliver a wide array of tree foods including fruits and leafy vegetables that are both cultivated and are remnants of natural forest (Table 2.1). When established in agroforestry systems with shade trees, food diversity and sustainability of tree crop systems increase. In Ethiopia, for example, the inclusion of fruit-bearing trees as shade in coffee plantations provides farmers with access to additional foods, such as mangoes, oranges, bananas and avocados, as well as firewood and timber (Muleta, 2007).

A small number of tropical food trees is widely cultivated globally as commodity crops (e.g., cocoa [*Theobroma cacao*], coffee [*Coffea* spp.] and oil palm [*Elaeis guineensis*]; Dawson et al., 2013; Dawson et al., 2014b) in a variety of production systems, some of which harbour high levels of tree diversity, especially smallholdings (Table 2.1). Tree foods are often rich sources of vitamins, minerals, proteins, fats and other nutrients (FAO, 1992; Ho et al., 2012; Leakey, 1999), although for many traditional and wild species such information is lacking or not reliable. A recent literature review on selected African indigenous fruit trees conducted by Stadlmayr et al. (2013), for example, clearly showed their high nutritional value, but also highlighted the huge variability and low quality of some of the data reported in the literature. Edible leaves of wild African trees such as baobab (*Adansonia digitata*) and tamarind (*Tamarindus indica*) are high in calcium and are sources of protein and iron (Kehlenbeck and Jamnadass, 2014). Fruits from trees such as mango (*Mangifera indica,* native to Asia, but widely introduced through the tropics) are high in provitamin A, but there is a huge variability of almost 12-fold among different cultivars, as indicated by the colour of the fruit pulp (Shaheen et al., 2013). A child's daily requirement for vitamin A can thus be met by around 25 g of a deep orange-fleshed mango variety, while 300 g of a yellow-fleshed variety would be required. As another example, the iron contents of dried seeds of the African locust bean (*Parkia biglobosa*) and raw cashew nut (*Anacardium occidentale*) are comparable with, or even higher than, that of chicken meat (FAO, 2012), although absorption of non-haem iron from plant sources is lower than from animal sources. Iron absorption is enhanced by the intake of vitamin C, which is found in high amounts in many tree fruits (WHO/FAO, 2004). Consumption of only 10 to 20 g of baobab fruit pulp (or a glass of its juice), for example, covers a child's daily vitamin C requirement. Increasing knowledge on the biochemical components of indigenous tree species that are not widely used in agriculture internationally remains an important area of research (Slavin and Lloyd, 2012; WHO/FAO, 2004).

Human foods from other (forest) sources

Bushmeat (wild meat), fish and insects can all be important food sources. Bushmeat is often the main source of animal protein available to forest and forest-boundary communities, serving as an important source of iron and fat, and diversifying diets (Golden et al., 2011; Wilkie et al., 2005). It plays a particularly important role in diet where livestock husbandry is not a feasible option and where wild fish are not available (Brashares et al., 2011; Elliott et al., 2002). The hunting of animals and eating of bushmeat also play special roles in the cultural and spiritual identity of indigenous peoples (Nasi et al., 2008; Sirén, 2012). For example, more than 580 animal species, distributed in 13 taxonomic categories, are used in traditional medicine in the Amazon region (Alves and Alves, 2011).

Consumption patterns for bushmeat can vary widely (Chardonnet, 1996; Fargeot and Dieval, 2000; Wilkie et al., 2005), but hunting has been estimated to provide 30 to 80 percent of the overall protein intake of rural households in parts of Central Africa and nearly 100 percent of animal protein (Koppert et al., 1996). Numerous studies in Latin America have shown the importance of bushmeat (Iwamura et al., 2014; Peres, 2001; Van Vliet et al., 2014; Zapata-Rios et al., 2009). In the Amazon, for example, rural consumption is believed to equal ~150,000 tonnes annually, equivalent to ~ 60 kg per person (Nasi et al., 2011).

In China, increasing affluence in major consumer markets has led to spiralling demand for many wild animals, a demand that is supported by improvements in transport infrastructure. Pangolins and turtles used for meat and in traditional Chinese medicine are the most frequently encountered mammals seized from illegal traders (TRAFFIC, 2008), with major markets also in Singapore and Malaysia. Bushmeat sales can constitute a significant source of revenue for rural communities, particularly where trade is driven by increased consumption in urban areas (Milner-Gulland and Bennett, 2003). Urban consumers may have a choice of several sources of animal protein but opt for bushmeat for reasons of preference or cost relative to alternatives (Wilkie et al., 2005). Surveys of bushmeat markets are a useful way to estimate the state of fauna and to infer the sustainability of hunting activities (Fa et al., 2015).

Table 2.1 Examples of tree-species-rich agroforests in Africa, Asia and Latin America, with information on tree uses and with particular reference to possible human food use. These case studies indicate that dozens and sometimes hundreds of tree species can be found in agroforestry landscapes in the tropics, with a wide range of species contributing directly to food production (*adapted from* Dawson et al., 2014b).

Reference	Location	Tree diversity	Tree uses
Das and Das (2005)	Barak Valley, Assam, India	87 tree species identified in agroforestry home gardens	Farmers indicated a mean of 8 species used as edible fruit per home garden, many of which were indigenous. Fruit trees were more dominant in smaller gardens. ~ 5 species per garden used for timber, 2 for woodfuel
Garen et al. (2011)	Los Santos and Rio Hato, Panama	99 tree species, 3/4 indigenous, utilised, planted and/or protected on farmers' land	~ 1/3 of species valued for human food. 27 mostly exotic fruits mentioned as planted. ~ 1/3 of species each valued for their wood or as living fences. > 60 % of species were assigned multiple uses
Kehlenbeck et al. (2011)	Surrounding Mount Kenya, Kenya	424 woody plant species, 306 indigenous, revealed in farm plots	Farmers indicated many species used for food. 7 of the 10 most common exotic species were planted, mainly for edible fruits/nuts. The most common indigenous species were used primarily for timber/firewood
Lengkeek et al. (2003)	East of Mount Kenya, Kenya	297 tree species, ~ 2/3 indigenous, revealed in smallholder farms	Farmers indicated that > 20 % of species yield fruits/nuts for human consumption. The most common exotic was coffee, then timber trees
Marjokorpi and Ruokolainen (2003)	Two areas of West Kalimantan, Indonesia	> 120 tree species identified in forest gardens, most species not planted	Farmers indicated ~ 30 % of species used for edible fruit, latex and in other non-destructive ways, ~ 50 % used for timber and in other destructive ways. Seedlings of unused trees removed around naturally-regenerating and intentionally-planted fruit/other useful trees
Philpott et al. (2008)	Bukit Barisan Selatan Park, Lampung province, Sumatra, Indonesia	92 and 90 trees species identified in coffee farm plots outside and inside the park, respectively	> 50 % of farmers grew a total of 17 other products in addition to coffee, including spices, timber and, most commonly, indigenous and exotic fruits. Farmers planting outside the park grew alternative tree products more often
Sambuichi and Haridasan (2007)	Southern Bahia, Brazil	293 tree species, 97 % indigenous, revealed in cacao plantation plots in forest understory	Many indigenous trees used for food. Seedlings favoured for retention during weeding were those providing edible fruit or good wood. The most abundant exotics were fruit species
Sonwa et al. (2007)	Yaoundé, Mbalmayo and Ebolowa sub-regions, Cameroon	206 mostly indigenous tree species revealed in cacao agroforestry plots	Farmers indicated 17 % of tree species used primarily for food, 2/3 of which were indigenous. 22 % of tree species primarily for timber, 8 % for medicine. Excluding cacao, the 3 most common species (2 indigenous) were used for food. Close to urban Yaoundé, the density of food trees was higher.

The value of fish as a nutritious food is well established (Kawarazuka and Béné, 2011). In many tropical forests, wild fish represent the main source of animal protein in the diet, outweighing the importance of bushmeat (cf. daSilva and Begossi, 2009 for the Amazon; Powell et al., 2010 for Laos; Wilkie et al., 2005 for Gabon). In the Rio Negro region of the Brazilian Amazon, for example, daSilva and Begossi (2009) found that fish caught in flooded forests and in forest rivers accounted for 70 percent of animal protein in the diet, excluding other aquatic species such as turtles. The importance of insects as a source of food has recently regained attention (FAO, 2013b). Insects are a cheap, available source of protein and fat, and to a lesser degree carbohydrate. Some species are also considered good sources of vitamins and minerals (Dunkel, 1996; FAO, 2013b; Schabel, 2010). Many forests and agroforests are managed by local communities to enhance edible insect supply (Johnson, 2010). For example, sago palms (*Metroxylon* spp.) are managed in forest-agriculture landscape mosaics in Papua New Guinea and eastern Indonesia to support grub production (Mercer, 1997). The global importance of insects as a food source is difficult to evaluate, as statistics are mostly restricted to a few specific studies. For example, a study of the Centre for Indigenous Peoples' Nutrition and Environment and FAO evaluated the nutritional and cultural importance of various traditional food items of 12 indigenous communities from different parts of the world, and found that leaf-eating and litter-feeding invertebrates provide many Amerindian groups with important foods that can be collected year-round (Kuhnlein et al., 2009).

Boy spear-fishing in riverine forest outside of Luang Prabang, Laos. Photo © Terry Sunderland

Tree products that support human food production and consumption

Trees provide animal fodder, enabling communities to keep livestock that provide them with nutritionally important milk and meat. They also provide green manure that replenishes soil fertility and supports annual crop production, as well as woodfuel that provides energy (Jamnadass et al., 2013). In the case of fodder production, for example, a recent initiative in East Africa involved more than 200,000 smallholder dairy farmers growing mostly introduced fodder shrubs (especially calliandra, *Calliandra calothyrsus*) as supplementary feed for their animals (Franzel et al., 2014). The typical increase in milk yield achieved enabled smallholders to raise extra revenue from milk sales of more than USD 100 per cow per year and allowed them to provide more milk more efficiently to urban consumers (Place et al., 2009). Such tree-and shrub-based practices for animal fodder production increase farmers' resilience to climate change (Dawson et al., 2014a). Many tree and other forest products are also used in ethnoveterinary treatments that support animal health and hence human food production (Dharani et al., 2014).

In the case of soil fertility replenishment, an analysis of more than 90 peer-reviewed studies found consistent evidence of higher maize yields in Africa from planting nitrogen-fixing green fertilisers, including trees and shrubs, to substitute for (or enhance) mineral fertiliser application, although the level of response varied by soil type and the particular management applied (Sileshi et al., 2008). A recent project in Malawi, for example, encouraged more than 180,000 farmers to plant fertiliser trees, leading to improvements in maize yields, more food secure months per year and greater dietary diversity (CIE, 2011). As well as increasing average yields, the planting of trees as green fertilisers in Southern Africa stabilised crop production in drought years and during other extreme weather events, and improved crop rain use efficiency (Sileshi et al., 2011; Sileshi et al., 2012), contributing to food security in the context of climate change in the region. Supporting the regeneration of natural vegetation in agroforestry systems also provides significant benefits for the production of staple crops, with farmer-managed natural regeneration (FMNR) of faidherbia (*Faidherbia albida*) and other leguminous trees in dryland agroforests (parklands) in semi-arid and sub-humid Africa being a good example. Supported in Niger by a policy shift that has awarded tree tenure to farmers, as well as by more favourable wetter weather, since 1986 FMNR is reputed to have led to the "regreening" of approximately 5 million hectares (Sendzimir et al., 2011). Improvements in sorghum and millet yields, and higher dietary diversity and household incomes, have resulted in some Sahelian locations (Place and Binam, 2013).

Traditional energy sources have received little attention in current energy debates, but firewood and charcoal are crucial for the survival and well-being of as many as two billion people, enabling them to cook food to make it safe for consumption and palatable, and to release the energy within it (Owen et al., 2013; Wrangham, 2009). In sub-Saharan Africa, for example, where perhaps 90 percent of the population relies on woodfuels for cooking (GEF 2013; IEA, 2006), the use of charcoal as a cooking fuel is still increasing rapidly, with the value of the charcoal industry there estimated at USD 8 billion in 2007 (World Bank, 2011). In Asia, even better-off rural households have often been observed to be highly dependent on woodfuels, as found by Narain et al. (2005) for India, the Government of Nepal (GN, 2004) for Nepal, and Chaudhuri and Pfaff (2002) for Pakistan. With the volatile and often high price of "modern" energy sources, this situation is unlikely to change for some time, a fact often neglected in policy discussions on "energy futures" in low-income nations, which place unrealistic emphasis on "more modern" energy sources, rather than attempting to make woodfuel production and use more efficient and sustainable (Iiyama et al., 2014a; Schure et al., 2013). Access to cooking fuel provides people with more flexibility in what they can eat, including foods with better nutritional profiles that require more energy to cook (Njenga et al., 2013). The cultivation of woodlots allows the production of wood that is less harmful when burnt (Tabuti et al., 2003), has higher energy content and requires less time for collection (freeing time for other activities; Thorlakson and Neufeldt, 2012). This is particularly beneficial for women, who do most of the woodfuel collection and the cooking, and whose health suffers most from cooking-smoke-related diseases (Bailis et al., 2005). Previously collected sources of fuel can then be used for other more beneficial purposes that support food production (e.g., not cutting fruit trees for fuel; Brouwer et al., 1997; Köhlin et al., 2011; Wan et al., 2011).

2.3.2 Dietary Choices, Access to Resources and Behavioural Change

Although trees and other forest plants can provide edible fruit, nuts and leaves, etc. that are often good potential sources of nutrients and are sometimes used in this regard (see examples earlier in this chapter), it does not follow that they are used by humans for food. In this sense, long lists of edible *non-timber forest products* (NTFPs) (Bharucha and Pretty, 2010) can sometimes be misleading, as the presence of wild food species in local forest and woodland landscapes does not necessarily mean that these are consumed. Termote et al. (2012) illustrated this point with a survey around the city of Kisangani in the Democratic Republic of the Congo, where a wide variety of wild food plants were found, but few contributed significantly to human diets, despite significant local dietary deficiencies. The real contribution of these foods to diets therefore needs to be assessed by measurements of intake (as noted in Section 2.2).

When there is availability but relatively low NTFP-food use in areas of dietary need, reasons can include the high labour costs involved in collection and processing, low yields, high phenotypic variability (with large proportions of non-preferred produce), and lack of knowledge in the community. Regarding the last point, in eastern Niger and northern Burkina Faso, for example, women prepare protein-rich condiments from the seeds of wild prosopis (*Prosopis africana*) and zanmné (*Acacia macrostachya*) trees, respectively, but women in other parts of the Sahel (where the same trees are found) are not aware of these food values and do not harvest or manage woodlands for them (Faye et al., 2011). Research suggests that knowledge on the use of such products is often higher among indigenous peoples than among immigrant communities, with knowledge being lost due to social change and "modernisation" (Kuhnlein et al., 2009; Moran, 1993). Within communities, cultural perceptions on who should eat particular foods, and when, are also important (Balée, 2013; Hladik et al., 1993; Keller et al., 2006; Lykke et al., 2002). Differences arise between genders and age groups with respect to specialised knowledge and preferences in tree use (Daniggelis, 2003). This is illustrated by the different relative use values assigned to plant products by different-aged respondents in the Yuracaré and Trinitario communities in the Bolivian Amazon, where older people generally had more recall on uses for particular categories of plant, but both young and old people assigned high use values to food products (higher than respondents in their mid-years; Thomas, 2008).

From the above discussion it is evident that the relationship between the availability of food and its consumption is often complex, and simple surveys of absence/presence are therefore not in themselves adequate for understanding diets (Webb and Kennedy,

2012). When collection costs, low yields and high proportions of non-preferred produce are factors inhibiting the use of wild sources, domestication to increase productivity, quality and access can play an important role (Dawson et al., 2014b). This is exemplified by improvements in the performance of wild African fruit trees being brought into cultivation in participatory domestication programmes in the Central African region (Jamnadass et al., 2011; Tchoundjeu et al., 2010). The option of cultivation also helps address the complex threats to the use of wild stands through a combination of over-harvesting, deforestation, the conflicting use of resources and restricted (or uncontrolled) access to forests (Dawson et al., 2013; FAO, 2010; Vinceti et al., 2013). The conventional wisdom that cultivation will support the maintenance of wild stands for conservation purposes and provide sustainable access for wild harvesters (rather than cultivators) is, however, not widely supported (Dawson et al., 2013).

When bringing trees from the wild into cultivation, an important aspect is to increase yields: if indigenous trees are perceived as relatively unproductive and can only be produced inefficiently, agricultural landscapes are likely to be dominated by staple crops, with agro-biodiversity (and hence, likely, dietary diversity) reduced (Sunderland, 2011). Since many tree species are essentially undomesticated, large increases in yield and quality are often available through selection, supporting cultivation; for example, this is the case for allanblackia (*Allanblackia* spp.), described further in Box 2.2 (Jamnadass et al., 2010). Lack of knowledge on appropriate tree management, however, can be a major limitation (Jamnadass et al., 2011). Increases in efficiency are important for markets, since price to the consumer is a significant factor influencing diet (Glanz et al., 2005; Ruel et al., 2005; Story et al., 2008). Where limited access to extant forest foods is a major issue, approaches that support access such as the development of community-based *forest management* plans can be beneficial (Schreckenberg and Luttrell, 2009), but wider efforts are required to include all significant stakeholders, and in particular women (Agarwal, 2001; Mitra and Mishra, 2011).

Household decision-making regarding food use and practice, mostly made by women, is influenced by levels of knowledge on nutrition (FAO, 1997; Jamnadass et al., 2011). Translating the harvest and cultivation of tree and other forest foods into improved dietary intakes therefore involves making nutrition education and behavioural-change communication to women a high priority (McCullough et al., 2004). There is, for example, a need to understand how best to educate on the benefits of eating fruit, how to prepare nutritious foods, and how to access them (Hawkes, 2013; Jamnadass et al., 2011). Children can also be effective agents of change in societies, so teaching them about agriculture and nutrition is a wise investment (Sherman, 2003). In Kenya, for example, the "Education for Sustainable Development" initiative included a "Healthy Learning" programme targeted at school children that resulted in attitudinal and behavioural changes in communities (Vandenbosch et al., 2009). Counselling to change feeding behaviours is important (Waswa et al., 2014), within the appropriate context of culture and knowledge (Bisseleua and Niang, 2013; Smith, 2013). The education of men should also not be neglected, since they often have most control over household incomes, and need to be aware of the importance of diverse cropping systems and the spending of income on healthy foods (Fon and Edokat, 2012).

2.4 The Indirect Roles of Forests and Tree-based Systems

2.4.1 Income and other Livelihood Opportunities

Income from non-timber forest products

Local communities derive income from timber and non-timber products in forests. In this subsection, the focus is on the latter, although research in the countries of the Congo Basin, as well as in Indonesia, Ecuador and elsewhere, shows that there is a large and vibrant –and largely informal – domestic timber sector that supports the livelihoods of hundreds of thousands of local forest users (Cerutti and Lescuyer, 2011; Lescuyer et al., 2011). In many countries, however, laws for timber extraction were designed largely around large-scale export-oriented forestry operations rather than to sustain healthy small-scale domestic markets, which can be criminalised, generating large revenues in bribes for unscrupulous state officials (Cerutti et al., 2013). There are in turn, some encouraging efforts to change forest and resource *governance* rules to favour strengthened local rights (Campese et al., 2009).

In addition to providing food directly, a multitude of NTFPs harvested from natural, incipiently- and/or semi-domesticated forests and woodlands provide a range of resources that are used by harvesters directly for other purposes, or are sold for income that can be used to purchase a variety of products, including food. The increased demand for forest products in low-income nations, prompted by population growth and urbanisation, provides particular opportunities to enhance rural livelihoods (Arnold et al., 2006). Difficulties in adequately quantifying NTFP value, however, include the multiplicity of products, informal trade and bartering that occur in unmonitored local markets, direct household provisioning without products entering markets at all, and the fact that wild-harvested resources have been excluded from many large-scale rural household surveys (Angelsen et al., 2011; Shackleton et al., 2007; Shackleton et al., 2011). The heterogeneity of challenges to harness the income- and livelihood-generating opportunities from these tree products include the diversity of markets and of market structures of which they are part (Jamnadass et al., 2014).

Despite difficulties in quantification, some overall estimates of value have been attempted. Pimentel et al. (1997), for example, estimated very approximately that USD 90 billion worth of food and other NTFPs were harvested annually from forests and trees in developing countries. FAO's latest (2010) Global Forest Resources Assessment (FRA) provided more recent estimates (based on 2005 figures), with worldwide values given of USD 19 billion and 17 billion annually for non-wood forest product- and woodfuel-removals, respectively. The data compiled for the FRA were, however, acknowledged to be far from complete (one problem is that, when they do report value for NTFPs, many countries only do so for the "top" few species of commercial importance; FAO, 2010). A good illustration of the discrepancy between current estimates of importance comes from comparing the value of woodfuel reported for Africa (most woodfuel is harvested from naturally-regenerating rather than planted sources in the continent) in the 2010 FRA (USD 1.4 billion annually) with the World Bank's (2011) much higher estimate of the value of the charcoal industry in the sub-Saharan region (USD 8 billion annually; quoted in Section 2.3; see also FAO, 2014). There is also some confusion regarding the meaning of the term "income" in estimates: some studies use it to mean the cash made from selling products; perhaps more commonly, however, the term is used in the sense of the "environmental income" from the diversity of goods provided "freely" by the environment, which includes the often higher value of subsistence extraction (Angelsen et al., 2014).

In recent years, more appropriate and systematic methods have been used to quantify the value of such products, including by the Poverty Environment Network (PEN), which compiled a comparative socio-economic data set from 8,000 households in 24 low-income tropical nations, focusing on tropical forest use and poverty alleviation (PEN, 2015; Wunder et al., 2014). The results of PEN revealed that, for the surveyed communities, environmental income constituted 28 percent of total household income, around three-quarters of which came from forests (with the highest proportion coming from forests in Latin America; Angelsen et al., 2014). According to the PEN analysis, across all sampled communities the major products and their contributions to forest income were woodfuel (firewood and charcoal, 35 percent), food (30 percent) and structure/fibre products (25 percent). There is variation between geographic regions in the importance of particular products to surveyed communities, with foods for example, being more important from forest sources in Latin America than in Africa, and the reverse being true for woodfuel. The PEN data also indicated that lower income classes were proportionally more dependent on NTFPs, partly because they have less access to private resources, although better-off households earned more in absolute terms (Angelsen et al., 2014; Wunder et al., 2014).

Carrying bushmeat in Vietnam. Photo © Terry Sunderland

A wide range of other studies have also indicated an important role for NTFPs in supporting rural peoples' livelihoods (Table 2.2). NTFPs are a common "safety net" for rural households in response to shocks and as gap-filling to seasonal shortfalls, and in some instances allow asset accumulation and provide a pathway out of poverty (Angelsen and Wunder, 2003; Mulenga et al., 2012; Shackleton and Shackleton, 2004).

The involvement of women, who have limited access to land and capital resources, in NTFP trade can have positive effects on intra-household equity (e.g., Kusters et al., 2006; Marshall et al., 2006). However, connecting such data with food consumption – through direct provisioning or through sales that are used to support food purchase and dietary diversity – is a different matter, and much less information is available (Ahmed, 2013). Given that much of the collection of NTFPs is done by women and children, they suffer more when access to resources is restricted or if resources are depleted (Agarwal, 2013).

As noted above and as is evident from Table 2.2, woodfuel is an important NTFP in many locations, which allows the preparation of food (Section 2.3). In contrast to subsistence firewood collection, traditionally handled by women and children, charcoal production is mainly an activity undertaken by men (Ingram et al., 2014), although the growing participation of women has been reported in some locations, such as in Zambia and northern Tanzania (Butz, 2013; Gumbo et al., 2013). Who benefits most from production depends on the specific context (Butz, 2013; Khundi et al., 2011; Schure et al., 2014; Zulu and Richardson, 2013). Charcoal production provides a good illustration of some of the dilemmas for intervention in NTFP harvest and trade since it is often based on unsustainable practices that are sometimes illegal (Mwampamba et al., 2013). Its value chain is generally affected by a complex and multi-layered regulatory context that is unclear for stakeholders (Iiyama et al., 2014b; Sepp, 2008). Interventions have rarely been effective, with economic rents accruing to the transport/wholesale stages of the value chain, as well as in bribes to those engaged in the illicit licence trade (Naughton-Treves et al., 2007). Partly as a result, producer margins are often low (Mwampamba et al., 2013).

Commercialising the wild harvest of NTFPs has been widely promoted as a conservation measure, based on the assumption that an increase in resource value is an incentive for collectors to manage forests and woodlands more sustainably (FAO, 2010). Experience shows, however, that the concept of commercialisation and conservation proceeding in tandem is often illusory (Belcher and Schreckenberg, 2007), as more beneficial livelihood outcomes are generally associated with more detrimental environmental outcomes (Kusters et al., 2006). The harvest of fruit from the argan tree (*Argania spinosa*), endemic to Morocco, is a good illustration of the dilemmas involved. The oil extracted from the kernels of argan fruit is one of the most expensive edible oils (as well as being used for cosmetic purposes) in the world and development agencies have widely promoted a "win-win" scenario for rural livelihoods and argan forest health based on further commercialisation (Lybbert et al., 2011). As Lybbert et al., showed, however, while the booming oil export market has benefited the local economy, it has also contributed to forest *degradation*. Thus, although the commercialisation of NTFP harvesting can contribute to livelihoods, not too much should be expected from it in terms of supporting sustainability, even if measures to engage in cultivation are taken (see Section 2.3; Dawson et al., 2013).

Table 2.2 Case studies indicating the proportional contribution of non-timber forest products to household budgets. The examples show that the scale of the contribution varies widely, depending on context and wealth group, with often higher proportional contributions to poorer households.

Reference	Location	Land use type	% household income **	Further information
Shackleton et al. (2007)	South Africa	Natural forest	20	
Appiah et al. (2007)	Ghana	Natural forest	38	
Kamanga et al. (2009)	Malawi	Forest, farmland	15 (17 P, 7 W)	Woodfuel, fodder, etc.
Babulo et al. (2009)	Northern Ethiopia	Natural forest	27	Woodfuel, farm implements, construction materials, wild foods, medicines
Yemiru et al. (2010)*	Southern Ethiopia	Forests (participatory management)	(53 P, 23 W)	
FAO (2011)	Mozambique	Natural forest	30	Woodfuel, fruit, mushrooms, insects, honey, medicines
FAO (2011)	Sahel	Parkland, savannah woodland	80	Shea nut
Mulenga et al. 2011	Zambia	Natural forest	32	Woodfuel, wild honey, mushrooms, ants, caterpillars
Heubach et al. (2011)	Northern Benin	Natural forest	39	
Adam and Pretzsch (2010)	Sudan	Savannah woodland	54	Ziziphus fruits
Ingram et al. (2012)	Congo Basin	Natural forest	47	
Pouliot (2012)	Burkina Faso	Parkland, forest	28 (43 P, 18 W)	Shea nut, woodfuel, locust bean pod, baobab fruit and leaves, fodder, thatching grass
Pouliot and Treue (2013)*	Ghana, Burkina Faso	Grassland, bushland, farmland, forest	Ghana (45 P, 20 W); Burkina Faso (42 P, 17 W)	Woodfuel, wild foods, fodder, construction materials, medicines
Bwalya (2013)	Zambia	Natural forest, woodland	30	Honey, mushrooms, tubers, berries, woodfuel, construction poles
Kar and Jacobson (2012)	Bangladesh	Forest-adjacent hilly areas	(16 P, 9 W)	Bamboo, wild vegetables, broom grass
Vedeld et al. (2004)	Review of 54 studies in 17 countries		20, ~ half as cash income	Woodfuel, wild foods, animal fodder, etc.

* Studies conducted under the Poverty Environment Network (PEN).

** Average for the sample, and/or (in parentheses) the range of contribution between poorer (P) and wealthier (W) groups. Values normally expressed in terms of environmental income.

Income from cultivated tree crops

Examples from Africa of widely-traded agroforestry tree foods that support farmers' incomes and consumers' choices include the indigenous semi-domesticated and widely cultivated fruit safou (*Dacryodes edulis*, Schreckenberg et al., 2006), the indigenous incipient domesticated njansang (*Ricinodendron heudelotii*, Ndoye et al., 1998) and exotic mango. New domestic markets for fruit are developing in Africa as a result of recent investments by Coca Cola, Del Monte and others to source produce locally for juice manufacture, and also to meet growing demand from population growth and increased urbanisation (Ferris et al., 2014). Worldwide, products supplied from tree-crop systems are fundamental raw materials underpinning the development of small scale to multibillion dollar industries. Coffee and cocoa are the most demanded tree crop commodities, particularly in the developed world, by beverage- and confectionery-producing giants such as Mars Inc., Nestlé and Cadbury, among others.

Women have particular opportunities to earn income from fruit and vegetable production because of their traditional involvement in harvesting and processing (Kiptot and Franzel, 2011), thereby supporting the expenditure of a greater proportion of the family income on food, although men may "co-opt" tree-based enterprises when they become more profitable (Jamnadass et al., 2011). Women are also more likely to grow a wider range of trees in the farm plots they control, including food trees (FAO, 1999).

There are still glaring gaps in the knowledge and efforts to realise the full potential of indigenous food trees, specifically in terms of production and trade status, and in the operation of value chains (Jamnadass et al., 2011). Big challenges to market engagement are the perishability of many fruits, combined with the geographic distance to larger market centres and the lack of suitable infrastructure, lack of market information, and value chains biased against small producers (Gyau et al., 2012). In addition to foods, the production of timber and other agroforestry tree products (AFTPs) for markets also provide incomes for food purchase. The high commercial value of timber planting in smallholdings pan-tropically is confirmed by the partial economic data available for the sector (e.g., for teak [*Tectona grandis*] in Indonesia see Roshetko et al., 2013; for acacia in Vietnam [*Acacia mangium* and *A. auriculiformis*] see Fisher and Gordon, 2007; Harwood and Nambiar, 2014). Many trees are also cultivated to provide medicines from bark, leaves, roots, etc., which are sold to support incomes and are used for self-treatment, supporting the health of communities along with the provision of healthy foods (Muriuki et al., 2012); however markets remain largely informal (McMullin et al., 2012; McMullin et al., 2014).

Market data recorded for agroforestry tree products are relatively sparse, but information on export value globally is quantified for major tree commodity crops such as palm oil, coffee, rubber (from *Hevea brasiliensis*), cocoa and tea (primarily from *Camellia sinensis*). Each of these crops is grown to a significant extent by smallholders, as illustrated in Indonesia where, in 2011, small farms were estimated to contribute 42 percent, 96 percent, 85 percent, 94 percent and 46 percent of the country's total production area for palm oil, coffee, rubber, cocoa and tea, respectively (GI, 2015). Unlike Indonesia, many countries do not formally differentiate between smallholder and larger-scale plantation production, but more than 67 percent of coffee produced worldwide is estimated to be from smallholdings (ICO, 2015), while the figure is 90 percent for cocoa (ICCO, 2015). Although in the 20th century there was a general transition from plantations to smallholder production for a number of tree crops, in some regions this may now be being reversed (Byerlee, 2014).

Moabi seeds contain highly valuable oil which is used for cooking, traditional healing and cosmetics. Photo © Terry Sunderland

Taken together, the current annual export value of the above five tree commodity crops is tens of billions of USD, while other cultivated tree crops (such as avocados, cashews, coconuts, mangoes and papayas) also provide additional valuable contributions (Figure 2.3; FAO, 2015). Total production of these crops and their export value have grown in recent decades, with FAOSTAT data showing that export values have increased at a rate roughly four times faster than that of production. Less clear is the proportion of the export value that accrues to smallholder producers, but often production constitutes a considerable proportion of farm takings. It is estimated that cocoa accounts for 80 percent of smallholders' incomes in Bolivia, while in Ghana it provides livelihoods for over 700,000 farmers (Kolavalli and Vigneri, 2011).

There is a danger that the planting of some tree commodities will result in the conversion of natural forest – which contains important local foods – to agricultural land, and a risk that food crops will be displaced from farmland in a trend towards the growing of monocultures (e.g., oil palm, the cultivation of which has led to the wide-scale loss of forest and *agrobiodiversity*; Danielsen et al., 2009). Although it has often been suggested that intensive monocultures raise productivity and therefore reduce the amount of forested land that needs to be cut for crop cultivation (leaving forest food sources intact), there are few quantitative data to support the notion that "*land sparing*" is more effective than "*land sharing*" as a conservation strategy (Balmford et al., 2012; Tscharntke et al., 2012; see discussion in Chapter 5).

Fig. 2.3 Global export values of a range of tree commodity crops over a twenty year period, 1991 to 2010

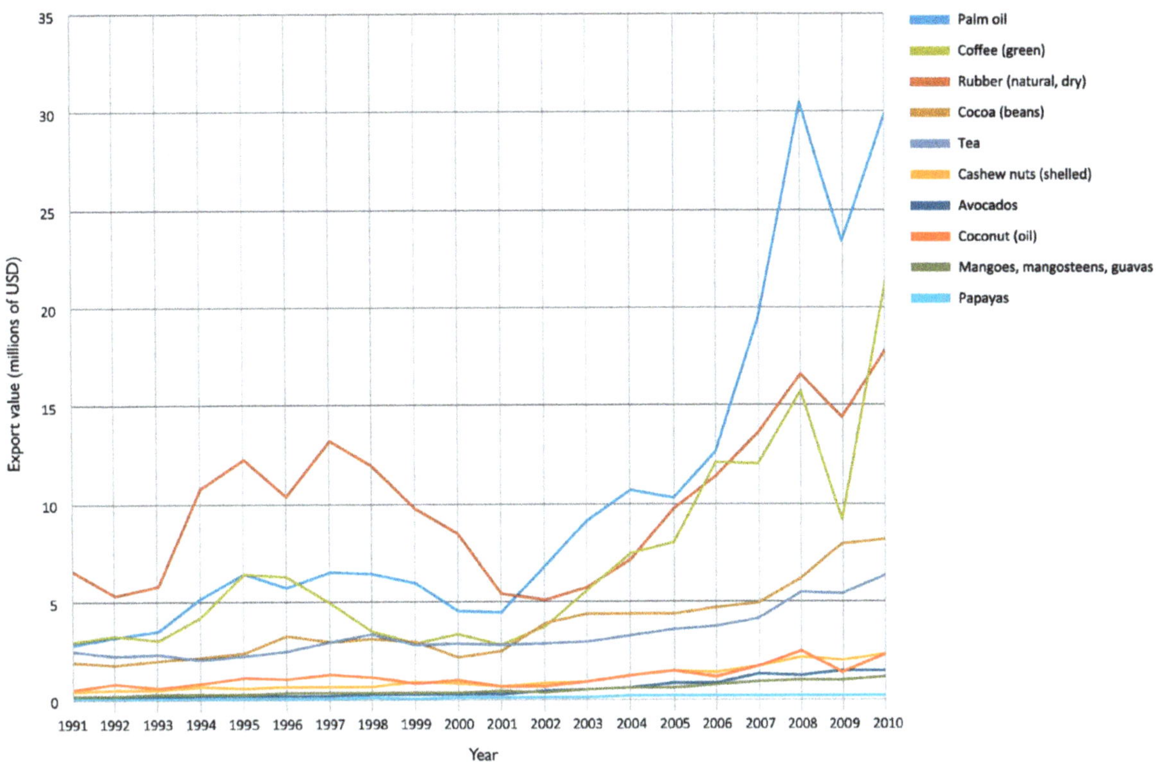

Data were extracted from FAO (2015) and are combined figures for all nations providing information. Data for mangoes, mangosteens and guava are reported together. Given values include re-exports (i.e., import into one nation followed by export to another). Some commodities, such as coffee, cocoa and coconut, are exported in more than one form and total export values are therefore higher than those shown here (for each of these crops only the most important form by export value is given). The graph shows that there was a significant increase in export value for crops during the decade leading up to 2010, but that value was volatile. The most notable feature over the period was a sharp rise in palm oil export value. Note that local trade can also be significant for many of these products

There is an important opportunity to diversify risks associated with the reliance on a few cash tree crops into other tree crops whose domestic production and export markets are growing steadily and rapidly, while also meeting food security and nutritional needs of the growing population. For example, currently, the global supply of fruits and vegetables falls, on average, 22 percent short of population need according to nutrition recommendations, while low income countries fall on average 58 percent short of need (Siegel et al., 2014). Although tree crop cultivation provides opportunities for farmers to diversify and minimise risk, especially for products that can be consumed by the family as well as sold (Jamnadass et al., 2011), buying food using the income received from a single commodity cash crop can lead to food insecurity for individual farm households when payments are one-off, delayed or volatile in value. Similarly, individual countries can become too dependent on one or a few commodities, with significant fluctuations in GDP, dependent on unpredictable world prices (Jamnadass et al., 2014). Monocultures of tree commodities also reduce resilience to shocks such as drought, flood and, often (although not always), the outbreak of pests and diseases. As a result, tree commodity crops are sometimes viewed sceptically within agricultural production-based strategies to improve nutrition (FAO, 2013a). For farmers who have too little land to cultivate enough food to directly meet their needs, however, income from tree commodity crops may be the only way to obtain sufficient food (Arnold, 1990).

2.4.2 Provision of Ecosystem Services

The Millennium Ecosystem Assessment (MA, 2005) provided a comprehensive overview of ecosystem services and much literature has been written on the subject. Here we provide a brief overview of key ecosystem services from forests and tree-based systems, and their roles in food security and nutrition.

Forests, agroforests and – to a certain extent – plantations, provide important ecosystem services including: soil, spring, stream and watershed protection; microclimate regulation; biodiversity conservation; and pollination, all of which ultimately affect food and nutritional security (Garrity, 2004; Zhang et al., 2007). Multiple ecosystem service scan generally be fund in any single forest fragment (see Box 2.3). Forest users and farmers can be encouraged to preserve and reinforce these functions by payments for ecosystem services (PES), but more important in determining their behaviour is the direct products and services they receive from trees (Roshetko et al., 2007). Neglect of this fact by PES schemes has led to sub-optimal results (Roshetko et al., 2015). Opportunities for ecological intensification (see Chapter 5) and for the better provision of environmental services to support food security vary by stage of the forest-tree landscape continuum (van Noordwijk et al., 2014 and see Chapter 3).

Forests, woodlands and trees elsewhere in landscapes play a vital role in controlling water flows, and preventing soil erosion and nutrient leaching, all of which are critical functions for food production systems (Bruinsma, 2003). At the same time, green manures in agroforestry systems maintain and enhance soil fertility, supporting crop yields when external fertiliser inputs are not available or are unaffordable (see Section 2.3; Garrity et al., 2010; Sanchez, 2002). Nitrogen-fixing trees have in particular received considerable attention for their ability to cycle atmospheric nitrogen in cropping systems (Sileshi et al., 2008; Sileshi et al., 2011; Sileshi et al., 2012). Microclimate regulation by trees in agroforestry systems, such as through the provision of a canopy that protects crops from direct exposure to the sun (reducing evapotranspiration), from extreme rainfall events and from high temperatures, can also promote more resilient and productive food-cropping systems (Pramova et al., 2012). In Sahelian zones with long dry seasons, for example, trees provide an environment for the cultivation of nutritious leafy vegetables and pulses (Sendzimir et al., 2011).

Box 2.3 Forest fragments modulate ecosystem services

Mitchell et al. (2014) provide empirical evidence that forest fragments influence the provision of multiple ecosystem service indicators in adjacent agricultural fields. Their study looked simultaneously at six ecosystem services (crop production, pest regulation, decomposition, carbon storage, soil fertility and water quality regulation) in soya bean fields at different distances from adjacent forest fragments that differed in isolation and size across an agricultural landscape in Quebec, Canada. The study showed significant effects of distance-from-forest, fragment isolation and fragment size on crop production, insect pest regulation, and decomposition. Distance-from-forest and fragment isolation had unique influences on service provision for each of the ecosystem services measured. For example, pest regulation was maximised adjacent to forest fragments (within 100 m), while crop production was maximised at intermediate distances from forest (150 m to 300 m). As a consequence, landscape multifunctionality depended on landscape heterogeneity: the range of field and forest fragment types present. The study also observed strong negative and positive relationships between ecosystem services that were more prevalent at greater distances from forest.

Forests, and frequently agroforests, are centres of plant and animal biodiversity, protecting species and the genetic variation that is found with them, which may be essential for future human food security (Dawson et al., 2013). As already noted in Section 2.3, as well as being sources of existing and "new" foods, many already cultivated tree species have their centres of genetic diversity within forests, and these resources may be crucial for future crop improvement. A good example is coffee, an important beverage globally, which is found wild in Ethiopian montane forests. These forests are under significant threat from agricultural expansion (Labouisse et al., 2008) and climate change (Davis et al., 2012). Economic "option value" analysis of wild coffee stands for breeding purposes – to increase yields, improve disease resistances and for a lower caffeine content in the cultivated crop – shows just how important it is to implement more effective conservation strategies for Ethiopian forests (Hein and Gatzweiler, 2006; Reichhuber and Requate, 2007).

Pollination is one of the most studied ecosystem services, with perhaps the most comprehensive reviews of animal pollination and how it underpins global food production being that of Klein et al. (2007). A diversity of trees in forests and in farmland can support populations of pollinator species such as insects and birds that are essential for the production of important human foods, including fruits in both forest and farmland, and a range of other important crops in farmland (Garibaldi et al., 2013; Hagen and Kraemer, 2010; for the specific case of coffee, see Ricketts et al., 2004; Priess et al., 2007). For communities living in or around forests, pollination is therefore a crucial ecosystem service (Adams, 2012). Of course, forests and trees in agroforests provide important habitat for a range of other fauna that include the natural predators of crop pests (as well as sometimes being hosts for the crop pests themselves; Tscharntke et al., 2005).

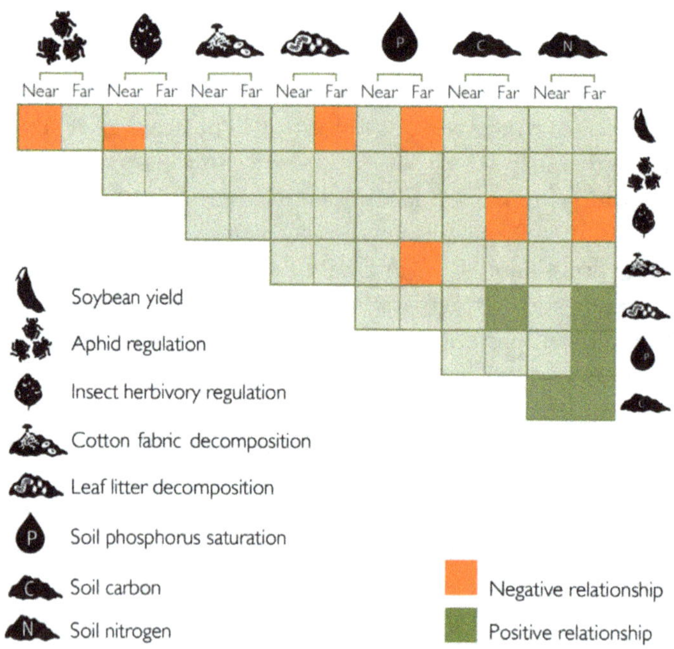

Fig. 2.4 Effects of distance-from-forest on pair-wise Spearman rank relationships between ecosystem service indicators. ***Source:*** **Mitchell et al., 2014**

2.5 Conclusions

Foods provided by forests and tree-based systems

There is increasing evidence of the importance of forests and tree-based systems for supporting food production and contributing to dietary diversity and quality, addressing nutritional shortfalls. By targeting particular species for improved harvest and/or cultivation, more optimal "portfolios" of species could be devised that best support communities' nutrition year-round. An overall increase in the production through cultivation of a wide range of foods, including tree fruits and vegetables, is required to bridge consumption shortfalls. There is much further potential for the domestication of currently little-researched indigenous fruit trees to bring about large production gains, although more information is needed on the nutritional value of many of these species. Trees also provide other important products (e.g., fodder, green fertiliser, fuel) that support food production and use.

Dietary choices, access to resources and behavioural change

Dietary choices are complex and depend on more than just what potential foods are available to communities in their environments. Rather than assumptions based on availability, assessments of actual diet through dietary diversity studies and other related estimators are therefore crucial. Then, the reasons behind current limitations in usage can be explored and possibly addressed. There are multiple targets to improve food choices, with women and children being key targets for education.

Income and other livelihood opportunities

NTFPs and AFTPs, including tree commodity crops within agroforestry systems, are important sources of revenue to local people and governments, which can support food supply. More is known about the economic value of tree commodity crops than of other products, but recent initiatives have provided a clearer picture of the "environmental income" from NTFPs (though not necessarily for AFTPs). Only limited information is available on how cash incomes from these resources are spent with regard to promoting food and nutritional security, and there are clear dangers in relying on cash incomes from single commodity crops.

Provision of ecosystem services

Forests and tree-based production systems provide valuable ecosystem services that support staple crop production and that of a wider range of edible plants. Many tree species that are important crops globally require pollinators to produce fruit. The presence of these pollinators is supported by forests and diverse cropping systems. More is known about the environmental service provisioning of tropical humid forests than of dry forests (Blackie et al., 2014).

Outstanding gaps

Pineapple – here in a homegarden in Cuba – is rich in manganese and vitamin C. Photo © Stephanie Mansourian

The value of the "hidden harvest" of edible forest foods, and the cultivation of trees by smallholders, is evident from this chapter. To maximise future potential, greater attention from both the scientific and the development communities is required. In particular, the development of a supportive policy framework requires proper attention to both the forestry and agriculture sectors in tandem. For this to take place, a better quantification of the relative benefits received by rural communities from different tree production categories is required, supported by an appropriate typology for characterisation (de Foresta et al., 2013). Despite recent advances such as PEN (2015), data are still required to quantify roles in supporting food and nutritional security that include dietary diversity measurements.

Policies that support communities' access to forest and that encourage the cultivation of tree products are required. Required reforms include more favourable land tenure arrangements for smallholders, in how farmers obtain tree planting material, and in the recognition of agroforestry as a viable investment option for food production (Jamnadass et al., 2013). Research should support food tree domestication options appropriate for meeting smallholders' needs. Emphasis should be placed on mixed agroforestry production regimes that can help to avoid many of the negative effects described in Section 2.4, by combining tree commodities in diverse production systems with locally-important food trees, staple crops, vegetables and edible fungi. Such regimes include shade coffee and shade cocoa systems (Jagoret et al., 2011; Jagoret et al., 2012; SCI, 2015), which increase or at least do not decrease commodity yields and profitability (Clough et al., 2011). Such systems have often been practised traditionally, but are now being actively encouraged through schemes such as certification by some international purchasers of tree commodity crops (Millard, 2011).

To support diverse production systems, genetic selection for commodity crop cultivars that do well under shade may be of particular importance (Mohan Jain and Priyadarshan, 2009). This may require returning to wild genetic resources still found in shaded, mixed-species forest habitats, reinforcing the value of their conservation. Not all tree commodities are, however, amenable to production in diversified systems; for example, oil palm is not well suited (Donald, 2004). There are also opportunities to develop valuable new tree commodities that are compatible with other crops and that therefore support more agro-biodiversity. Further research is also required to assess the complementarity and resilience of different crops in agroforestry systems under climate change, in the context also of other global challenges to food and nutritional security.

The development of "nutrient-sensitive" value chains is also needed, which means improving nutritional knowledge and awareness among value-chain actors and consumers, focusing on promoting the involvement of women, and considering markets for a wider range of tree foods. By promoting tree food processing and other value additions, the non-farm rural economy can also be stimulated. As highlighted elsewhere in this publication, however, more research is required to understand the economic, environmental and other trade-offs for the different sectors of rural societies when the harvesting of NTFPs is commercialised or they are planted (and perhaps are converted to new commodity crops; Dawson et al., 2014b), as the benefits and costs for different members of society vary. For example, wild harvesters without access to farmland can be disadvantaged when NTFPs become cultivated as AFTPs (Page, 2003). More work is therefore needed to ensure equitable relationships between the different participants in market supply chains (Marshall et al., 2006).

References

Adam, Y.O. and Pretzsch, J., 2010. Contribution of local trade in Ziziphusspina-christi L. fruits to rural household's economy in Rashad Locality: Sudan. *Forestry Ideas* 1: 19-27. http://www.qucosa.de/fileadmin/data/qucosa/documents/7148/PhD thesis 22.07.11.pdf

Adams, W.M., 2012. Feeding the next billion: Hunger and conservation. *Oryx* 46: 157-158. http://dx.doi.org/10.1017/s0030605312000397

Agarwal, B., 2001. Participatory exclusions, community forestry, and gender: An analysis for South Asia and a conceptual framework. *World Development* 29: 1623-1648. http://dx.doi.org/10.1016/s0305-750x(01)00066-3

Agarwal, B., 2013. *Gender and Green Governance: The Political Economy of Women's Inclusion in Community Forestry Institutions*. Oxford: Oxford University Press.

Appiah, M., Blay, D., Damnyag, L., Dwomoh, F.K., Pappinen, A. and Luukkanen, O., 2007. Dependence on forest resources and tropical deforestation in Ghana. *Environment, Development and Sustainability* 11: 471-487. http://dx.doi.org/10.1007/s10668-007-9125-0

Ahmed, M.A., 2013. Contribution of non-timber forest products to household food security: The case of Yabelo Woreda, Borana Zone, Ethiopia. *Food Science and Quality Management* 20. http://www.iiste.org/Journals/index.php/FSQM/article/view/8019

Alves, R.N. and Alves, H.N., 2011. The faunal drugstore: Animal-based remedies used in traditional medicines in Latin America. *Journal of Ethnobiology and Ethnomedicine* 7: 9. http://dx.doi.org/10.1186/1746-4269-7-9

Angelsen, A., Jagger, P., Babigumira, R., Belcher, B., Hogarth, N., Bauch, S., Börner, B., Smith-Hall, C. and Wunder, S., 2014. Environmental income and rural livelihoods: A global-comparative analysis. *World Development* 64(1): S12-S28. http://dx.doi.org/10.1016/j.worlddev.2014.03.006

Angelsen, A. and Wunder, S., 2003. *Exploring the Forest-poverty Link: Key Concepts, Issues and Research Implications*. Occasional Paper No. 40. Bogor: Center for International Forestry Research (CIFOR). http://dx.doi.org/10.17528/cifor/001211

Angelsen, A., Wunder, S., Babigumira, R., Belcher, B., Börner, J. and Smith-Hall, C., 2011. *Environmental Incomes and Rural Livelihoods: A Global-comparative Assessment*. Occasional Paper. 4th Wye Global Conference, Rio de Janeiro, 9 to 11 November 2011. Wye City Group on statistics on rural development and agriculture household income. Rio de Janeiro and Rome: The Brazilian Institute of Geography and Statistics and FAO. http://www.fao.org/fileadmin/templates/ess/pages/rural/wye_city_group/2011/documents/session4/Angelsen_Wunder_Babigumira_Belcher_Birner__Smith-Hall-Paper.pdf

Arimond, M., Wiesmann, D., Becquey, E., Carriquiry, A., Daniels, M.C., Deitchler, Fanou-Fogny, N., Joseph, N.J., Kennedy, G., Martin-Prevel, Y. and Torheim, L., 2010. Simple food group diversity indicators predict micronutrient adequacy of women's diets in 5 diverse: Resource-poor settings. *Journal of Nutrition* 140: 2059S-2069S. http://dx.doi.org/10.3945/jn.110.123414

Arnold, J.E.M., 1990. Tree components in farming systems. *Unasylva* 160: 35-42. http://www.fao.org/docrep/t7750e/t7750e06.htm

Arnold, J.E.M., Kohlin, G. and Persson, R., 2006. Woodfuels, livelihoods, and policy interventions: Changing perspectives. *World Developmen*t 34: 596-611. http://dx.doi.org/10.1016/j.worlddev.2005.08.008

Babulo, B., Muys, B., Nega, F., Tollens, E., Nyssen, J., Dekkers, J. and Mathijs, E., 2009. The economic contribution of forest resources to rural livelihoods in Tigray, Northern Ethiopia. *Forest Policy and Economics* 11: 109-117. http://dx.doi.org/10.1016/j.forpol.2008.10.007

Balée, W., 2013. *Cultural Forests of the Amazon: A Historical Ecology of People and their Landscapes.* Tuscaloosa: University of Alabama Press.

Bailis, R., Ezzati, M. and Kammen, D.M., 2005. Mortality and greenhouse gas impacts and petroleum energy futures in Africa. *Science* 308(5718): 98-103. http://dx.doi.org/10.1126/science.1106881

Balmford, A., Green, R. and Phalan, B., 2012. What conservationists need to know about farming. *Proceedings of the Royal College of London* 279: 2714-2724. http://dx.doi.org/10.1098/rspb.2012.0515

Belcher, B. and Schreckenberg, K., 2007. Commercialisation of non-timber forest products: A reality check. *Development Policy Review* 25: 355-377. http://dx.doi.org/10.1111/j.1467-7679.2007.00374.x

Belcher, B., Ruiz Pérez, M. and Achdiawan, R., 2005. Global patterns and trends in the use and management of commercial NTFPs: Implications for livelihoods and conservation. *World Development* 9: 1435-1452. http://dx.doi.org/10.1016/j.worlddev.2004.10.007

Bharucha, Z. and Pretty, J. 2010. The roles and values of wild foods in agricultural systems. *Philosophical Transactions of the Royal Society B: Biological Sciences* 365(1554): 2913-2926. http://dx.doi.org/10.1098/rstb.2010.0123

Bhutta, Z.A., Das, J.K., Rizvi, A., Gaffey, M.F., Walker, N., Horton, S., Webb, P., Lartey, A. and Black, R.E., 2013. Evidence-based interventions for improvement of maternal and child nutrition: What can be done and at what cost? *The Lancet* 382(9890): 452-477. http://dx.doi.org/10.1016/s0140-6736(13)60996-4

Bisseleua, H.B.D. and Niang, A. I., 2013. Lessons from Sub-Saharan Africa. Delivery Mechanisms for Mobilizing Agricultural Biodiversity for Improved Food and Nutrition Security. In: *Diversifying Food and Diets: Using agricultural biodiversity to improve nutrition and health*, edited by J. Fanzo, D. Hunter, T. Borelli, and F. Mattei. Oxford and New York: Routledge, 111-121. http://dx.doi.org/10.4324/9780203127261

Black, R.E., Victora, C.G., Walker, S.P., Bhutta, Z.A., Christian, P., de Onis, M., Ezzati, M., Grantham-McGregor, S., Katz, J., Martorell, R. and Uauy, R., 2013. Maternal and child undernutrition and overweight in low-income and middle-income countries. *The Lancet* 382: 427-451. http://dx.doi.org/10.1016/s0140-6736(13)60937-x

Blackie, R., Baldauf, C., Gautier, D., Gumbo, D., Kassa, H., Parthasarathy, N., Paumgarten, F., Sola, P., Pulla, S., Waeber, P. and Sunderland, T., 2014. *Tropical Dry Forests: The State of Global Knowledge and Recommendations for Future Research.* Discussion Paper 2. Bogor: CIFOR. http://www.cifor.org/publications/pdf_files/WPapers/DPBlackie1401.pdf

Boeing, H., Bechthold, A., Bub, A., Ellinger, S., Haller, D., Kroke, A., Leschik-Bonnet, E., Müller, M.J., Oberritter, H., Schulze, M.S., Stehle, P., and Watzl, B., 2012. Critical review: Vegetables and fruit in the prevention of chronic diseases. *European Journal of Nutrition* 51: 637-663. http://dx.doi.org/10.1007/s00394-012-0380-y

Brashares, J., Goldena, C., Weinbauma, K., Barrett, C. and Okello, G., 2011. Economic and geographic drivers of wildlife consumption in rural Africa. *Proceedings of the National Academy of Sciences of the USA* 108: 13931-13936. http://dx.doi.org/10.1073/pnas.1011526108

Brouwer, I.D., Hoorweg, J.C. and van Liere, M.J., 1997. When households run out of fuel: Responses of rural households to decreasing fuelwood availability, Ntcheu District, Malawi. *World Development* 25: 255-266. http://dx.doi.org/10.1016/s0305-750x(96)00100-3

Bruinsma, J., 2003. *World Agriculture: Towards 2015/2030. An FAO perspective.* London: Earthscan Publications Ltd. http://www.fao.org/fileadmin/user_upload/esag/docs/y4252e.pdf

Bryce J., Coitinho, D., Darnton-Hill I., Pelletier, D. and Pinstrup-Andersen, P., 2008. Maternal and child under-nutrition: Effective action at national level. *The Lancet* 371: 510-526. http://dx.doi.org/10.1016/s0140-6736(07)61694-8

Burlingame, B. and Dernini, D., 2012. *Sustainable Diets and Biodiversity Directions and Solutions for Policy, Research and Action*. FAO Nutrition and Consumer Protection Division Proceedings of the International Scientific Symposium Biodiversity and Sustainable Diets United Against Hunger 3-5 November 2010, Rome: FAO. http://www.fao.org/docrep/016/i3004e/i3004e.pdf

Butz, R.J., 2013. Changing land management: A case study of charcoal production among a group of pastoral women in northern Tanzania. *Energy for Sustainable Development* 17: 138-145. http://dx.doi.org/10.1016/j.esd.2012.11.001

Bwalya, S.M., 2013. Household dependence on forest income in rural Zambia. *Zambia Social Science Journal* 2: 6. http://scholarship.law.cornell.edu/cgi/viewcontent.cgi?article=1021&context=zssj

Byerlee, D., 2014. The fall and rise again of plantations in tropical Asia: History repeated? *Land* 3: 574-597. http://dx.doi.org/10.3390/land3030574

Byron, N. and Arnold, J.E.M., 1997. *What Futures for the People of the Tropical Forests*. CIFOR Working Paper 19. Bogor: CIFOR. http://dx.doi.org/10.17528/cifor/000079

Campese, J., Sunderland, T.C.H., Greiber, T. and Oviedo, G., 2009. *Rights Based Approaches: Exploring Issues and Opportunities for Conservation*. Bogor: CIFOR. http://www.cifor.org/publications/pdf_files/Books/BSunderland0901.pdf

Cerutti, P.O. and Lescuyer, G., 2011. *The Domestic Market for Small-scale Chainsaw Milling in Cameroon: Present Situation, Opportunities and Challenges*. CIFOR Occasional Paper 61. Bogor: CIFOR. http://dx.doi.org/10.17528/cifor/003421

Cerutti, P.O., Tacconi, L., Lescuyer, G. and Nasi, R., 2013. Cameroon's hidden harvest: Commercial chainsaw logging, corruption and livelihoods. *Society and Natural Resources* 26: 539-553. http://dx.doi.org/10.1080/08941920.2012.714846

CGIAR, 2014. *Research Program on Agriculture for Nutrition and Health*. http://www.a4nh.cgiar.org/

Chardonnet P. (ed.), 1996. *Faune sauvage africaine: La ressource oubliée*. Luxembourg: Commission européenne.

Chaudhuri, S. and Pfaff, A., 2002. *Economic Growth and the Environment: What Can We Learn from Household Data?* Working Paper 2002, Department of Economics, Columbia University, USA. http://hdl.handle.net/10022/AC:P:367

CIE, 2011. *Evaluation of ICRAF's Agroforestry Food Security Programme (AFSP) 2007-2011*. Final report submitted to IRISH AID. Lilongwe: Center for Independent Evaluations.

Clement, C.R. and Junqueira, A.B., 2010. Between a pristine myth and an impoverished Future. *Biotropica* 42: 534-536. http://dx.doi.org/10.1111/j.1744-7429.2010.00674.x

Clement, C.R., 1989. A center of crop genetic diversity in western Amazonia. *BioScience* 39: 624-631. http://dx.doi.org/10.2307/1311092

Clement, C.R., 1999. 1492 and the loss of Amazonian crop genetic resources. I. The relation between domestication and human population decline. *Economic Botany* 52 (2): 188-202. http://dx.doi.org/10.1007/bf02866498

Colfer, C.J.P., 2008. *Human Health and Forests: Global Overview of Issues, Practice and Policy*. London: Earthscan Publications Ltd. http://dx.doi.org/10.5860/choice.46-1459

Clough, Y., Barkmann, J., Juhrbandt, J., Kessler, M., Wanger, T.C., Anshary, A., Buchori, D., Cicuzza, D., Darras, D., Dwi Putra, D., Erasmi, S., Pitopang, R., Schmidt, C.,Schulze, C.H., Seidel, D., Steffan-Dewenter, I., Stenchly, K., Vidal, S., Weist, M.,Wielgoss, A.C. and Tscharntke, T., 2011. Combining high biodiversity with high yields in tropical agroforests. *Proceedings of the National Academy of Sciences of the USA* 108: 8311-8316. http://dx.doi.org/10.1073/pnas.1016799108

Danielsen, F., Beukema, H., Burgess, N.D., Parish, F., Brühl, C.A., Donald, P.F., Murdiyarso, D., Phalan, B., Reijnders, L., Struebig, M. and Fitzherbet, E.B., 2009. Biofuel plantations on forested lands: Double jeopardy for biodiversity and climate. *Conservation Biology* 23: 348-358. http://dx.doi.org/10.1111/j.1523-1739.2008.01096.x

Daniggelis, E., 2003. Women and 'Wild' Foods: Nutrition and Household Security among Rai and Sherpa Forager Farmers in Eastern Nepal. In: *Women and Plants: Gender Relations in Biodiversity Management and Conservation*, edited by P.L. Howard. New York and London: Zed Books and St. Martin's Press.

Das, T. and Das, A.K., 2005. Inventorying plant biodiversity in home gardens: A case study in Barak Valley: Assam, North East India. *Current Science* 89: 155-163. http://www.iisc.ernet.in/currsci/jul102005/155.pdf

daSilva, A. and Begossi, A., 2009. Biodiversity, food consumption and ecological niche dimension: A study case of the riverine populations from the Rio Negro, Amazonia, Brazil. *Environment Development and Sustainability* 11: 489-507. http://dx.doi.org/10.1007/s10668-007-9126-z

Davis, A.P., Gole, T.W., Baena, S. and Moat, J., 2012. The impact of climate change on indigenous arabica coffee (*Coffea arabica*): Predicting future trends and identifying priorities. *Public Library of Science One* 7: e47981. http://dx.doi.org/10.1371/journal.pone.0047981

Dawson, I., Carsan, S., Franzel, S., Kindt, R., van Breugel, P., Graudal, L., Lillesø, J.P., Orwa, C. and Jamnadass, R., 2014a. *Agroforestry, Livestock, Fodder Production and Climate Change Adaptation and Mitigation in East Africa: Issues and Options*. ICRAF Working Paper No. 178. Nairobi: ICRAF. http://www.worldagroforestry.org/downloads/Publications/PDFS/WP14050.pdf

Dawson, I.K., Leakey, R., Clement, C.R., Weber, J., Cornelius, J.P., Roshetko, J.M., Vinceti, B., Kalinganire, A., Tchoundjeu, Z., Masters, E. and Jamnadass, R., 2014b. The management of tree genetic resources and the livelihoods of rural communities in the tropics: Non-timber forest products, smallholder agroforestry practices and tree commodity crops. *Forest Ecology and Management* 333: 9-21. http://dx.doi.org/10.1016/j.foreco.2014.01.021

Dawson, I.K., Guariguata, M.R., Loo, J., Weber, J.C., Lengkeek, A., Bush, D., Cornelius. J., Guarino, L., Kindt, R., Orwa, C., Russell, J. and Jamnadass, R., 2013. What is the relevance of smallholders' agroforestry systems for conserving tropical tree species and genetic diversity in circa situm, in situ and ex situ settings: A review. *Biodiversity and Conservation* 22: 301-324. http://dx.doi.org/10.1007/s10531-012-0429-5

de Foresta, H., Somarriba, E., Temu, A., Boulanger, D., Feuilly, H. and Gauthier, M., 2013. *Towards the Assessment of Trees outside Forests*. FAO Resources Assessment Working Paper No. 183. Rome: FAO. http://www.fao.org/docrep/017/aq071e/aq071e00.pdf

Dharani, N., Yenesew, A., Ermais, B., Tuei, B. and Jamnadass, R., 2014. *Traditional Ethnoveterinary Medicine in East Africa: A Manual on the Use of Medicinal Plants*, edited by I.K. Dawson. Nairobi: ICRAF.

Donald, P.F., 2004. Biodiversity impacts of some agricultural commodity production systems. *Conservation Biology* 18: 17-38. http://dx.doi.org/10.1111/j.1523-1739.2004.01803.x

Dunkel, D., 1996. Nutritional values of various insects per 100 grams. *The Food Insect Newsletter* 9: 1-8.

Ecker, O., Breisinger, C. and Pauw, K., 2011. *Growth is Good but is not Enough for Improving Nutrition*. Paper prepared for the 2020 Conference: Leveraging Agriculture for Improving Nutrition and Health: New Delhi, India, February 10-12. https://www.microlinks.org/sites/microlinks/files/resource/files/Ecker et al - Growth is Good but is not enough to improve nutrition - 2011.pdf

Elliott, J., Grahn, R., Sriskanthan, G. and Arnold, C., 2002. *Wildlife and Poverty Study*. London: Department for International Development. http://www.eldis.org/vfile/upload/1/document/0708/DOC11657.pdf

Fa, J. E., Olivero, J., Farfán, M. Á., Márquez, A. L., Duarte, J., Nackoney, J., Hall, A., Dupain, J., Seymour, S., Johnson, P. J., Macdonald, D. W., Real, R. and Vargas, J. M. 2015. Correlates of bushmeat in markets and depletion of wildlife. *Conservation Biology*. http://dx.doi.org/10.1111/cobi.12441

FAO, 2015. FAOSTAT Statistical Database, 2015. http://faostat.fao.org

FAO, 2014. *State of the World's Forests: Enhancing the Socioeconomic Benefits from the Forests*. Rome: FAO. http://www.fao.org/3/a-i3710e.pdf

FAO, 2013a. *Synthesis of Guiding Principles on Agriculture Programming for Nutrition*. Rome: FAO. http://www.fao.org/docrep/017/aq194e/aq194e.pdf

FAO, 2013b. *Edible Insects: Future Prospects for Food and Feed Security*. FAO Forestry Paper. Rome: FAO. http://www.fao.org/docrep/018/i3253e/i3253e.pdf

FAO, 2013c. *The State of Food and Agriculture: Better Food Systems for Better Nutrition*. Rome: FAO. http://www.fao.org/docrep/018/i3300e/i3300e.pdf

FAO, 2012. *The West African Food Composition Table*. Rome: FAO. http://www.fao.org/infoods/infoods/tables-and-databases/africa/en/

FAO, 2011. *Forests for Improved Food Security and Nutrition Report*. Rome: FAO. http://www.fao.org/docrep/014/i2011e/i2011e00.pdf

FAO, 2010. *Global Forest Resources Assessment: Progress Towards Sustainable Forest Management*. Rome: FAO. http://www.fao.org/docrep/013/i1757e/i1757e09.pdf

FAO, 2009. *Declaration of the World Summit on Food Security*. WSFS 2009/2. Rome: FAO. http://www.fao.org/fileadmin/templates/wsfs/Summit/Docs/Final_Declaration/WSFS09_Declaration.pdf

FAO, 1999. *Agroforestry Parklands in Sub-Saharan Africa*. FAO Conservation Guide No. 34. Rome: FAO. http://www.fao.org/docrep/005/x3940e/x3940e00.htm

FAO, 1997. *Agriculture Food and Nutrition for Africa: A Resource Book for Teachers of Agriculture*. Food and Nutrition Division. Rome: FAO. http://www.fao.org/docrep/w0078e/w0078e00.HTM

FAO, 1992. *Forest, Trees and Food*. Rome: FAO. http://www.fao.org/docrep/006/u5620e/U5620E00.HTM

FAO, IFAD and WFP, 2013. *The State of Food Insecurity in the World 2013. The Multiple Dimensions of Food Security*. Rome: FAO. http://www.fao.org/docrep/018/i3434e/i3434e.pdf

FAO, WFP and IFAD, 2012. *The State of Food Insecurity in the World 2012. Economic Growth is Necessary but not Sufficient to Accelerate Reduction of Hunger and Malnutrition*. Rome: FAO. http://www.fao.org/docrep/016/i3027e/i3027e.pdf

Fargeot, C. and Dieval, S., 2000. La consommation de gibier à Bangui, quelques données économiques et biologiques. *Canopée* 18: 5-7.

Faye, M.D., Weber, J.C., Abasse, T.A., Boureima, M., Larwanou, M., Bationo, A.B., Diallo, B.O., Sigué, H., Dakouo, J.M., Samaké, O. and Diaté, D.S., 2011. Farmers' preferences for tree functions and species in the West African Sahel forests. *Trees and Livelihoods* 20: 113-136. http://dx.doi.org/10.1080/14728028.2011.9756702

Ferris, S., Robbins, P., Best, R., Seville, D., Buxton, A., Shriver, J. and Wei, E., 2014. *Linking Smallholder Farmers to Markets and the Implications for Extension and Advisory Services.* MEAS Discussion Paper 4. Washington DC: CRS and USAID. http://agrilinks.org/sites/default/files/resource/files/MEAS Discussion Paper 4 - Linking Farmers To Markets - May 2014.pdf

Fischler, C., 1988. Food, self and identity. *Social Science Information* 27: 275-92. http://dx.doi.org/10.1177/053901888027002005

Fisher, H. and Gordon, J., 2007. *Improved Australian Tree Species for Vietnam.* ACIAR Impact Assessment Series Report No. 47. Canberra: Australian Centre for International Agricultural Research. http://aciar.gov.au/files/node/2677/ias47_pdf_52502.pdf

Fon, D. and Edokat, T., 2012. Marginalization of women's role in sub-Saharan Africa towards crop production: A review. *Agricultural Science Research Journal* 2(9): 499-505. http://www.resjournals.com/journals/agricultural-science-research-journal/AGRIC 2012 SEPT/Fon and Edokat.pdf

Franzel, S., Carsan, S., Lukuyu, B., Sinja, J. and Wambugu, C., 2014. Fodder trees for improving livestock productivity and smallholder livelihoods in Africa. *Current Opinion in Environmental Sustainability* 6: 98-103. http://dx.doi.org/10.1016/j.cosust.2013.11.008

Frison, E.A., Cherfas, J. and Hodgkin, T., 2011. Agricultural biodiversity is essential for a sustainable improvement in food and nutrition security. *Sustainability* 3: 238-53. http://dx.doi.org/10.3390/su3010238

Garen, E.J., Saltonstall, K., Ashton, M.S., Slusser, J.L., Mathias, S. and Hall, J.S., 2011. The tree planting and protecting culture of cattle ranchers and small-scale agriculturalists in rural Panama: Opportunities for reforestation and land restoration. *Forest Ecology and Management* 261: 1684-1695. http://dx.doi.org/10.1016/j.foreco.2010.10.011

Garibaldi, L.A., Steffan-Dewenter, I., Winfree, R., Aizen, M.A., Bommarco, R. and Cunningham, S.A., Kremen, C., Carvalheiro, L.G., Harder, L.D., Afik, O., Bartomeus, I., Benjamin, F., Boreux, V., Cariveau, D., Chacoff, N.P., Dudenhöffer, J.H., Freitas, B.M., Ghazoul, J., Greenleaf, S., Hipólito, J., Holzschuh, A., Howlett, B., Isaacs, R., Javorek, S.K., Kennedy, C.M., Krewenka, K.M., Krishnan, S., Mandelik, Y., Mayfield, M.M., Motzke, I., Munyuli, T., Nault, B.A., Otieno, M., Petersen, J., Pisanty, G., Potts, S.G., Rader, R., Ricketts, T.H., Rundlöf, M., Seymour, CL., Schüepp, C., Szentgyörgyi, H., Taki H, Tscharntke, T., Vergara, C.H., Viana, BF., Wanger, T.C., Westphal, C., Williams, N. and Klein, A.M., 2013. Wild pollinators enhance fruit set of crops regardless of honeybee abundance. *Science* 339(6127): 1608-1611. http://dx.doi.org/10.1126/science.1230200

Garrity, D., Akinnifesi, F., Ajayi, O., Weldesemayat, S., Mowo, J., Kalinganire, A. and Bayala, J., 2010. Evergreen Agriculture: A robust approach to sustainable food security in Africa. *Food Security* 2: 197-214. http://dx.doi.org/10.1007/s12571-010-0070-7

Garrity, D.P., 2004. Agroforestry and the achievement of the Millennium Development Goals. *Agroforestry Systems* 61: 5-17. http://dx.doi.org/10.1007/978-94-017-2424-1_1

GEF, 2013. *Africa Will Import – not Export – Wood.* Washington DC: Global Environment Facility. http://www.globalenvironmentfund.com/wp-content/uploads/2013/05/GEF_Africa-will-Import-not-Export-Wood1.pdf

GI, 2015. Government of Indonesia. http://www.pertanian.go.id

Glanz, K., Sallis, J.F., Saelens, B.E. and Frank, L.D., 2005. Healthy nutrition environments: Concepts and measures. *American Journal of Health Promotion* 19: 330-333. http://dx.doi.org/10.4278/0890-1171-19.5.330

GN (Government of Nepal), 2004. *Nepal Living Standards Survey 2003-2004.* Statistical Report. Kathmandu: CBS, National Planning Commission Secretariat, Government of Nepal.

Golden, C.D., Fernald, L.C.H, Brashares, J.S., Rasolofoniaina, B.J.R. and Kremen, C., 2011. Benefits of wildlife consumption to child nutrition in a biodiversity hotspot. *Proceedings of the National Academy of Sciences of the USA* 108: 19653-19656. http://dx.doi.org/10.1073/pnas.1112586108

Grivetti, L.E. and Ogle, B.M., 2000. Value of traditional foods in meeting macro- and micronutrient needs: The wild plant connection. *Nutrition Research* Reviews 13: 31-46. http://dx.doi.org/10.1079/095442200108728990

Gumbo, D.J., Moombe, K.B., Kandulu, M.M., Kabwe, G., Ojanen, M., Ndhlovu, E. and Sunderland, T.C.H., 2013. *Dynamics of the Charcoal and Indigenous Timber Trade in Zambia: A Scoping Study in Eastern, Northern and Northwestern Provinces'* Occasional Paper. Bogor: CIFOR. http://dx.doi.org/10.17528/cifor/004113

Gyau, A., Takoutsing, B., De Grande, A. and Franzel, S., 2012. Farmers motivation for collective action in the production and marketing of kola in Cameroon. *Journal of Agriculture and Rural Development in the Tropics and Sub Tropics* 113: 43-50. http://dx.doi.org/10.5539/jas.v4n4p117

Hagen, M. and Kraemer, M., 2010. Agricultural surroundings support flower-visitor networks in an Afrotropical rain forest. *Biological Conservation* 143: 54-63. http://dx.doi.org/10.1016/j.biocon.2010.03.036

Harwood, C.E. and Nambiar, E.K.S., 2014. Productivity of acacia and eucalypt plantations in South East Asia: Trends and variations. *International Forestry Review* 16: 249-260. http://dx.doi.org/10.1505/146554814811724766

Hawkes, C., 2013. *Promoting Healthy Diets through Nutrition Education and Changes in the Food Environment: An International Review of Actions and their Effectiveness.* Nutrition Education and Consumer Awareness Group, Rome: FAO. http://www.fao.org/3/a-i3235e.pdf

Hein, L. and Gatzweiler, F., 2006. The economic value of coffee (*Coffea arabica*) genetic resources. *Ecological Economics* 60: 176-185. http://dx.doi.org/10.1016/j.ecolecon.2005.11.022

Herforth, A. and Dufour, C., 2013. *Key Recommendations for Improving Nutrition through Agriculture: Establishing a Global Consensus SCN News.* Rome: FAO. http://www.unscn.org/files/Publications/SCN_News/SCNNEWS40_final_standard_res.pdf

Heubach, K., Wittig, R., Nuppenau, E. and Hahn, K., 2011. The economic importance of non-timber forest products for livelihood maintenance of rural West African communities: A case study from northern Benin. *Ecological Economics* 70: 1991-2001. http://dx.doi.org/10.1016/j.ecolecon.2011.05.015

Hladik, C. M., Hladik, A., Linares, O., Pagezy, H., Semple, A. and Hadley, M. (Eds.), 1993. *Tropical Forests, People and Food: Biocultural Interactions and Applications to Development.* Carnforth, UK: Parthenon Publishing Group.

Ho, S-T., Tung, Y-T., Chen, Y-L., Zhao, Y-Y., Chung, M-J. and Wu, J-H., 2012. Antioxidant Activities and Phytochemical Study of Leaf Extracts from 18 Indigenous Tree Species in Taiwan. *Evidence-Based Complementary and Alternative Medicine.* Article ID 215959: 8. http://dx.doi.org/10.1155/2012/215959

ICCO, 2015. International Cacao Organization. http://www.icco.org

Ickowitz, A., Powell, B., Salim, M.A. and Sunderland, T.C.H., 2014. Dietary quality and tree cover in Africa. *Global Environmental Change* 24: 287. 294. http://dx.doi.org/10.1016/j.gloenvcha.2013.12.001

ICO, 2015. International Coffee Organization. http://www.ico.org

IEA, 2006. *World Energy Outlook*. Paris: IEA/OECD. http://dx.doi.org/10.1787/weo-2006-en

Iiyama, M., Neufeldt, H., Dobie, P., Jamnadass, R., Njenga, M. and Ndegwa, G., 2014a. The potential of agroforestry in the provision of sustainable woodfuel in sub-Saharan Africa. *Current Opinion in Environmental Sustainability* 6: 138-147. http://dx.doi.org/10.1016/j.cosust.2013.12.003

Iiyama, M., Chenevoy, A., Otieno, E., Kinyanjui, T., Ndegwa, G., Vandenabeele, J. and Johnson, O., 2014b. *Achieving Sustainable Charcoal in Kenya: Harnessing the Opportunities for Crosssectoral Integration*. Nairobi, Kenya: ICRAF-SEI Technical Brief. http://www.sei-international.org/mediamanager/documents/Publications/ICRAF-SEI-2014-techbrief-Sustainable-charcoal.pdf

Ingram, V., NdumbeNjie, L. and EwaneElah, M., 2012. Small scale, high value: Gnetumafricanum and buchholzianum value chains in Cameroon. *Small-scale Forestry* 11(4): 539-566. http://dx.doi.org/10.1007/s11842-012-9200-8

Ingram, V., Schure, J., Tieguhong, J.C., Ndoye, O., Awono, A. and Iponga, D.M., 2014. Gender implications of forest product value chains in the Congo Basin. *Forest, Trees and Livelihoods* 23(1-2): 67-86. http://dx.doi.org/10.1080/14728028.2014.887610

Iwamura, T., Lambin, E. F., Silvius, K. M., Luzar, J. B. and Fragoso, J. M. V., 2014. Agent-based modeling of hunting and subsistence agriculture on indigenous lands: Understanding interactions between social and ecological systems. *Environmental Modelling & Software*, 58: 109-127. http://dx.doi.org/10.1016/j.envsoft.2014.03.008

Jagoret, P., Michel-Dounias, I. and Malézieux, E., 2011. Long-term dynamics of cocoa agroforests: A case study in central Cameroon. *Agroforestry Systems* 81: 267-278. http://dx.doi.org/10.1007/s10457-010-9368-x

Jagoret, P., Michel-Dounias, I., Snoeck, D., Todem Ngnogué, H. and Malézieux, E., 2012. Afforestation of savannah with cocoa agroforestry systems: A small-farm innovation in central Cameroon. *Agroforestry Systems* 86: 493-504. http://dx.doi.org/10.1007/s10457-012-9513-9

Jamnadass, R., Langford, K., Anjarwalla, P. and Mithöfer, D., 2014. Public–Private Partnerships in Agroforestry. In: *Encyclopedia of Agriculture and Food Systems*, edited by N. Van Alfen Vol. 4, San Diego: Elsevier, 544-564. http://dx.doi.org/10.1016/b978-0-444-52512-3.00026-7

Jamnadass, R., Dawson, I.K., Anegbeh, P., Asaah, E., Atangana, A., Cordeiro, N., Hendrickx, H., Henneh, S., Kadu, C.A.C., Kattah, C., Misbah, M., Muchugi, A., Munjuga, M., Mwaura, L., Ndangalasi, H.J., Sirito Njau, C., Kofi Nyame, S., Ofori, D., Peprah, T., Russell, J., Rutatina, F., Sawe, C., Schmidt, L., Tchoundjeu, Z. and Simons, T., 2010. Allanblackia, a new tree crop in Africa for the global food industry: Market development, smallholder cultivation and biodiversity management. *Forests, Trees and Livelihoods* 19: 251-268. http://dx.doi.org/10.1080/14728028.2010.9752670

Jamnadass, R., Place, F., Torquebiau, E., Malézieux, E., Iiyama, M., Sileshi, G.W., Kehlenbeck, K., Masters, E., McMullin, S. and Dawson, I.K., 2013. Agroforestry for food and nutritional security. *Unasylva* 241(64): 2. http://www.fao.org/forestry/37082-04957fe26afbc90d1e9c0356c48185295.pdf

Jamnadass, R.H., Dawson, I.K., Franzel, S., Leakey, R.R.B., Mithöfer, D., Akinnifesi, F.K. and Tchoundjeu, Z., 2011. Improving livelihoods and nutrition in sub-Saharan Africa through the promotion of indigenous and exotic fruit production in smallholders 'agroforestry systems: A review. *International Forest Review* 13: 338-354. http://dx.doi.org/10.1505/146554811798293836

Johns, T. and Eyzaguirre, P., 2006. Linking biodiversity: Diet and health in policy and practice. *Proceedings of the Nutrition Society* 65: 182-189. http://dx.doi.org/10.1079/pns2006494

Johnson, D.V., 2010. The contribution of edible forest insects to human nutrition and to forest management: Current status and future potential. In: *Forest Insects as Food: Humans Bite Back*, edited by P.B. Durst, D.V. Johnson, R.N. Leslie and K. Shono. Proceedings of a workshop on Asia-Pacific resources and their potential for development, February 2008. Food and Agriculture Organization of the United Nations, Regional Office for Asia and the Pacific, Chiang Mai, Thailand. http://www.fao.org/docrep/013/i1380e/I1380e00.pdf

Johnson, K.B., Jacob, A. and Brown, M.W., 2013. Forest cover associated with improved child health and nutrition: Evidence from Malawi demographic and health survey and satellite data. *Global Health: Science and Practice* 1: 237- 248. http://dx.doi.org/10.9745/ghsp-d-13-00055

Kamanga, P., Vedeld, P. and Sjaastad, E., 2009. Forest incomes and rural livelihoods in Chiradzulu District, Malawi. *Ecological Economics* 68: 613-624. http://dx.doi.org/10.1016/j.ecolecon.2008.08.018

Kar, S.P. and Jacobson, M.G., 2012. NTFP income contribution to household economy and related socio-economic factors: Lessons from Bangladesh. *Forest Policy and Economics* 14: 136-142. http://dx.doi.org/10.1016/j.forpol.2011.08.003

Kawarazuka, N. and Béné, C., 2011. The potential role of small fish species in improving micronutrient deficiencies in developing countries: Building evidence. *Public Health Nutrition* 14: 1927-1938. http://dx.doi.org/10.1017/s1368980011000814

Keatinge, J.D.H., Waliyar, F., Jamnadass, R.H., Moustafa, A., Andrade, M., Drechsel, P., Hughes, J.A., Kadirvel, P. and Luther, K., 2010. Relearning old lessons for the future of food—by bread alone no longer: Diversifying diets with fruit and vegetables. *Crop Science* 50: S51–S62. https://dl.sciencesocieties.org/publications/cs/articles/50/Supplement_1/S-51

Kehlenbeck, K. and Jamnadass R., 2014. Chapter 6.2.1 Food and Nutrition – Fruits, Nuts, Vegetables and Staples from Trees. In: *Treesilience: An Assessment of the Resilience Provided by Trees in the Drylands of Eastern Africa*, edited by J. De Leeuw, M. Njenga, B. Wagner, M., Iiyama. Nairobi, Kenya: ICRAF, 166. http://www.worldagroforestry.org/downloads/Publications/PDFS/B17611.pdf

Kehlenbeck, K., Kindt, R., Sinclair, F.L., Simons, A.J. and Jamnadass, R., 2011. Exotic tree species displace indigenous ones on farms at intermediate altitudes around Mount Kenya. *Agroforestry Systems* 83: 133-147. http://dx.doi.org/10.1007/s10457-011-9413-4

Keller, G.B., Mndiga, H. and Maass, B., 2006. Diversity and genetic erosion of traditional vegetables in Tanzania from the farmer's point of view. *Plant Genetic Resources* 3: 400-413. http://dx.doi.org/10.1079/pgr200594

Kennedy, G.L., Pedro, M. R., Sehieri, C., Nantel, G. and Brouwer, I., 2007. Dietary diversity score is a useful indicator of micronutrient intake in non-breast-feeding Filipino children. *Journal of Nutrition* 137: 472-477. http://jn.nutrition.org/content/137/2/472.full

Khare, R.S., 1980. Food as nutrition and culture: Notes towards an anthropological methodology. *Social Science Information* 19: 519-542. http://dx.doi.org/10.1177/053901848001900303

Khundi, F., Jagger, P., Shively, G. and Sserunkuuma, D., 2011. Income, poverty and charcoal production in Uganda. *Forest Policy and Economics* 13: 199-205. http://dx.doi.org/10.1016/j.forpol.2010.11.002

Kiptot, E. and Franzel, S., 2011. *Gender and Agroforestry in Africa: Are Women Participating?* ICRAF Occasional Paper No. 13. Nairobi: ICRAF. http://dx.doi.org/10.5716/op16988

Klein, A.M., Vaissière, B.E., Cane, J.H., Steffan-Dewenter, I., Cunningham, S.A., Kremen, C. and Tscharntke, T., 2007. Importance of pollinators in changing landscapes for world crops. *Proceedings of the Royal Society B: Biological Sciences* 274: 303-313. http://dx.doi.org/10.1098/rspb.2006.3721

Köhlin, G., Sills, E. O., Pattanayak, S. K. and Wilfong, C., 2011. *Energy, Gender and Development. What Are the Linkages? Where is the Evidence?* Social Development World Bank Policy Research Working Paper 1-63. http://dx.doi.org/10.1596/1813-9450-5800

Kolavalli, S. and Vigneri, M., 2011. Cocoa in Ghana: Shaping the success of an economy. In: *Yes Africa Can: Success Stories from a Dynamic Continent*, edited by P. Chuhan-Pole and M. Angwafo. Washington DC: The World Bank. http://dx.doi.org/10.1596/978-0-8213-8745-0

Koppert, G.J.A., Dounias. E., Froment, A. and Pasquet, P., 1996. Consommation alimentaire dans trois populations forestières de la région côtière du Cameroun: Yassa, Mvae et Bakola. In: *L'alimentation en forêt tropicale. Interactions bioculturelles et perspectives de développement*, edited by C.M. Hladik, A. Hladik and H. Pagezy. Paris: UNESCO, 477-496. https://halshs.archives-ouvertes.fr/hal-00586904/document

Kuhnlein, H., Erasmus, B. and Spigelski, D., 2009. *Indigenous Peoples' Food Systems: The Many Dimensions of Culture, Diversity and Environment for Nutrition and Health*. Rome: FAO. ftp://ftp.fao.org/docrep/fao/012/i0370e/i0370e.pdf

Kuhnlein, H.V. and Receveur, O., 1996. Dietary change and traditional food systems of Indigenous Peoples. *Annual Review of Nutrition* 16: 417-442. http://dx.doi.org/10.1146/annurev.nu.16.070196.002221

Kusters, K., Achdiawan, R., Belcher, B. and Ruiz Pérez, M., 2006. Balancing development and conservation? An assessment of livelihood and environmental outcomes of nontimber forest product trade in Asia, Africa, and Latin America. *Ecology and Society* 11(2): 20. http://www.ecologyandsociety.org/vol11/iss2/art20/

Labouisse, J., Bellachew, B., Kotecha, S. and Bertrand, B., 2008. Current status of coffee (*Coffea arabica* L.) genetic resources in Ethiopia: Implications for conservation. *Genetic Resources and Crop Evolution* 55: 1079-1093. http://dx.doi.org/10.1007/s10722-008-9361-7

Leakey, R.R., 1999. Potential for novel food products from agroforestry trees: A review. *Food Chemistry* 66: 1-14. http://dx.doi.org/10.1016/s0308-8146(98)00072-7

Lengkeek, A.G., Carsan, S. and Jaenicke, H., 2003. A wealth of knowledge: How farmers in Meru, central Kenya, manage their tree nurseries. In: *Diversity Makes a Difference: Farmers Managing Inter- and Intra-Specific Tree Species Diversity in Meru Kenya*, edited by A.G., Lengkeek. PhD Thesis, Wageningen: Wageningen University.

Lescuyer, G., Cerutti, P.O., Ndotit Manguiengha, S. and Bilogo bi Ndong, L., 2011. *The Domestic Market for Small-scale Chainsaw Milling in Gabon Present Situation: Opportunities and Challenges*. Occasional Paper 65. Bogor: CIFOR. http://dx.doi.org/10.17528/cifor/003421

Lock, J., Agras, W.S., Bryson, S. and Kraemer, H., 2005. A comparison of short- and long-term family therapy for adolescent anorexia nervosa. *Journal of the American Academy of Child and Adolescent Psychiatry* 44: 632-639. http://dx.doi.org/10.1097/01.chi.0000161647.82775.0a

Lybbert, T.J., Aboudrare, A., Chaloud, D., Magnan, N. and Nash, M., 2011. Booming markets for Moroccan argan oil appear to benefit some rural households while threatening the endemic argan forest. *Proceedings of the National Academy of Sciences USA*. 108: 13963-13968. http://dx.doi.org/10.1073/pnas.1106382108

Lykke, A.M., Mertz, O. and Ganaba, S., 2002. Food consumption in rural Burkina Faso. *Ecology of Food Nutrition* 41: 119-153. http://dx.doi.org/10.1080/03670240214492

MA (Millennium Ecosystem Assessment), 2005. *Ecosystems and Human Well-being: Synthesis*. Washington DC: Island Press. http://www.millenniumassessment.org/documents/document.356.aspx.pdf

Marjokorpi, A. and Ruokolainen, K., 2003. The role of traditional forest gardens in the conservation of tree species in West Kalimantan: Indonesia. *Biodiversity and Conservation* 12: 799-822. http://dx.doi.org/10.1023/a:1022487631270

Marshall, D., Schreckenberg, K. and Newton, A.C., 2006. *Commercialization of Non-timber Forest Products: Factors Influencing Success. Lessons Learned from Mexico and Bolivia and Policy Implications for Decision-makers.* Cambridge: UNEP World Conservation Monitoring Centre. http://dx.doi.org/10.5962/bhl.title.44910

McCullough, F.S.W., Yoo, S. and Ainsworth, P., 2004. Food choice: Nutrition education and parental influence on British and Korean primary school children. *International Journal of Consumer Studies* 28: 235-244. http://dx.doi.org/10.1111/j.1470-6431.2003.00341.x

McMullin, S., Phelan, J., Jamnadass, R., Iiyama, M., Franzel, S. and Nieuwenhuis, M., 2012. Trade in medicinal tree and shrub products in three urban centres in Kenya. *Forests, Trees and Livelihoods* 21(3): 188-206. http://dx.doi.org/10.1080/14728028.2012.733559

McMullin, S., Nieuwenhuis, M. and Jamnadass. R., 2014. Strategies for sustainable supply of and trade in threatened medicinal tree species: A case study of Genus Warburgia. *Ethnobotany Research and Applications* 12: 671-683. http://journals.sfu.ca/era/index.php/era/article/viewFile/1000/650

Mercer, C.W.L., 1997. Sustainable production of insects for food and income by New Guinea villagers. *Ecology of Food and Nutrition* 36: 151-157. http://dx.doi.org/10.1080/03670244.1997.9991512

Millard, E., 2011. Incorporating agroforestry approaches into commodity value chains. *Environmental Management* 48: 365-377. http://dx.doi.org/10.1007/s00267-011-9685-5

Milner-Gulland, E.J. and Bennett, E.L., 2003. Wild meat: The bigger picture. *Trends in Ecology and Evolution* 18: 351-357. http://dx.doi.org/10.1016/s0169-5347(03)00123-x

Mitra, A. and Mishra, D.K., 2011. Environmental resource consumption pattern in rural Arunachal Pradesh. *Forest Policy and Economics* 13: 166-170. http://dx.doi.org/10.1016/j.forpol.2010.11.001

Mitchell, M.G.E., Bennett, E.M. and Gonzalez, A., 2014. Forest fragments modulate the provision of multiple ecosystem services. *Journal of Applied Ecology* 51: 909-918. http://dx.doi.org/10.1111/1365-2664.12241

Mohan Jain, S. and Priyadarshan, P.M., 2009. Breeding Plantation Tree Crops: Tropical Species. New York: *Springer Science & Business Media.* http://dx.doi.org/10.1007/978-0-387-71201-7

Moran, E. F., 1993. Managing Amazonian Variability with Indigenous Knowledge. In: *Tropical Forests, People and Food. Biocultural Interactions and Applications to Development*, edited by C.M. Hadlik, A. Hladik, O.F. Linares, H. Pagezy, A. Semple and M. Hadley. Vol. 765: 753-766. Indiana: Anthropological Centre for Training and Research on Global Environmental Change Indiana University. https://halshs.archives-ouvertes.fr/hal-00586888/document

Mulenga, B.P., Richardson, R.B. and Tembo, G., 2012. *Non-timber forest products and rural poverty alleviation in Zambia.* Working Paper No. 62. http://fsg.afre.msu.edu/zambia/wp62.pdf

Mulenga, B.P., Richardson, R.B., Mapemba, L. and Tembo, G., 2011. *The Contribution of Non-timber Forest Products to Rural Household Income in Zambia.* Food Security Research Project Lusaka, Zambia. Working Paper No. 54. http://ageconsearch.umn.edu/bitstream/109887/2/wp54.pdf

Muleta, D., 2007. *Microbial Inputs in Coffee (Coffea arabica L.) Production Systems, Southwestern Ethiopia: Implications for Promotion of Biofertilizers and Biocontrol Agents.* PhD thesis, Swedish University of Agricultural Sciences/Uppsala. http://pub.epsilon.slu.se/1657/1/Am_Thesis_template_LC_PUBL..pdf

Muriuki, J., Franzel, F., Mowo, J., Kariuki P. and Jamnadass, R., 2012. Formalisation of local herbal product markets has potential to stimulate cultivation of medicinal plants by smallholder farmers in Kenya. *Forests, Trees and Livelihoods* 21: 114-127. http://dx.doi.org/10.1080/14728028.2012.721959

Mwampamba, T.H., Ghilardi, A., Sander, K. and Chaix, K.J., 2013. Dispelling common misconceptions to improve attitudes and policy outlook on charcoal in developing countries. *Energy for Sustainable Development* 17: 158-170. http://dx.doi.org/10.1016/j.esd.2013.01.001

Narain, U.S., Gupta, S. and Veld, K.V., 2005. *Poverty and the Environment: Exploring the Relationship between Household Incomes, Private Assets and Natural Assets.* Discussion Paper 05-18. Washington DC: Resources for the Future. http://dx.doi.org/10.2139/ssrn.850071

Nasi, R., Taber, A. and Van Vliet, N., 2011. Empty forests, empty stomachs: Bushmeat and livelihoods in the Congo and Amazon Basins. *International Forestry Review* 13(3): 355-368. http://dx.doi.org/10.1505/146554811798293872

Nasi, R., Brown, D., Wilkie, D., Bennett, E., Tutin, C., van Tol, G. and Christophersen, T., 2008. *Conservation and Use of Wildlife-based Resources: The Bushmeat Crisis.* Technical Series No. 33. Secretariat of the Convention on Biological Diversity: Montreal, and Centre for International Forestry Research: Bogor, Indonesia. https://www.cbd.int/doc/publications/cbd-ts-33-en.pdf

Naughton-Treves, L., Kammen, D.M. and Chapman, C., 2007. Burning biodiversity: Woody biomass use by commercial and subsistence groups in western Uganda's forests. *Biological Conservation* 134: 232-241. http://dx.doi.org/10.1016/j.biocon.2006.08.020

Ndoye, O., Ruiz Pérez, M. and Eyebe, A., 1998. *The Markets of Non-timber Forest Products in the Humid Forest Zone of Cameroon.* Paper 22c. London: Rural Development Forestry Network. http://www.odi.org/sites/odi.org.uk/files/odi-assets/publications-opinion-files/1168.pdf

Njenga. M., Yonemitsu, A., Karanja, N., Iiyama, M., Kithinji, J., Dubbeling M., Sundberg, C. and Jamnadass, R., 2013. Implications of charcoal briquette produced by local communities on livelihoods and environment in Nairobi, Kenya. *International Journal of Renewable Energy Development* 2(1): 19-29. http://dx.doi.org/10.14710/ijred.2.1.19-29

Ogle, B.M., Hung, P.H. and Tuyet, H.T., 2001. Significance of wild vegetables in micronutrient intakes of women in Vietnam: An analysis of food variety. *Asia Pacific Journal of Clinical Nutrition* 10: 21-30. http://dx.doi.org/10.1046/j.1440-6047.2001.00206.x

Owen, M., Van der Plas, R. and Sepp, S., 2013. Can there be energy policy in Sub-Saharan Africa without biomass? *Energy Sustainable Development* 17: 146-152. http://dx.doi.org/10.1016/j.esd.2012.10.005

Padoch, C. and Sunderland T., 2013. Managing landscapes for greater food security and livelihoods. *Unasylva* 241(64): 3-13.

Page, B., 2003. The political ecology of *Prunus africana* in Cameroon. *Area* 35: 357-370. http://dx.doi.org/10.1111/j.0004-0894.2003.00187.x

PEN, 2015. Poverty Environment Network. *A Comprehensive Global Analysis of Tropical Forests and Poverty.* http://www1.cifor.org/pen

Peres, C.A., 2001. Synergistic effects of subsistence hunting and habitat fragmentation on Amazonian forest vertebrates. *Conservation Biology* 15: 1490-1505. http://dx.doi.org/10.1046/j.1523-1739.2001.01089.x

Philpott, S.M., Arendt, W.J., Armbrecht, I., Bichier, P., Diestch, T.V., Gordon, C., Greenberg, R., Perfecto, I., Reynoso-Santos, R., Soto-Pinto, L., Tejeda-Cruz, C., Williams-Linera, G., Valenzuelaa, J. and Zolotoff, J.M., 2008. Biodiversity loss in Latin American coffee landscapes: Review of the evidence on ants, birds, and trees. *Conservation Biology* 22: 1093-1105. http://dx.doi.org/10.1111/j.1523-1739.2008.01029.x

Pimentel, D., McNair, M., Buck, L., Pimentel, M. and Kamil, J., 1997. The value of forests to world food security. *Human Ecology* 25: 91-120. http://dx.doi.org/10.1023/a:1021987920278

Pinstrup-Andersen, P., 2013. Special debate section, can agriculture meet future nutrition challenges? *European Journal of Development Research* 25(1): 5-12. http://dx.doi.org/10.1057/ejdr.2012.44

Pinstrup-Andersen, P., 2009. Food security: Definition and measurement. *Food Security* 1: 5-7. http://dx.doi.org/10.1007/s12571-008-0002-y

Place, F. and Binam, J.N., 2013. *Economic Impacts of Farmer Managed Natural Regeneration in the Sahel.* End of project technical report for the Free University Amsterdam and IFAD. Nairobi: ICRAF.

Place, F., Roothaert, R., Maina, L., Franzel, S., Sinja, J. and Wanjiku, J., 2009. *The Impact of Fodder Trees on Milk Production and Income among Smallholder Dairy Farmers in East Africa and the Role of Research.* ICRAF Occasional Paper No. 12. Nairobi: ICRAF. http://impact.cgiar.org/pdf/217.pdf

Pouliot, M., 2012. Contribution of "Women's Gold" to West African livelihoods: The case of Shea (Vitellaria paradoxa) in Burkina Faso. *Economic Botany* 66(3): 237-248. http://dx.doi.org/10.1007/s12231-012-9203-6

Pouliot, M. and Treue, T., 2013. Rural people's reliance on forests and the non-forest Environment in West Africa: Evidence from Ghana and Burkina Faso. *World Development* 43: 180-193. http://dx.doi.org/10.1016/j.worlddev.2012.09.010

Powell, B., Ickowitz, A., McMullin, S., Jamnadass, R., Padoch, C., Pinedo-Vasquez, M. and Sunderland, T., 2013. *The Role of Forests, Trees and Wild Biodiversity for Improved Nutrition: Sensitivity of Food and Agriculture Systems.* Expert background paper for the International Conference on Nutrition 2. FAO: Rome.

Powell, B., Watts, J., Boucard, A., Urech, Z., Feintrenie, L., Lyimo, E., Asaha, S. and Sunderland-Groves, J., 2010. The role of wild species in the governance of tropical forested landscapes. In: *Collaborative Governance of Tropical Landscapes*, edited by C.J.P. Colfer and J.L. Pfund. London: Earthscan Publications Ltd. Chapter 7. http://dx.doi.org/10.4324/9781849775601

Powell, L.M., Schermbeck, R.M., Szczypka, G., Chaloupka, F.J. and Braunschweig, C.L., 2011. Trends in the nutritional content of television food advertisements seen by children in the United States: Analyses by age, food categories, and companies. *Arch Pediatrics and Adolescent Medicine* 165(12): 1078-1086. http://dx.doi.org/10.1001/archpediatrics.2011.131

Pramova, E., Locatelli, B., Djoudi, H. and Somorin, O.A., 2012. Forests and trees for social adaptation to climate variability and change. *Wiley Interdisciplinary Reviews: Climate Change* 3(6): 581-596. http://dx.doi.org/10.1002/wcc.195

Priess, J.A., Mimler, M., Klein, A.M., Schwarze, S., Tscharntke, T. and Steffan-Dewenter, I., 2007. Linking deforestation scenarios to pollination services and economic returns in coffee agroforestry systems. *Ecological Applications* 17: 407-417. http://dx.doi.org/10.1890/05-1795

Reichhuber, A. and Requate, T., 2007. *Alternative Use Systems for the Remaining Cloud Forest in Ethiopia and the Role of Arabica Coffee: A Cost-benefit Analysis.* Economics Working Paper No. 7, 2007. Kiel: Department of Economics, Christian-Albrechts-Universität. http://dx.doi.org/10.1016/j.ecolecon.2012.01.006

Ricketts, T.H., Daily, G.C., Ehrlich, P.R. and Michener, C.D., 2004. Economic value of tropical forest to coffee production. *Proceedings of the National Academy of Sciences of the USA* 101: 12579-12582. http://dx.doi.org/10.1073/pnas.0405147101

Roshetko, J., Dawson, I.K., Urquiola, J., Lasco, R., Leimona, B., Graudal, L., Bozzano, M., Weber, J.C. and Jamnadass, R., 2015. Tree planting for environmental services: Considering source to improve outcomes (in preparation).

Roshetko, J.M., Rohadi, D., Perdana, A., Sabastian, G., Nuryartono, N., Pramono, A.A., Widyani, N., Manalu, P., Fauzi, M.A., Sumardamto, P. and Kusumowardhani, N., 2013. Teak agroforestry systems for livelihood enhancement, industrial timber production, and environmental rehabilitation. *Forests, Trees, and Livelihoods* 22(4) http://dx.doi.org/10.1080/14728028.2013.855150

Roshetko, J.M., Lasco, R.D. and Delos Angeles, M.S., 2007. Smallholder agroforestry systems for carbon storage. *Mitigation and Adaptation Strategies for Global Climate Change* 12: 219-242. http://dx.doi.org/10.1007/s11027-005-9010-9

Ruel, M.T. and Alderman, H., 2013. Nutrition-sensitive interventions and programmes: How can they help to accelerate progress in improving maternal and child nutrition? *The Lancet* 382: 536-551. http://dx.doi.org/10.1016/s0140-6736(13)60843-0

Ruel, M.T., 2003. Operationalizing dietary diversity: A review of measurement issues and research priorities. *Journal of Nutrition* 133: 3911S-3926S. PMID: 14672290

Ruel, M.T., Minot, N. and Smith, L., 2005. *Patterns and Determinants of Fruit and Vegetable Consumption in Sub-Saharan Africa: A Multi-country Comparison.* Geneva: World Health Organization and Washington: IFPRI. http://www.who.int/dietphysicalactivity/publications/f%26v_africa_economics.pdf

Sambuichi, R.H.R. and Haridasan, M., 2007. Recovery of species richness and conservation of native Atlantic forest trees in the cacao plantations of southern Bahia in Brazil. *Biodiversity and Conservation* 16: 3681-3701. http://dx.doi.org/10.1007/s10531-006-9017-x

Sanchez, P.A., 2002. Soil fertility and hunger in Africa. *Science* 295: 2019-2020. http://dx.doi.org/10.1126/science.1065256

Schabel, H.G., 2010. Forests insects as food: A global review. In: *Forest Insects as Food: Humans Bite Back*, edited by P.B. Durst, D.V. Johnson, R.N. Leslie and K. Shono. Proceedings of a workshop on Asia-Pacific resources and their potential for development, 19-21 February 2008: 37-64. http://www.fao.org/docrep/013/i1380e/I1380e00.pdf

Schreckenberg, K. and Luttrell, C., 2009. Participatory forest management: A route to poverty reduction? *International Forestry Review* 11: 221-238. http://dx.doi.org/10.1505/ifor.11.2.221

Schreckenberg, K., Awono, A., Degrande, A., Mbosso, C., Ndoye, O. and Tchoundjeu, Z., 2006. Domesticating indigenous fruit trees as a contribution to poverty reduction. *Forests, Trees and Livelihoods* 16: 35-51. http://dx.doi.org/10.1080/14728028.2006.9752544

Schure, J., Ingram, V., Assembe-Mvondo, S., Mvula-Mampasi, E., Inzamba, J. and Levang, P. 2013. La Filière Bois Énergie des Villes de Kinshasa et Kisangani. In: *Quand la ville mange la forêt* edited by J.-N. Marien, E. Dubiez, D. Louppe and A. Larzilliere. Versailles: Editions QUAE, 27-44.

Schure, J., Levang, P. and Wiersum, K.F., 2014. Producing woodfuel for urban centres in the Democratic Republic of Congo: A path out of poverty for rural households. *World Development* 64: 80-90. http://dx.doi.org/10.1016/j.worlddev.2014.03.013

SCI, 2015. Sustainable Cocoa Initiative. http://cocoasustainability.com

Sendzimir, J., Reij, C.P. and Magnuszewski, P., 2011. Rebuilding resilience in the Sahel: Regreening in the Maradi and Zinder regions of Niger. *Ecology and Society* 16(3): 1. http://dx.doi.org/10.5751/es-04198-160301

Sepp, S., 2008. *Shaping Charcoal Policies: Context, Process and Instruments as Exemplified by Country Cases.* Bonn: GTZ. https://energypedia.info/images/1/1e/Shaping_charcoal_policies.pdf

Shackleton, C. and Shackleton, S., 2004. The importance of non-timber forest products in rural livelihood security and as safety nets: A review of evidence from South-Africa. *Southern Africa Journal of Science* 100: 658-664. http://pdf.wri.org/ref/shackleton_04_the_importance.pdf

Shackleton, C.M., Shackleton, S.E., Buitenb, E. and Bird, N., 2007. The importance of dry woodlands and forests in rural livelihoods and poverty alleviation in South Africa. *Forest Policy and Economics* 9(5): 558-577. http://dx.doi.org/10.1016/j.forpol.2006.03.004

Shackleton, S., Paumgarten, F., Kassa, H., Husseelman, M. and Zida, M., 2011. Opportunities for enhancing poor women's socioeconomic empowerment in the value chains of three African non timber forest products. *International Forestry Review* Special Issue 13. http://dx.doi.org/10.1505/146554811797406642

Shaheen, N., Rahim, A.T., Abu Torab, M.A.R., Mohiduzzaman, M.D., Banu, C.P., Bari, M.D.L., Tukan, A.B., Mannan, M.A., Bhattacharjee, L., and Stadlmayr, B., 2013. *Food Composition Table for Bangladesh*. Dhaka: University of Dhaka, Bangladesh.

Shepard, G.H. and Ramirez, H., 2011. Made in Brazil: Human dispersal of the Brazil nut (*Bertholletiaexcelsa*, Lecythidaceae) in ancient Amazonia. *Economic Botany* 65: 44-65. http://dx.doi.org/10.1007/s12231-011-9151-6

Sherman, J., 2003. From nutritional needs to classroom lessons: Can we make a difference? *Food, Nutrition and Agriculture* 33: 45-51. ftp://ftp.fao.org/docrep/fao/006/j0243m/j0243m06.pdf

Siegel, K.R., Ali, M.K., Srinivasiah, A., Nugent, R.A. and Narayan, K.M.V., 2014. Do we produce enough fruits and vegetables to meet global health need? *PLoS ONE* 9(8): e104059. http://dx.doi.org/10.1371/journal.pone.0104059

Sileshi, G., Akinnifesi, F.K., Ajayi, O.C. and Place, F., 2008. Meta-analysis of maize yield response to planted fallow and green manure legumes in sub-Saharan Africa. *Plant and Soil* 307: 1-19. http://dx.doi.org/10.1007/s11104-008-9547-y

Sileshi, G.W., Akinnifesi, F.K., Ajayi, O.C. and Muys B., 2011. Integration of legume trees in maize-based cropping systems improves rain-use efficiency and yield stability under rain-fed agriculture. *Agricultural Water Management* 98: 1364-1372. http://dx.doi.org/10.1016/j.agwat.2011.04.002

Sileshi, G.W., Debusho, L.K. and Akinnifesi, F.K., 2012. Can integration of legume trees increase yield stability in rainfed maize cropping systems in Southern Africa? *Agronomy Journal* 104: 1392-1398. http://dx.doi.org/10.2134/agronj2012.0063

Sirén, A., 2012. Festival hunting by the kichwa people in the Ecuadorian amazon. *Journal of Ethnobiology* 32: 30-50. http://dx.doi.org/10.2993/0278-0771-32.1.30

Slavin, J.L. and Lloyd, B., 2012. Health benefits of fruits and vegetables. *Advances in Nutrition* 3(4): 506-16. http://dx.doi.org/10.3945/an.112.002154

Smith, I.F. 2013. Sustained and integrated promotion of local, traditional food systems for nutrition security. In: *Diversifying Food and Diets: Using Agricultural Biodiversity to Improve Nutrition and Health Issues in Agricultural Biodiversity*, edited by J. Fanzo, D. Hunter, T. Borelli and F. Mattei. London: Earthscan Publications Ltd. 122-139. http://www.bioversityinternational.org/uploads/tx_news/Diversifying_food_and_diets_1688_02.pdf

Sobal, J., Bisogni, C.A., and Jastran, M., 2014. Food choice is multifaceted: Contextual, dynamic, multilevel, integrated, and diverse. *Mind, Brain, and Education* 8: 6-12. http://dx.doi.org/10.1111/mbe.12044

Sonwa, D.S., Nkongmeneck, B.A., Weise, S.F., Tchatat, M., Adesina, A.A. and Janssens, M.J.J., 2007. Diversity of plants in cocoa agroforests in the humid forest zone of Southern Cameroon. *Biodiversity Conservation* 16: 2385-2400. http://dx.doi.org/10.1007/s10531-007-9187-1

Stadlmayr, B., Charrondière, U.R., Eisenwagen, S., Jamnadass, R. and Kehlenbeck, K., 2013. Review: Nutrient composition of selected indigenous fruits from sub-Saharan Africa. *Journal of the Science of Food and Agriculture* 93: 2627-2636. http://dx.doi.org/10.1002/jsfa.6196

Story, M., Kaphingst, K.M., Robinson-O'Brien, R. and Glanz, K., 2008. Creating healthy food and eating environments: Policy and environmental approaches. *Annual Review of Public Health* 29: 253-272. http://dx.doi.org/10.1146/annurev.publhealth.29.020907.090926

Sunderland, T.C.H., 2011. Food security: Why is biodiversity important? *International Forestry Review* 13: 265-274. http://dx.doi.org/10.1505/146554811798293908

Tabuti, J.R.S., Dhillion, S.S. and Lye, K.A., 2003. Firewood use in Bulamogi County, Uganda: Species selection, harvesting and consumption patterns. *Biomass and Bioenergy* 25: 581-596. http://dx.doi.org/10.1016/s0961-9534(03)00052-7

Tchoundjeu, Z., Degrande, A., Leakey, R.R.B., Nimino, G., Kemajou, E., Asaah, E., Facheux, C., Mbile, P., Mbosso, C., Sado, T., and Tsobeng, A., 2010. Impacts of participatory tree domestication on farmer livelihoods in West and Central Africa. *Forests, Trees and Livelihoods* 19: 217-234. http://dx.doi.org/10.1080/14728028.2010.9752668

Termote, C., BwamaMeyi, M., Dhed'aDjailo, B., Huybregts, L., Lachat, C., Kolsteren, P. and Van Damme, P., 2012. A biodiverse rich environment does not contribute to better diets. A case study from DR Congo. *Plos One* 7: e30533. http://dx.doi.org/10.1371/journal.pone.0030533

Thomas, E., 2008. *Quantitative Ethnobotanical Research on Knowledge and Use of Plants for Livelihood among Quechua: Yuracaré and Trinitario Communities in the Andes and Amazon Regions of Bolivia.* PhD thesis, Ghent University: Ghent, Belgium. 10.13140/2.1.3577.0249

Thorlakson, T. and Neufeldt, H., 2012. Reducing subsistence farmers' vulnerability to climate change: Evaluating the potential contributions of agroforestry in western Kenya. *Agriculture and Food Security* 1: 15 http://dx.doi.org/10.1186/2048-7010-1-15

TRAFFIC, 2008. *What is Driving the Wildlife Trade: A Review of Expert Opinion on Economic and Social Drivers of the Wildlife Trade and Trade Control Efforts in Cambodia, Indonesia, Lao PDR and Vietnam.* East Asia and Pacific Region Sustainable Development Discussion Papers. Washington, DC: World Bank.

Tscharntke, T., Clough, Y., Wanger, T.C., Jackson, L., Motzke, I., Perfecto, I., Vandermeer, J., and Whitbread, A., 2012. Global food security, biodiversity conservation and the future of agricultural intensification. *Biological Conservation* 151(1): 53-59. http://dx.doi.org/10.1016/j.biocon.2012.01.068

Tscharntke,T., Klein, A., Kruess, M., Steffan-Dewenter, A. and Thies, C., 2005. Landscape perspectives on agricultural intensification and biodiversity: Ecosystem service management. *Ecology Letters* 8: 857-874. http://dx.doi.org/10.1111/j.1461-0248.2005.00782.x

Turner, W.R., Brandon, K., Brooks, T.M., Gascon, C., Gibbs, H.K., Lawrence, K., Mittermeier, R.A. and Selig, E.R., 2012. Global Biodiversity Conservation and the Alleviation of Poverty. *BioScience* 62: 85-92. http://dx.doi.org/10.1525/bio.2012.62.1.13

UN-SCN., 2010. *Sixth Report on the World Nutrition Situation: Progress in Nutrition*. Geneva, Switzerland: United Nations, Standing Committee on Nutrition and International Food Policy Research Institute. http://www.unscn.org/files/Publications/RWNS6/report/SCN_report.pdf

Van Noordwijk, M., Bizard, V., Wangpakapattanawong, P., Tata, H.L., Villamor, G.B. and Leimona, B., 2014. Tree cover transitions and food security in Southeast Asia. *Global Food Security* 3: 200-208. http://dx.doi.org/10.1016/j.gfs.2014.10.005

van Vliet, N., Quiceno-Mesa, MP., Cruz-Antia, D., Johnson Neves de Aquino, L., Moreno, J. and Nasi, R., 2014. The uncovered volumes of bushmeat commercialized in the Amazonian trifrontier between Colombia, Peru & Brazil. *Ethnobiology Conservation* 3: 7. http://dx.doi.org/10.15451/ec2014-11-3.7-1-11

Vandenbosch T., Ouko B.A., Guleid N.J., Were R.A., Mungai P.P., Mbithe D.D., Ndanyi M., Chesumo. J., Laenen-Fox L., Smets K., Kosgei C.K. and Walema B., 2009. Integrating food and nutrition: Health, agricultural and environmental education towards education for sustainable development: Lessons learned from the Healthy Learning Programme in Kenya. In: *Sustainable Initiatives in the Decade of Education for Sustainable Development: Where are We?* Proceedings of the July 2009 Environmental Education Association of Southern Africa (EEASA) conference.

Vedeld, P., Angelsen, A., Sjaastad, E. and Kobugabe Berg, G., 2004. *Counting on the Environment: Forest Incomes and the Rural Poor.* Environment Economics Series No. 98. World Bank, Washington DC, USA. http://www-wds.worldbank.org/servlet/WDSContentServer/WDSP/IB/2004/09/30/000090341_20040930105923/Rendered/PDF/300260PAPER0Counting0on0ENV0EDP0198.pdf

Vinceti, B., Termote, C., Ickowitz, A., Powell, B., Kehlenbeck, K. and Hunter, D., 2013. Strengthening the contribution of forests and trees to sustainable diets: Challenges and opportunities. *Sustainability* 5: 4797-4824.

Von Grebmer, Saltzman, K., Birol, A., Wiesmann, E., Prasai, D., Yin, N., Yohannes, S., Menon, Y., Thompson, P. and Sonntag, A., 2014. *2014 Global Hunger Index: The Challenge of Hidden Hunger.* Bonn, Washington, DC and Dublin: Welthungerhilfe, International Food Policy Research Institute and Concern Worldwide. http://dx.doi.org/10.2499/9780896299580

Wan, M., Colfer, C.J.P. and Powell, B., 2011. Forests, women and health: Opportunities and challenges for conservation. *International Forestry Review* Special Issue, 13. http://dx.doi.org/10.1505/146554811798293854

Waswa, F., Kilalo, C.R.S. and Mwasaru, D.M., 2014. *Sustainable Community Development: Dilemma of Options in Kenya.* Basingstoke: Palgrave Macmillan. http://dx.doi.org/10.1057/9781137497413

Webb, P. and Kennedy, E., 2012. *Impacts of Agriculture on Nutrition: Nature of the Evidence and Research Gaps.* Nutrition CRSP Research Brief No. 4. Boston: Tufts University. https://agrilinks.org/sites/default/files/resource/files/13_Webb_FNB_v35_n1_FINAL-PROOF_CE-+PW1.pdf

Webb Girard, A., Self, J.L., McAuliffe, C. and Oludea, O., 2012. The Effects of Household Food Production Strategies on the Health and Nutrition Outcomes of Women and Young Children: A Systematic Review. *Paediatric and Perinatal Epidemiology* 26 (Suppl. 1): 205-222. http://dx.doi.org/10.1111/j.1365-3016.2012.01282.x

WHO, 2004. *Global Strategy on Diet, Physical Activity and Health.* Geneva: WHO. http://www.who.int/dietphysicalactivity/strategy/eb11344/strategy_english_web.pdf

WHO, 2014. *Healthy Diet.* WHO Fact Sheets. http://www.who.int/mediacentre/factsheets/fs394/en/

WHO/FAO, 2004. *WHO/FAO Expert Consultation on Vitamin and Mineral Requirements in Human Nutrition* (2nd edition). Geneva: WHO.

Wilkie, D., Starkey, M., Abernethy, K., Nstame, E., Telfer, P. and Godoy, R., 2005. Role of prices and wealth in consumer demand for bushmeat in Gabon: Central Africa. *Conservation Biology* 19: 268-274. http://dx.doi.org/10.1111/j.1523-1739.2005.00372.x

World Bank, 2008. *Implementation, Completion and Results Report (TF-50612) on a Grant in the Amount of SDR 3.7 Million Equivalent (US $ 4.5 million) to Centro Agronomico Tropical De Investigacion Y Ensenanza (CATIE) for the Integrated Silvopastoral Approaches to Ecosystem Management Project in Columbia, Costa Rica, and Nicaragua.* Washington DC: World Bank.

World Bank, 2011. *Wood-based Biomass Energy Development for Sub-Saharan Africa: Issues and Approaches.* Washington, DC 20433, USA: The International Bank for Reconstruction and Development. http://siteresources.worldbank.org/EXTAFRREGTOPENERGY/Resources/717305-1266613906108/BiomassEnergyPaper_WEB_Zoomed75.pdf

Wrangham, E., 2009. *Catching Fire: How Cooking Made us Human.* New York: Basic Books.

Wunder, S., Borner, J., Shively, J. and Wyman, M., 2014. Safety nets, gap filling and forests: A global-comparative perspective. *World Development* 64(1): S29-S42. http://dx.doi.org/10.1016/j.worlddev.2014.03.005

Yemiru, T., Roos, A., Campbell, B.M. and Bohlin, F., 2010. Forest incomes and poverty alleviation under participatory forest management in the Bale highlands, Southern Ethiopia. *International Forestry Review* 12(1): 66-77. http://dx.doi.org/10.1505/ifor.12.1.66

Zapata-Rios, G., Urgiles, C. and Suarez, E., 2009. Mammal hunting by the Shuar of the Ecuadorian Amazon: Is it sustainable? *Oryx* 43: 357-385. http://dx.doi.org/10.1017/s0030605309001914

Zhang, W., Ricketts, T.H., Kremen, C., Carney, K. and Swinton, S.M., 2007. Ecosystem services and dis-services to agriculture. *Ecological Economics* 64: 253-260. http://dx.doi.org/10.1016/j.ecolecon.2007.02.024

Zulu, L.C. and Richardson R.B., 2013. Charcoal: Livelihoods and poverty reduction: Evidence from sub-Saharan Africa. *Energy for Sustainable Development* 17: 127-137. http://dx.doi.org/10.1016/j.esd.2012.07.007

3. The Historical, Environmental and Socio-economic Context of Forests and Tree-based Systems for Food Security and Nutrition

Coordinating lead author: *John A. Parrotta*

Lead authors: *Jennie Dey de Pryck, Beatrice Darko Obiri, Christine Padoch, Bronwen Powell and Chris Sandbrook*

Contributing authors: *Bina Agarwal, Amy Ickowitz, Katy Jeary, Anca Serban, Terry Sunderland and Tran Nam Tu*

Forests and tree-based systems are an important component of rural landscapes, sustaining livelihoods and contributing to the food security and nutritional needs of hundreds of millions of people worldwide. Historically, these systems developed under a wide variety of ecological conditions, and cultural and socio-economic contexts, as integrated approaches that combined management of forest and agricultural areas to provide primarily for the needs of producers and their local communities. Today they serve food and nutrition demands of growing global populations, both urban and rural. Population increase, globalisation, deforestation, land degradation, and ever-increasing demand and associated conflict for land (including forest) resources are placing pressure on these lands. Farmers have been encouraged to intensify food production on existing agricultural lands, by modifying some traditional practices (such as agroforestry) or abandoning others (such as shifting cultivation) that evolved over centuries to cope with biophysical constraints (e.g. limited soil fertility, climate variability) and changing socio-economic conditions. This chapter provides an overview of forests and tree-based systems and their role in enhancing food security and nutrition for rural communities and those served through the marketplace. The variability and viability of these management systems are considered within and across geographical regions and agro-ecological zones. Also discussed is the role of the social, cultural and economic contexts in which these systems exist, with a focus on three factors that affect the socio-economic organisation of forests and tree-based systems, namely: land and tree tenure and governance, human capital (including knowledge and labour) and financial capital (including credit). How these biophysical and socio-economic conditions and their complex interactions influence food security and nutrition outcomes, particularly for vulnerable segments of the population (i.e., the poor, women and children), are of particular concern.

 http://dx.doi.org/10.11647/OBP.0085.03

3.1 Introduction

Forests[1] and trees outside of forests have ensured the *food security* and *nutrition* of human populations since time immemorial. Throughout the world, forests and associated *ecosystems* have been managed to enhance their production of a vast array of wild, semi-domesticated and domesticated foods, including fruits, nuts, tubers, leafy vegetables, mushrooms, honey, insects, game animals, fish and other wildlife (discussed in detail in Chapter 2). The development and spread of crop agriculture and animal husbandry over the past few centuries, and particularly since the early 20th century, has diminished dependence on forests for food security and nutrition in many societies, particularly those relying primarily on staple crops. Nonetheless *forests and tree-based systems* – which generally co-exist in the *landscape* with other land management practices – continue to play a very important role for food security and nutrition, often complementing other food production systems, particularly on lands unsuited to other forms of agriculture due to soil productivity constraints.

The earth's diverse forest ecosystems and the human cultures associated with them through the course of history have produced a vast array of *food systems* connected to forests and trees. These forests and tree-based systems are based on the traditional wisdom, knowledge, practices and technologies of societies, developed and enriched through experimentation and adaptation to changing environmental conditions and societal needs over countless generations (Altieri, 2002; Berkes et al., 2000; Colfer et al., 2005; Galloway-McLean, 2010; Parrotta and Trosper, 2012). Traditional forest-related knowledge and farmer innovation have played a critical role in the development of highly diverse, productive and sustainable food production systems within and outside of forests (Anderson, 2006; Kuhnlein et al., 2009; Posey, 1999; Turner et al., 2011). Starting early in the 20th century, when anthropologists began documenting the ethnobotany and food production systems of indigenous and local communities worldwide, these forests and tree-based systems and the *traditional knowledge* upon which they are based have been "rediscovered" by a broader audience within the (formal) scientific community, principally among agricultural scientists and ecologists.

A number of inter-related factors continue to drive the general shift from forests and tree-based systems towards intensive agriculture (discussed in detail in Chapter 4). These include, among others, population growth, urbanisation, and the progressive movement from subsistence to market-driven economies and food production systems required to serve growing numbers of consumers globally. The resultant increased demand for staples and other food crops has led to expansion of mechanised agriculture and livestock production into forests and woodlands. This has frequently included introduction of crop and livestock species and production technologies developed under very different environmental and socio-cultural

1 All terms that are defined in the glossary (Appendix 1), appear for the first time in italics in a chapter.

conditions. It should be noted, however that in some regions such as Amazonia, urbanisation has increased the demand for, and production of, foods from forests and tree-based systems (Padoch et al., 2008).

Deforestation continues unabated in many parts of the world, in large part the result of agricultural expansion and cattle ranching (particularly in Latin America) (FAO, 2010), driven notably by urbanisation and globalisation of agricultural trade (c.f. De Fries et al., 2010; Rudel et al., 2009). Further, an increasing proportion of the world's remaining forests have been degraded both structurally and functionally. The drivers of *forest degradation* include unsustainable *forest management* for timber, fuelwood, wildlife and other *non-timber forest products*, overgrazing of livestock within forests, and uncontrolled human-induced fires, exacerbated in many regions by a number of factors, including *climate change* (Chazdon, 2014; Cochrane, 2003; ITTO, 2002; Thompson et al., 2012) and changing rural demographics (c.f. Uriarte et al., 2012).

These trends are not encouraging, particularly in light of extensive and ongoing *land degradation*, i.e., the long-term decline in ecosystem function and productivity caused by disturbances from which land cannot recover unaided. Land degradation currently affects hundreds of millions of hectares of agricultural lands and forests and woodlands, and an estimated 1.5 billion people who live in these landscapes (Zomer et al., 2009). Land degradation is the long-term result primarily of poor agricultural management (both historic and ongoing) associated with the expansion of extensive and intensive agricultural production practices into lands that are only marginally suitable for such activities. Without adequate organic or fossil fuel-derived fertilisers or other agricultural inputs (e.g. irrigation, pesticides, etc.) agricultural productivity typically declines in such areas, jeopardising food security for producers and those who depend on them.

In this chapter, we provide an overview of forests and tree-based systems and their role in enhancing food security and nutrition in rural communities. Our discussion includes not only management of forests, woodlands, agroforests and *tree crops* for direct food provisioning, but also the management of forested landscapes for the conditions they create that in turn affect other agricultural systems. The continuum of systems included in our analysis covers *managed forests* to optimise yields of wild foods and fodder, *shifting cultivation*, a broad spectrum of *agroforestry* practices, and single-species tree crop production (see Figure 3.1). We consider the variability and applicability of these management systems within and across geographical regions and biomes (agro-ecological zones). The social, cultural and economic contexts in which these systems exist and how they determine food security and nutrition outcomes are of particular concern. We therefore focus (in Section 3.4) on four factors that affect the socio-economic organisation of forests and tree-based systems, namely: land and tree *tenure* and *governance*; gender relations; human capital (including labour); and financial capital (including credit).

Fig. 3.1 The forest-tree-landscape continuum.
Photo 1 © Terry Sunderland, Photo 2 © Miguel Pinedo-Vasquez, Photo 3 © Liang Luohui, Photo 4 © PJ Stephenson

3.2 Forests and Tree-based Systems: An Overview

3.2.1 Historical Overview and the Role of Traditional Knowledge

Most of the forest and tree-based systems found in the world today have deep historical roots, developed and enriched over generations through experimentation and adaptation to changing environmental conditions and societal needs. While the scientific community, development economists and policymakers have generally disregarded and under-valued local and indigenous knowledge, such knowledge and associated management practices continue to serve communities living in or near forests in meeting their food security, nutrition and other health needs (Altieri, 2004; Cairns, 2007; Cairns, 2015; Johns, 1996; Kuhnlein et al., 2009; Parrotta and Trosper, 2012).

Traditional knowledge includes such things as weather forecasting, the behaviour, ecological dynamics, and health values of countless forest food species. It has been used to develop techniques for modifying habitats (as discussed in Section 3.2.2), enhance soil fertility, manage water resources, in the breeding of agricultural crops, domesticated trees and animals, and management of habitats and species assemblages to increase their production of food, fodder, fuel, medicine and other purposes (c.f., Altieri, 2004; Feary et al., 2012; Lim et al., 2012; Oteng-Yeboah et al., 2012; Parrotta and Agnoletti, 2012; Pinedo-Vasquez et al., 2012; Ramakrishnan et al., 2012).

An often-cited example of the sucessful application of traditional knowledge on a massive scale is the re-greening of the Sahel in Burkina Faso, Mali and Niger (Reij,

2014) where hundreds of thousands of poor farmers have turned millions of acres of what had become semi-desert by the 1980s into more productive land. Traditional knowledge regarding shea nut (from the shea tree, *Vitellaria paradoxa*) harvesting and processing among women engaged in shea butter production in Ghana and Burkina Faso has led to local selection of trees for desired fruit and nut traits and culling of other trees for fuel or construction. This is enabling the expansion of intensively-managed shea parklands to meet growing export markets (Carney and Elias, 2014).

The local and indigenous knowledge that underpins traditional forest- and tree-based systems is eroding in most parts of the world (Collings, 2009; Maffi, 2005; Parrotta and Trosper, 2012) as a result of a number of pressures, notably shifts to a market-based economy, cultural homogenisation, and dramatic changes in governance arrangements related to forest lands and trees outside of forests in favour of state (or colonial) ownership and control (Garcia Latorre and Garcia Latorre, 2012; Jarosz, 1993; United Nations, 2009). Development and conservation policies that discourage the traditional forest management practices that have historically ensured food security within indigenous and local communities have inevitably led to the loss of the traditional knowledge underpinning these practices (Collings, 2009; Parrotta and Trosper, 2012).

There is, however, a growing recognition of the value of traditional knowledge and innovation underpinning the management of forests and tree-based systems by indigenous and local communities worldwide. Beyond its importance for food security and nutrition, the forested landscapes that traditional management practices have produced can be appreciated for their provision of *ecosystem services* (including carbon sequestration), as well as conservation of biological and cultural diversity (Cairns, 2015; De Foresta and Michon, 1997; Fox et al., 2000; Palm et al., 2005; Swift et al., 1996).

Only recently have the scientific community and decision-makers in dominant societies begun to appreciate the limitations of land use policies and the often unsustainable agricultural intensification practices that they have encouraged (c.f. Altieri, 2002; Sanchez, 1995). Part of this reassessment is a growing awareness of the value of forest-based food production systems and the traditional knowledge and wisdom that underpins them. Today, an increasing number of scientists in universities, research organisations and networks are involved in efforts to better understand and apply knowledge of forests and tree-based systems to help farmers and communities to maintain, further develop, and extend the use of these management practices to meet current and emerging challenges (such as land and forest degradation, climate change adaptation, and market changes). A useful framework for evaluating sustainability issues associated with these systems and the roles that *agroecology*, traditional knowledge and farmer innovation can all play in understanding and enhancing the *resilience* of forests and tree-based systems is presented in Figure 3.2 (Altieri, 2004).

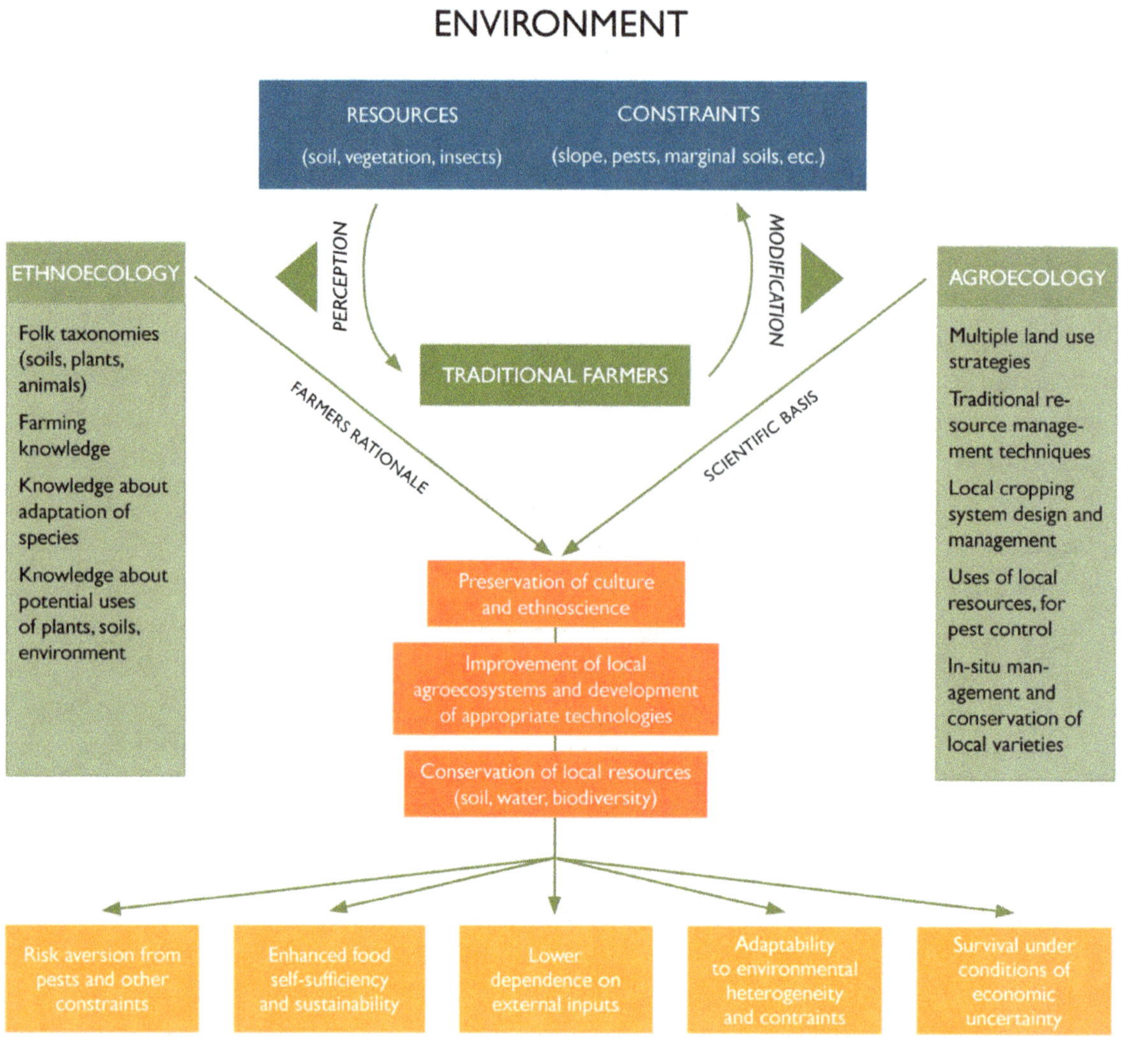

Fig. 3.2 Agroecology and ethnoecology are complementary approaches for understanding and systematising the ecological rationale inherent in traditional agriculture and enhancing sustainability of forest and tree-based systems. *Source:* Altieri (2004)

3.2.2 Managed Forests, Woodlands and Parklands

People living in and near forests have, for millennia, been altering forests in many ways and on many levels. Although precise estimates are difficult to obtain, as many as 1.5 billion people are thought to be dependent on forests (Chao, 2012; Agrawal et al., 2013). Paleobotanical research in New Guinea by Hladik et al. (1993) has shown that people as early as the late Pleistocene (30,000-40,000 years ago) were manipulating the forest by trimming, thinning and ring-barking in order to increase the natural stands of taro, bananas and yams. Throughout the world, people have changed the diversity and density of edible plant and animal species, modified the structure of forest stands and populations of food trees, made gaps in forests to plant crops in temporary clearings, introduced new species, burned understories, transplanted seedlings, changed watercourses, and substantially altered the nutritional, economic and *biodiversity* value of many if not most, forests we see today (c.f. Boerboom and Wiersum, 1983; Sauer, 1969; Wiersum, 1997).

Fire is probably the most frequently cited and most effective management tool that past generations as well as today's small farmers wield for changing and enriching forests and other areas with food and other useful plants. Fire is still widely used in shifting cultivation (or swidden) systems to temporarily increase soil fertility (through release of nutrients from standing vegetation), and in the management of both forests and grasslands around the world to enhance game production. Fire not only affects standing vegetation but also the soils upon which those forests stand and thus their potential productivity when cleared and planted to crops (Blate, 2005; Hammond et al., 2007; Hecht, 2009; McDaniel et al., 2005; Nepstad et al., 2001).

Many forms of traditional and contemporary forest management for food (including the creation of multi-storied agroforests, the planting of diverse forest gardens or the management of shifting cultivation fallows for food) have remained, with few exceptions, either invisible to researchers and planners or condemned by governments and conservationists (Hecht et al., 2014). Even the many contributions that woodlands make to agricultural production outside of forests have been largely overlooked (Foli et al., 2014).

There is little doubt that many of the forests that are now found throughout the tropics and elsewhere show the marks of management by people whether in the past or present (Balée, 2006). Often different types and patterns of forest manipulation have been superimposed in complex patterns whose histories and even purposes are not easily deciphered or understood. These patterns of forest disturbance, management, or manipulation continue to be developed and adapted to emerging needs and changing environmental and socio-economic conditions (Pinedo-Vasquez et al., 2012; Hecht et al., 2014). Rural communities living in and near forests around the globe and throughout history, and belonging to various communities, have not only enhanced the nutritional and economic value of their environments by increasing the supply of plant-based foods, they have also changed – and often increased – the availability of favoured animal species. Simple categories of hunting, gathering and agriculture, simply do not fit the realities of many of these *livelihood* strategies, while "forest management" does not adequately describe the multifaceted nature of these processes and practices. Some examples are outlined in Box 3.1.

The examples cited above give only a glimpse of how tropical forests have been and continue to be managed for food in complex and subtle ways that defy conventional categorisation. Even these few examples, however, challenge the ahistorical view held by many that old forests, particularly those of the tropics are "primordial" (Balée, 2006; Denevan, 1992) and question the facile dichotomisation of forests into "pristine" and "degraded".

3.2.3 Shifting Cultivation Systems

Shifting cultivation, also known as swidden (or, more pejoratively, "slash-and-burn"), encompasses a highly diverse range of land use practices that human societies worldwide have used to manage forests for food over the past 10,000 years. Shifting cultivation is practised in a variety of landscapes, from steeply sloped hilly areas to flat lands and low-lying valleys, and in a variety of ecosystems ranging from tropical moist forests to dry tropical forests and savannahs, grasslands, and seasonal floodplains (Thrupp et al., 1997). Until the 19th and even into the 20th century, shifting cultivation was common in the temperate zones of the Mediterranean and Northern Europe as well as in the southwestern and northeastern pine woodlands of North America (Dove, 1983; Dove et al., 2013; Warner, 1991). Currently, shifting cultivation is practised in over 40 countries in tropical regions of Africa, South and Southeast Asia, and Latin America under a variety of environmental, social and political conditions (Mertz, 2009). It remains the dominant form of agriculture in many rural upland areas where it contributes to the creation of complex landscapes and livelihoods (Mertz et al., 2008; Raintree and Warner, 1986; Spencer, 1966).

Box 3.1 Contemporary examples of forest management systems employed to enhance food security and nutrition in Southeast Asia and Amazonia

The "Forest Gardens" of West Kalimantan

On the island of Borneo there are significant forest stands that resemble "natural" forests but are in fact largely planted and are all heavily managed by farmers. A good example of such forests are the forest gardens that are commonly termed "tembawang" across the interior of the island. These complex forest gardens are largely found in what were once village sites and were originally formed by planting fruit trees and other trees around houses, by preserving useful species that came up spontaneously and by periodically weeding the areas selectively. When villages moved to other sites the gardens remained and grew, exhibiting an impressive tree diversity. For example in the village of Tae, an area of just one-fifth of a hectare was found to contain 224 trees belonging to 44 different species; 30 of which produce edible fruits, leaves or other edible products (Padoch and Peters, 1993; Padoch and Peluso, 1996). The most important fruits commonly found in tembawang include the especially prized durian (*Durio zibethinus*), as well as langsat (*Lansium domesticum*), jackfruit (*Artocarpus heterophyllus*), rambutan (*Nephelium lappaceum*), mangosteen (*Garcinia mangostana*), sugar palm (*Arenga* spp.) and the illipe nut (*Shorea macrophylla*) which produces an edible oil that also has industrial uses.

Managed forests of the Amazon estuary

The fruit of the açai palm (*Euterpe oleracea*) in the forests of the Amazon estuary has long been a staple of rural diets in Amazonian Brazil. It has recently also become an important source of cash, as consumption of the nutrient-rich açai fruit – once almost exclusively a local, rural food – has expanded to urban areas and into markets well beyond Amazonia. It is now highly prized and sold processed into a variety of products in North America, Europe and elsewhere (Brondizio, 2008; Brondizio et al., 2002; Padoch et al., 2008). The application of diverse management and planting practices and strategies is increasingly transforming the tidally-flooded forests of the estuary and beyond into açaí agroforests, locally called "açaizais" (Hiraoka, 1994; Brondizio, 2008). Açai agroforests include stands under different types and intensities of management, with varying population densities, structures, species diversity and composition. These practices range from selective weeding of existing açai-rich stands to further increase the production of the palm fruit, to enrichment planting and management of shifting cultivation fallows in the area. Often açai is not the only product that açai forest managers seek to promote, as açaizais contain other useful products including timbers, game and other fruits. Brondizio (2008) suggests that " ...while at the plot level one may observe a decline in tree species diversity in managed açaizais (avg 17 species) when compared to unmanaged floodplain forest (average 44 species), a broader landscape view (combining data from plots in different parts of the landscape) shows an increase of [native and exotic] tree species diversity (total 96 species)."

Building upon the management of others in the Amazon

Amazonian forests far from the estuary also abound in patches and plots that stand out from surrounding forests because of their richness in fruits and other foods. Many of these forest patches are almost certainly remnants of gardens, perhaps not unlike Borneo's tembawang, that may have once been intensively managed but have since been largely abandoned. Other food-rich plots scattered throughout Amazonia include planted or protected vegetation along footpaths and rivers that are periodically manipulated by passersby, including indigenous groups that continue to seasonally trek following the changing availability of animals or fish, as well as other forest travellers or migrants (Alexiades, 2009; Anderson and Posey, 1989; Kerr and Posey, 1984; Rival, 2002). Many of these patches are further enriched and casually maintained by fruit harvesters, who often take the time to do some selective weeding, cut back intruding vines, or occasionally transplant new seedlings. In Brazil and Peru most of these forests are named after their most abundant and valuable tree species. In the Peruvian Amazon, zapotales (rich in the zapote fruit (*Quararibea cordata*)) are frequently found along paths used for centuries by indigenous and non-indigenous people. The exact origin of these stands is unknown, but many are believed to have originated centuries ago, and been maintained up to this day either intentionally or accidentally by people dispersing the seeds (while eating or processing food), protecting the seedlings and juveniles in the forests through selective weeding, and occasionally by transplanting seedlings from forests to the edges of pathways, agricultural fields or fallows. People not only value zapotes as a tasty fruit, but also as an attractor of game animals ranging from monkeys to tapirs.

While the importance of shifting cultivation for food security and nutrition in many tropical regions is indisputable, the numbers of people who depend on shifting cultivation and the land areas involved remain unclear. This is due to a general lack of useful demographic data, ethnographic studies, and explicit knowledge about the location and intensity of these practices, a failure of land cover/land use maps to identify these practices from the global to the sub-national scale (Mertz et al., 2009a; Padoch et al., 2007; Schmidt-Vogt et al., 2009). Earlier empirically-based assessments have yielded estimates of the numbers of people dependent on shifting cultivation ranging from 40 to more than 500 million worldwide (Russell, 1988; Goldammer, 1988; Kleinman et al., 1996; Sanchez et al., 2005). A more systematic study by Mertz et al. (2009a) provided conservative estimates of between 14 and 34 million people engaged in shifting cultivation in nine countries in Southeast Asia alone. Similarly, accurate estimates of land areas involved in shifting cultivation are also lacking, although it can be assumed that they include a significant proportion of the 850 million hectares of tropical *secondary forests* in Africa, Latin America and Asia (Mertz et al., 2008). There is a clear need for further research to provide more accurate estimates of shifting cultivator populations and land areas involved using a combination of remote sensing data, ethnographic studies and special information databases. Promising steps are being taken by scientists in this direction, for example by Hett et al. (2012) in their work in northern Laos.

These management systems usually begin with the formation of a gap in the forest, frequently a secondary forest. The forest gaps or clearings made by shifting cultivators may range from several hectares in size, especially in Southeast Asia when several households choose to farm contiguously, to only a few square metres. This phase of the cycle which usually, but not always, involves the use of fire, and creates a space to plant agricultural crops ranging from the dryland rice and vegetable combinations frequent in montane zones of Southeast Asia (Cairns, 2007; Conklin, 1957; Condominas, 1977; Padoch et al., 2007; Mertz et al., 2009b), to assemblages of cassava, banana, and a variety of tubers and herbs representative of Amazonian fields (Denevan et al., 1984; Denevan and Padoch, 1987; Padoch and de Jong, 1992). The *agrobiodiversity* of some of these systems is extremely high (Rerkasem et al., 2009). For example, the pioneering study of shifting cultivation fields in the Philippines by the Hanunoo people of Mindoro Island (Conklin, 1957) found over 280 types of food crops and 92 recognised rice varieties, with several dozen usually showing up in any particular field. Intensive cropping of annual species usually lasts for only a year or two after which management generally becomes less intensive, allowing for a more or less spontaneous or natural vegetation to gradually dominate the site.

In the past, the change in types or intensity of management was commonly characterised as "abandonment" of the field; more recently there has been considerable recognition that much of the "natural" or "forest" fallow can be and often is manipulated or managed by shifting cultivators for a variety of economic and food products (Cairns, 2007; Alcorn, 1981; Denevan and Padoch, 1987; Colfer et al., 1997; Colfer, 2008a; Padoch and de Jong, 1992). The "less intensive management" phase, or fallow, often relies heavily on the regrowth of forest vegetation for the provision of many of the environmental qualities necessary for efficient food production, including restoration of soil fertility and structure. The accumulation of biomass in the regrowing vegetation and the suppression of pests, diseases and weeds make agricultural production, especially in the tropics, a difficult and labour-demanding activity. Fallows or young regrowth also often feature many useful species that households collect and rely upon for food and the preparation of food. Thus shifting cultivation is increasingly seen and described as a complex and dynamic form of "swidden-fallow agroforestry" (Denevan and Padoch, 1987).

The complexity of alternating forest and field phases is further enhanced by other practices that result in the mixture of planted and spontaneous vegetation in swidden fields. When fields are first cleared, any useful tree species found in the plot are generally spared, left standing, and even protected from fire. These plants, frequently fruit trees, then become integral parts of the field together with planted crops and any spontaneous vegetation that survives weeding and further fires. "Selective weeding" is the norm; plants valuable for food or other purposes are again spared while those that are not valued are cut and removed. Especially in the later stages of the "fallow" phase, spontaneous or forest vegetation tends to predominate in shifting cultivators' fields, the boundaries between forests and fields disappear, although the food value of these plots is often far higher than that of less "disturbed" forests (Rerkasem et al., 2009). Many areas of regrowth in these systems continue to be heavily managed for economic and other products, including such nutritionally valuable resources as bushmeat (Wadley and Colfer, 2004). "Garden hunting" is often carried out in shifting cultivation fields and fallows that can be rich in animals (Linares, 1976; Hiraoka, 1995) as they are attracted by the fruits that are frequently planted or spared. In summary, many shifting cultivation landscapes are largely forests that have been enriched with crops and a broad array of species by diverse management practices that are often applied iteratively and are difficult to classify or even see.

The dynamics of shifting cultivation have changed over time, and in some regions these changes have been rapid particularly since the mid-20th century. Many shifting cultivators have intensified their land use practices over time, including through the introduction of new crops and technologies that are not always well-suited to local agroecological conditions. While such changes can sometimes increase the cultivators' immediate incomes, the agricultural results have often been adverse or unsustainable, especially if unsuitable land is overused or inappropriate inputs or crops are used. These changes have often resulted in instabilities in previously well-adapted shifting cultivation and resource use, jeopardising their ecological and in some cases economic sustainability (Raintree and Warner, 1986; Warner, 1991). For example, shortened cropping cycles or other management practices have in many situations contributed to soil fertility and productivity declines (Borggaard et al., 2003; Cairns and Garrity, 1999; Ramakrishnan, 1992). Destabilisation of traditional shifting cultivation systems is usually the result of a combination of socioeconomic and political changes, demographic pressures, and biophysical factors that force cultivators to change their practices (Table 3.1). Factors that commonly contribute to these changes include government restrictions of forest use, changes in land tenure systems, demographic pressures including large-scale migration and resettlements, and policies that promote cash crop production (Nair and Fernandes, 1984).

While such unstable conditions are not found in all shifting cultivation systems, they have reinforced negative perceptions of shifting cultivators and their practices (Fox et al., 2009; Mertz et al., 2009b). Arguments typically used to condemn shifting cultivation have included its low productivity, negative impacts on soils, hydrology and biodiversity conservation. However, broad generalisations regarding shifting cultivation are not helpful and obscure the fact that environmental impacts of shifting cultivation are diverse, and depend not only on farmers' management practices, but the environmental, social, economic and political contexts in which they occur (c.f., Thrupp et al., 1997; Lambin et al., 2001). Efforts to ameliorate the perceived shortcomings or negative impacts of shifting cultivation can be counter-productive, particularly in relation to food security and nutrition. For example, recent studies on land use change in the Lao People's Democratic Republic (also see Chapter 5), found that policies aimed at increasing forest cover, protecting wildlife, and promoting more intensive, commercial farming have had significant negative impacts on the well-being of rural community members and especially on their ability to adapt to change and respond to a variety of "shocks" that economic and environmental change may bring (Hurni et al., 2013; Castella et al., 2013).

Table 3.1 Causes of destabilisation and degradation in shifting cultivation systems (*adapted from* Thrupp et al., 1997)

Outcomes of Destabilisation and Degradation	Proximate Causes	Underlying Causes
• Shortening or ceasing fallows • Over-exploitation of land/ soils • Declining soil fertility • Decreasing yields • Increasing deforestation • Loss of biodiversity	• Development of roads and other infrastructure • Expansion of monoculture agriculture and timber industries • Scarcity of land and other resources available to cultivators • Changing demographic trends, e.g. migration and population growth • Lack of alternatives for production and income for rural people • Resettlement of new groups in frontier areas • Lack of access to stable markets for shifting cultivators	• Inequitable political-economic structures affecting use of resources • International/national economic policies, especially trade liberalisation, structural adjustment • Disrespect for, or neglect of, the rights of shifting cultivators • Lack of knowledge of environmental factors in agriculture • Lack of sustained economic development and employment for poor • Lack of political commitment for poverty alleviation • Inadequate attention to social needs in environmental policies

A growing body of research indicates that in many areas where shifting cultivation is still practised, particularly where traditional knowledge regarding fallow management is well-developed and applied, these systems can be managed sustainably – without undermining soil fertility and jeopardising productivity – while conserving biodiversity and maintaining provision of an array of forest ecosystem services (c.f. Cairns, 2007; Cairns, 2015; Colfer et al., 2015; Cramb, 1993; Finegan and Nasi, 2004; Kleinman et al., 1996; Mertz et al., 2008; Palm et al., 2005; Parrotta and Trosper, 2012; Ramakrishnan, 1992; Swift et al., 1996). With respect to efforts to mitigate climate change through REDD+ programmes, it is important to note that while the secondary forest-dominated landscapes created through shifting cultivation do not store as much carbon as *primary forests*, their carbon sequestration potential is far greater than those dominated by alternative agricultural or single species tree crop management systems (c.f. Bruun et al., 2009; Chazdon, 2014; Martin et al., 2013). Such findings have important implications for REDD+ policies and programmes, particularly where they may exclude shifting cultivation areas (and their practitioners) from REDD+ funding consideration, or use REDD+ policies as a lever to eradicate shifting cultivation practices (Angelsen, 2008; Brown et al., 2011; Ziegler et al., 2012).

Finally, although shifting cultivation is a prominent feature of food production in forested areas in many tropical regions, the food values of forest mosaics that result from shifting cultivation systems have to date been little researched as they fall between conventional "farm" and "forest" categories. Shifting cultivation landscapes are often "illegible" to outsiders (Scott, 1999), are frequently devalued and labelled "degraded". Yet what research there is suggests that these landscapes that harbour a great variety of plants and animals in fields and food-rich fallows and forests, and create multiple and diverse "edges", have been the larders of human communities around the globe and throughout millennia (Andrade and Rubio-Torgler, 1994). As shifting cultivation systems disappear around the world (van Vliet et al., 2012; Padoch et al, 2008), being replaced by other forms of production that yield more food calories per area, it is important to understand what is being lost in micronutrient output, food diversity and resilience to shocks when these practices vanish.

3.2.4 Agroforestry Systems

Agroforestry encompasses a vast array of food production systems in which woody perennials are deliberately integrated in spatial mixtures or temporal sequences with crops and/or animals on the same land unit. These systems involve careful selection of species and management of trees and crops to optimise productivity and positive interactions among their components and minimise the need for chemical fertilisers and other inputs to maintain their productivity.

Like managed forests and shifting cultivation systems, most agroforestry practices are based on the traditional knowledge of people in local and indigenous communities. A staggering variety of agroforestry systems have been developed and modified by farmers in tropical, subtropical and temperate regions worldwide over centuries, or even millennia in some regions. The systematic study of agroforestry by the scientific community, which began only a few decades ago, has sought to understand the accumulated knowledge and wisdom of agroforestry practitioners using established theoretical bases from ecology and agroecology. This knowledge is being used to promote and in some cases modify these traditional systems in ways that will enhance their applicability, relevance and adaptability to changing environmental, economic and social conditions (Sanchez, 1995).

Overview of agroforestry systems and their variability

Agroforestry systems are typically classified on the basis of their structure, i.e., the nature and spatial and/or temporal arrangement of tree and non-tree components. Three broad classes are generally distinguished, based on the inclusion of agricultural crops and/or livestock in these systems: "agrisilvicultural systems" involving combinations of agricultural crops and trees or shrubs; "silvopastoral systems" that include combinations of trees and pasture for grazing livestock; and "agrosilvopastoral systems" combining crops, pastures and trees (Nair, 1993).

Agrisilvicultural systems include a very diverse array of agroforestry subsystems and practices, all of which involve the cultivation and management of trees and/or shrubs for food and/or non-food values (such as soil conservation or providing shelter for crops), generally in combination with agricultural crops. These subsystems and practices include for example, improved fallows, multilayer tree gardens and alley cropping. In some cases agrisilvicultural systems also combine the production of timber with agricultural crops, as is the case with "Taungya" which was originally used to promote teak plantations by the British colonial government in Burma in the late 19th century and which is widely practised today thoughout much of the tropics. Other agrisilvicultural systems include different plantation crop combinations, notably for fuelwood but also homegardens with fruit trees.

Silvopastoral systems include plantation crops with pastures and animals; trees on rangeland or pastures; and protein banks, involving concentrated production of protein-rich tree fodder outside of grazing areas.

Agrosilvopastoral systems include homegardens with domesticated animals; multipurpose woody hedgerows, involving fast-growing and coppicing fodder trees and shrubs in woody hedges for browse, mulch, green manure and soil improvement; apiculture with trees; aquaforestry where selected trees and shrubs line fish ponds, and multipurpose woodlots.

Within and across these broad categories, agroforestry systems vary in the functional characteristics of their components (especially of their tree and shrub components), including both productive functions (food, fodder, fuelwood, timber and other non-timber forest products) as well as protective functions (windbreaks and shelterbelts, soil conservation and fertility improvement, moisture conservation, and shade for crops, livestock and people). Considerable variation exists within all categories of agroforestry systems with respect to management intensity and the level of inputs used (such as labour, fertilisers and other agricultural inputs) which affect their adoption by farmers (Bannister and Nair, 2003; Franzel, 1999; Mercer, 2004; Scherr, 1995; see also discussion below in 3.4.4). They also differ in the predominant end uses of their products – ranging from subsistence (directly contributing to household food security and nutrition) as in the case of homegardens, to predominantly commercial, as in the case of cocoa, coffee, tea, rubber and oil palm agroforestry systems.

Regional and global patterns in agroforestry practice

Agroforestry systems serve a major role in food security and nutrition for their practitioners (and consumers of commercialised products) within a number of agroecological zones on all continents although the exact extent of these practices is difficult to quantify (notably because of a lack of standardised definitions and procedures for delineating the zone of influence of trees in mixed tree/crop systems (Nair et al., 2009)). Of particular importance to this book are those regions where food security is considered to be a more significant challenge. These include extensive areas where agroforestry systems also have a long history, i.e., the majority of tropical and sub-tropical humid, sub-humid, semi-arid and highland regions. The prevalence of different agroforestry systems in these regions, and their actual or potential contributions to enhanced food security and nutrition, are influenced by climate, natural vegetation and soils, and dominant land use systems, as well as a host of other socio-economic factors (Nair, 1993).

In humid and sub-humid tropical lowland regions, agroforestry is practised extensively in Southeast and South Asia, Central and West Africa, and Central and South America. In these regions, agroforestry can help to reduce deforestation and forest degradation, and overcome productivity constraints on conventional agriculture related to soil degradation caused by unsustainable forest management, poorly managed shifting cultivation (including reduction of fallow lengths), overgrazing, soil acidity, low soil fertility and high rates of soil erosion (Nair, 1993).

Tropical and sub-tropical highlands (over 1000m in elevation) with agroforestry potential include humid and sub-humid regions in the Himalayan region, parts of southern India and Southeast Asia, the highlands of east and central Africa, Central America and the Caribbean, and the Andes. Dominant land uses in these regions include shifting cultivation, arable farming, plantation agriculture and forestry, and ranching (in Central and South America). Agricultural productivity and food security in these regions may be constrained by soil erosion, shortening of fallows in shifting cultivation, overgrazing, deforestation and forest degradation, and fodder and fuelwood shortages (Nair, 1993).

Semiarid and arid regions where agroforestry systems are common include the cerrado of South America, savannah and sub-Saharan zones of Africa, drier regions of the Mediterranean, North Africa and the Near East, and parts of South Asia (Nair, 1993).

Parklands, one of the most extensive farming systems in the tropics and the dominant farming systems in semi-arid West Africa, cover the vast majority of cultivated area in Sahelian countries. This includes an estimated 90 percent (5.1 million ha) of all agricultural lands in Mali (Cissé 1995; Boffa, 1999) where scattered multipurpose trees such as baobab (*Adansonia digitata* L.), detar (*Detarium microcarpum*), néré (*Parkia biglobosa*), tamarind (*Tamarindus indica*), shea tree or karité (*Vitellaria paradoxa*) and ber (*Ziziphus mauritiana*) are managed on farmlands.

A recent geospatial analysis by Zomer et al. (2014) estimated the extent and recent changes in agroforestry practices at a global scale, based on remote sensing-derived global datasets on land use, tree cover and population. Agroforestry systems (defined in their study as agricultural lands with greater than 10 percent tree cover) were found to comprise 43 percent (over 1 billion ha) of all agricultural land globally (Figure 3.3). These lands include 320 million ha in South America, 190 million ha in sub-Saharan Africa, and 130 million ha in Southeast Asia. In Central America, 96 percent of agricultural lands were classified as agroforestry, as were over 80 percent of agricultural lands in

Southeast Asia and South America. Globally, the amount of tree cover on agricultural land increased substantially between 2000 and 2010, with the area of >10 percent tree cover increasing from 40 to 43 percent (+82.8 million ha). The proportion of agricultural lands with varying levels of tree cover and proportions of people living in these landscapes in different regions of the world are presented in Table 3.2.

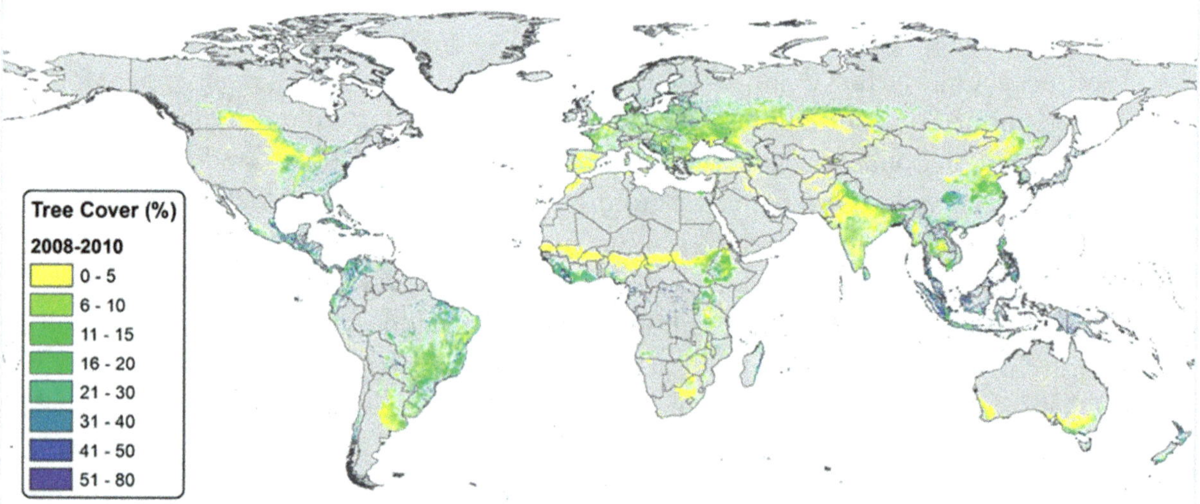

Fig. 3.3 Global estimates of tree cover (percent) on agricultural land in the years 2008-2010 (averaged). ***Source:*** **Zomer et al. (2014)**

Zomer et al. (2009) found a strong relationship between aridity and tree cover in Southeast Asia, Central America and South America, although there are many exceptions to this rule (i.e., high tree cover found in more arid zones and low tree cover found in more humid zones) that must be explained by other factors, such as tenure, markets or other policies and institutions that affect incentives for tree planting and management, as well as context-specific historical trends (Zomer et al., 2014; Zomer et al., 2007; Zomer et al., 2009). Further, although patterns in the relationship between tree cover and human population densities in agricultural landscapes exist within aridity classes and continents, these correlations are neither consistently positive nor negative except in the very low or high range of tree cover, and there appears to be no general trade-off between human population density and tree cover in these landscapes. Additional work is needed to refine estimates of land cover (versus land use) in agricultural landscapes and the extent of agroforestry practice in its varied forms, both at the global level and at finer spatial scales, as well as their relationship with factors other than climate and population density.

Table 3.2 Percentage of land area and population living in agricultural areas with greater than 10%, 20% and 30% tree cover in 2008-2010 (*adapted from* Zomer et al., 2014).

(% of all land area/persons in agricultural area)	>10% tree cover		>20% tree cover		>30% tree cover	
Region	% land area	% popu-lation	% land area	% popu-lation	% land area	% popu-lation
North America	42.4	66	26.3	46	15.5	30
Central America	96.1	95	79.0	78	54.8	54
South America	65.6	74	31.8	35	17.7	19
Europe	45.0	46	20.4	19	11.6	10
North Africa/Western Asia	11.0	13	5.5	4	3.3	2
sub-Saharan Africa	30.5	39	15.0	16	8.4	7
Northern and Central Asia	25.3	23	9.7	7	4.3	3
South Asia	27.7	34	7.8	8	3.6	2
Southeast Asia	79.6	73	62.9	46	49.9	30
East Asia	47.5	57	22.1	21	11.8	8
Oceania	33.3	80	23.8	67	17.0	52
Global average	***43.4***	***46***	***23.1***	***19***	***14.2***	***10***
Change since 2000-2002	+3.7	+5	*1.8*	+2	+*1.1*	+2

3.2.5 Single-species Tree Crop Production Systems

Single-species tree crop production systems can be found in forest and agricultural landscapes in tropical, sub-tropical and temperate regions worldwide. They involve a wide variety of designs and management practices that have evolved over time in response to local, regional and global commoditization of domesticated forest species.

The domestication of forest tree species is rooted in antiquity. Genetic selection, vegetative propagation (including grafting) and cultivation of tree crops such as date palm (*Phoenix dactilifera*), olive (*Olea europaea*), sycamore fig (*Ficus sycomorus*), pomegranate (*Punica granatum*), apple (*Malus* x *domestica*), pear (*Pyrus communis*), apricot (*P. armeniaca*), almond (*P. dulcis*), sweet cherry (*P. avium*), peach (*P. persica*), mango (*Mangifera indica*) before avocado (*Persea americana*) all date back 4,000 to 6,000 years (Janick, 2005). In the case of the common fig (*Ficus carica*) its domestication may have begun at the time when wild grains such as rice, wheat and other staple crops were first cultivated in North Africa and Southwest Asia 11,000-12,000 years ago (Kislev et al., 2006).

Worldwide, many hundreds of tree species are cultivated today by farmers for household and local consumption, a lesser number for sale in urban markets, and still fewer for international markets. These cultivated species include beverage and confectionery crops (e.g. coffee, cocoa, tea), fruits, oils (e.g. oil palm, coconut), staples (e.g. bananas, plantains, breadfruit, peach palm and sago palm), spices (e.g. cinnamon, clove) and nuts. The diversity of forest species cultivated by farmers in tropical and subtropical regions is impressive; an indicative list presented by Smith et al. (1992) of domesticated tropical moist and wet forest trees for their edible fruits or nuts includes over 170 species. Production from these tree crop systems contributes significantly to the food security and nutrition of farmers – either directly for their nutritional value, or indirectly by providing income, as discussed in Chapter 2.

Tree crop systems are managed on large, medium or small scales either as single-species or multi-strata systems with other woody or herbaceous species. They may also be intercropped in agroforestry systems with annual or perennial crops in temporal or spatial sequences. For example, coffee production in Ethiopia mainly involves agroforestry-based systems, although there are both natural coffee forests and single-species plantations (Muleta, 2007). Similarly, cocoa is cultivated under the canopy of shade trees in traditional agroforests, although single-species plantations are also cultivated (Obiri et al., 2007). Weeding, fertiliser application, pest and disease control, and branch pruning are among the cultural practices used in tree crop systems for enhancement of yield (Table 3.3).

The introduction of new hybrids of some species with large international markets has led to a rapid expansion in acreage in producing countries. A number of major tree crops are listed in the FAO database, FAOSTATS, on agricultural commodities traded globally. These include: cocoa (*Theobroma cacao*), coffee (*Coffea arabica, Coffea robusta*), tea (*Camellia sinensis*), oil palm (*Elaeis gineensis*), coconut (*Cocos nucifera*), date palm (*Phoenix dactylifera*), mango (*Mangifera indica*), avocados (*Persia americana*), orange, tangerine, lemon, grapefruit (*Citrus* spp.), shea (*Vitellaria paradoxa*), guava (*Psidium guajava*), fig (*Ficus carica*), banana and plantain (*Musa* spp.), apple (*Malus domestica*), peach, plum, and apricot (*Prunus* spp.), olive (*Olea europaea*), cashew (*Anacardium occidentale*), walnut (*Juglans* spp.) and hazelnut (*Corylus* spp.). Information on a number of these tree crop species, their management and contributions to food security and nutrition, are summarised in Table 3.3 (see also Chapter 2).

Production of some tree crops with major global markets has been organised on a large scale with smallholder participation, making significant contributions to local and national economies (Watson, 1990). While smallholder farmers typically earn the least profit margin in tree crop commodity value chains, single-species tree crop systems do create employment and income opportunities locally and internationally as well as improved trade and foreign exchange balances for producing nations. For example, Ethiopia, the oldest exporter of coffee in the world, is the largest coffee producer and exporter in Africa. The cultivation, processing, trading, transportation and marketing of coffee provide employment for 15 million Ethiopians who depend on the industry for at least a significant part of their livelihood on a subsistence basis or as a sole source of income. The industry plays a fundamental role in both the cultural and socio-economic life of the nation (Muleta, 2007). In Uganda the coffee industry employs over 5 million people and the sector contributes 20-30 percent of the country's foreign exchange earnings (Kiyingi and Gwali, 2013).

Climate change and its potentially devastating effects on crop production threaten the productivity of tree crop systems in many regions. For example, it is predicted that rising temperatures will dramatically reduce cocoa production between 2030 and 2050 in Côte d'Ivoire and Ghana, the world's first and second cocoa producers accounting for 53 percent of the world's cocoa output (CTA, 2012). This has necessitated a critical analysis of promising multi-purpose tree-based systems that have the potential for ensuring sustainable income and food security while mitigating climate change effects. Shade-grown cocoa and coffee are also being advocated in response to certification schemes and also the increasing demand for "specialty" products (Afari-Sefa et al., 2010; WOCAT, 2007). Generally, growing tree crops under the shade of upper canopy forest trees is considered to be more ecologically and economically sustainable than open-grown systems (WOCAT, 2007). However, the value of such systems for biodiversity conservation is very much context-specific, and has been questioned in the case of shade coffee (Tejada-Cruz et al., 2010).

Box 3.2 Shade-grown cocoa

Although it has been argued that the perennial nature of tree crop systems makes them inherently more sustainable and less environmentally damaging in comparison with annual food crop systems (Watson, 1990), their biodiversity impacts, particularly for the production of cocoa and coffee, have increased with the expansion of plantations in many producing countries. In the case of cocoa, the total area under cultivation worldwide increased by 3 million ha (4.4 million to 7.4 ha) in the last 50 years (Clough et al., 2010), contributing to the ongoing transformation of many lowland tropical forest landscapes in Latin America, Africa and Southeast Asia that began centuries ago (Schroth and Harvey, 2007). Expansion of cocoa farms accounts for much of the deforestation in lowland West Africa (Gockowski and Sonwa, 2011) where intact tropical forests have been converted for this purpose. This transformation has been expedited by the development and introduction of highly productive cocoa hybrid varieties that require little or no forest tree shade. However, since open-grown cocoa requires increased investments in agro-chemical inputs to support optimum productivity, it has a shorter productive period with deleterious effects on soil fertility and plantation health (Ruf and Schroth, 2004). In contrast, cocoa traditionally grown under filtered shade of forest trees often results in a multi-strata agroforestry system that is considered to be one of the best examples of permanent agriculture that preserves a forest environment and biodiversity (Ruf and Schroth, 2004; Rice and Greenberg, 2000). Under optimal soil conditions and rainfall regimes, shade grown cocoa may produce good yields for 60-100 years whereas optimum production may last for 20 or less years without shade (Ruf and Schroth, 2004; Obiri et al., 2007; Obiri et al., 2011).

Theobroma cacao (cocoa) pods. Photo © sarahemcc, Wikimedia

Table 3.3 Geographical distribution, management and nutritional values of selected tree crops with international markets.

Common (and scientific) name & centre of origin	Major producing countries	Establishment and management	Principle food uses and nutritional value	References
Sweet orange (*Citrus x sinensis*) Most likely Southeast Asia	Cultivated worldwide in tropical, subtropical and Mediterranean climates: Largest producer is Brazil, followed by USA, India, Mexico, China, Spain, Italy, Egypt, Iran, Indonesia, Turkey, Pakistan and South Africa.	Grown in agroforestry systems with food crops and in monocrop plantations; propagated from seeds/seedlings or vegetatively from grafted seedlings. Weed control, insect pest and disease control, fertiliser application, irrigation and branch pruning required to sustain productivity.	Fruit is eaten fresh, or processed for its juice or fragrant peel for marmalade. Orange juice is a rich source of vitamin C; the edible peel has significant contents of vitamin C, dietary fibre, total polyphenols, carotenoids, limonene and dietary minerals, such as potassium and magnesium.	Barros et al. (2012); FAOSTAT Statistical Database: http://faostat.fao.org/
Apple (*Malus domestica*) Central Asia in southern Kazakhstan, Kyrgyzstan, Tajikistan, and Xinjiang, China.	Cultivated worldwide in temperate and some subtropical regions. Largest producer is China, followed by USA, Turkey, Iran, Poland, Italy, France, India, Russia, Chile, Argentina and Brazil.	Grown in orchards and agroforestry systems. Generally propagated by grafting, although wild apples grow readily from seed. Apple trees highly susceptible to fungal and bacterial diseases and insect pests. Intensive programme of chemical sprays important to maintain high fruit quality, tree health, and high yields in commercial plantations.	Fruit often eaten fresh but also cooked in prepared foods (especially desserts) and drinks. Used for juice, vinegar and other beverages and confectionery. Fruit contains significant dietary fibre and modest vitamin C content, with otherwise a generally low content of essential nutrients compared to other fruits. Apple peels contain various phytochemicals with unknown nutritional value, including quercetin, epicatechin, and procyanidin B2.	Boyer and Liu (2004); FAOSTAT Statistical Database: http://faostat.fao.org/; Lauri et al. (2006); USDA Nutrient Database: http://ndb.nal.usda.gov/ndb/search/list
Mango (*Mangifera indica*) Tropical South and Southeast Asia	Cultivated throughout the tropics and subtropics. Largest producer is India, followed by China, Thailand, Indonesia, Pakistan, Mexico, Brazil, Philippines, Egypt, Kenya.	Grown in smallholder agroforestry systems and in large scale monocrop plantations; propagated from seeds, seedlings and grafted seedling. Weed control, insect pest and disease control, fertiliser application, irrigation and branch pruning required to sustain productivity.	Fruits are eaten fresh or used to prepare juices, smoothies, sherbets or other desserts. Also used (dried or fresh) in cooking, preparation of chutneys and preserves. Both green and ripe mango fruits are rich in carbohydrates, minerals and vitamin C. Fruits and sometimes leaves used as livestock fodder.	FAO (1982); FAOSTAT Statistical Database: http://faostat.fao.org/; Mukherjee (1972).
Avocado (*Persia americana*) Mexico and Central America	Cultivated worldwide in tropics, subtropical and Mediterranean climates. Largest producer is Mexico, followed by Indonesia, Chile, USA, Dominican Republic, Colombia, Brazil, Peru, China and Kenya.	Grown in agroforestry systems with food crops and in orchards; propagated from seeds/seedlings or asexually from grafted seedlings. Weed control, insect pest and disease control, fertiliser application, irrigation and branch pruning required to sustain productivity.	The fruit is eaten fresh and for preparation of various recipes; it is a major ingredient in vegetarian diets. A typical serving of avocado (100 g) is a very good source of several B vitamins and vitamin K, and a good source of vitamin C, vitamin E and potassium. Avocados also contain phytosterols and carotenoids, such as lutein and zeaxanthin, and diverse fats, mostly oleic acid but also palmitic acid and linoleic acid, among others.	Chen et al. (2008); Dreher and Davenport (2013); FAOSTAT Statistical Database: http://faostat.fao.org/; NutritionData.com. 2013.

Common (and scientific) name & centre of origin	Major producing countries	Establishment and management	Principle food uses and nutritional value	References
Common fig (*Ficus carica*) Middle East and western Asia	Cultivated in many temperate and subtropical countries worldwide, particularly in the Middle East and areas with a Mediterranean climate. Major producers include: Turkey, Egypt, Iran, Morocco, Algeria, Syria, USA, Greece, Spain, Afghanistan, Brazil, Tunisia and Italy.	Propagated from seeds, but more commonly by vegetative methods, i.e., cuttings, air-layering or grafting.	Figs are consumed fresh or dried and are often processed as a paste for pastries or canned. The fruit can be fermented and distilled into alcohol. Dried figs are a rich source (> 20% of the daily value) of dietary fibre and the essential mineral, manganese, while vitamin K and numerous other minerals are in moderate content. Figs contain diverse phytochemicals, including polyphenols such as gallic acid, chlorogenic acid, syringic acid, (+)-catechin, (−)-epicatechin and rutin.	FAOSTAT Statistical Database: http://faostat.fao.org/; Janick (2005); USDA Nutrient Database: http://ndb.nal.usda.gov/ndb/search/list; Veberic et al. (2008)
Cocoa (*Theobroma cacao*) Southeastern Mexico to Amazon Basin	Cultivated in humid tropics. Top 10 producing countries (in 2005): Cote d'Ivoire, Ghana, Indonesia, Nigeria, Brazil, Cameroon, Ecuador, Colombia, Mexico, Papua New Guinea.	Grown both in large agroindustrial plantations and by small producers, the bulk of production coming from millions of small producers; Planted under forest shade or in monocrop plantations; propagated from seeds/seedlings. Pruning, ferti liser application, pest, disease and weed control, pod harvesting and bean processing are main cultural practices for managing cocoa plantations.	Seeds or beans contain 40-50% fat as cocoa butter used for chocolate, cocoa mass and powder; pulp used for juice, smoothies, jelly and nata; fermented pulp distilled into alcoholic beverages.	CacaoNet. (2012); FAOSTAT Statistical Database: http://faostat.fao.org/; Figueira et al. (1993)
Coconut (*Cocos nucifera*) Asia-Pacific	Coastal regions throughout humid tropics and subtropics. Major producers include: Indonesia, Philippines, and India, followed by Brazil, Sri Lanka, Vietnam, Papua New Guinea, Mexico, Thailand, Malaysia and Tanzania.	Cultivated in a variety of agroforestry systems and in monocrop plantations; propagated from seeds (nuts) and seedlings.	Fruit (nut) contains water suspended in the endosperm with an outer hard shell (mesocarp) and fibrous husk(exocarp).Water from immature fruits consumed as a refreshing beverage rich in vitamins and trace minerals; Endosperm when mature contains 35-40% oil, 10% carbohydrate and 3% protein; oil extracted from dried endosperm (copra) used as a cooking oil, in margarine, cocoa butter, beverages and numerous non-food products; dried endosperm used in confectionery, cooking, and may be ground into flour for baking.	FAOSTAT Statistical Database: http://faostat.fao.org/; Opeke (1982); Parrotta (1993)

Table 3.3, cont. Geographical distribution, management and nutritional values of selected tree crops with international markets.

Common (and scientific) name & centre of origin	Major producing countries	Establishment and management	Principle food uses and nutritional value	References
Shea (*Vitellaria paradoxa*) Guinea and Sudan savannah zone from Senegal to Sudan, and to western Ethiopia and Uganda.	Managed in natural stands near human settlements, as well as planted throughout its African range. Major producers include Nigeria, Mali, Burkina Faso, Ghana, Côte d'Ivoire, Benin and Togo.	Throughout its range it is managed in natural stands and in agroforestry systems (parklands), either involving livestock and/or staple crop production. Control of bush fire, insects and parasites and drought are major management activities.	Shea butter extracted from the seeds widely used in cosmetics as a moisturiser, salve or lotion. It is one of the most important sources of vegetable oil in rural areas of the savannah zone of West Africa, used in food preparation in many African countries, and occasionally (mixed with other oils) in the chocolate industry as a substitute for cocoa butter. Shea butter is composed of five principal fats mostly stearic, oleic but also palmitic, linoleic, and arachidic; it is a rich source of vitamins A and E, and contains phenolic compounds known to have antioxidant properties.	Hall et al. (1996); Masters et al. (n.d.); PROTA database: http://database.prota.org/PROTAhtml/Vitellaria paradoxa_En.htm
Cashew (*Anacardium occidentale*) Northeastern Brazil	Cultivated in many tropical countries worldwide. Major producers include: India, Côte d'Ivoire, Brazil, Indonesia, Vietnam, Nigeria, Benin, Guinea-Bissau, Mozambique and Philippines.	Grown in smallholder agroforestry systems and in large scale monocrop plantations from seeds, seedlings and grafted seedling. Weeding, mulching of young plants, fertiliser application and pruning of branches enhance growth and yield.	Nut (kernel) eaten as a snack food or used in cooking. They are a rich source of protein, carbohydrate and fat and contains minerals such as Ca, P, Na, K, Mg, Fe, Cu, Zn and Mn. Cashew kernel lipids are rich in unsaturated fats, mainly oleic acid. It is also a good source of antioxidants. The spongy, juicy, pear shaped stalk (cashew apple) contains sugars, tannins, phenols, amino acids, ascorbic acid, riboflavin, minerals and fibre. It is used to prepare juices or distilled into a liqueur (feni);also used to prepare pickle and other food products.	FAOSTAT Statistical Database: http://faostat.fao.org/; Johnson (1973); Ohler (1979); Saroj and Rupa (2014)
Walnut (*Juglans regia*) Central Asia, including Uzbekistan, Kyrgyzstan, Tajikistan, Turkmenistan and southern Kazakhstan.	Largest producer is China, followed by Iran, USA, Turkey, Mexico, Ukraine, India, Chile, France and Romania.	Commonly propagated from seeds. Insect pest and disease control, fertiliser application, branch pruning required to sustain productivity.	Walnuts are eaten raw, toasted, pickled or cooked in various recipes; also processed for oil. 100 grams of walnuts contain 15.2 grams of protein, 65.2 grams of fat, and 6.7 grams of dietary fibre. They are rich in vitamins, particularly thiamine (B1), B6, folate (B9), and in trace metals, particularly manganese, but also magnesium, phosphorus, iron, and zinc. Unlike most nuts that are high in monounsaturated fatty acids.	FAOSTAT Statistical Database: http://faostat.fao.org/; Molnar et al. (2011); USDA Nutrient Database: http://ndb.nal.usda.gov/ndb/search/list

Common (and scientific) name & centre of origin	Major producing countries	Establishment and management	Principle food uses and nutritional value	References
Oil palm (*Elaeis guineensis*) Tropical West and Southwest Africa, between Angola and the Gambia.	Cultivated in many countries in the humid tropics outside of its African range since the mid-20th century when large-scale plantations were established in Malaysia. At present (2014-2015), Indonesia is the major producer, followed by Malaysia and Nigeria. Smaller producer countries include Thailand, Colombia, Benin, Cameroon, Kenya and Ghana. Largest importers of palm oil include India, the European Union and China.	Cultivated in mixed cropping with food crops in smallholder systems and increasingly in large scale monoculture plantations. Propagated by seeds. Weed control, insect pest and disease control, fertiliser application, irrigation and branch pruning required to sustain productivity.	An edible oil derived from the mesocarp (reddish pulp) of the fruit kernels is used for household cooking (especially in tropical Africa and Southeast Asia) and industrial food and non-food applications worldwide (e.g. margarine, cosmetics, soaps, toothpaste, waxes, lubricants and ink). From a nutritional and health perspective, palm oil has an especially high concentration of saturated fat, specifically of palmitic acid, as well as the monounsaturated oleic acid. While palm oil is an important source of calories and a food staple in poor communities, its overall health impacts, particularly in relation to cardiovascular disease, are controversial. Much of the palm oil that is consumed as food is to some degree oxidised rather than in the fresh state, and this oxidation appears to be responsible for the health risk associated with consuming palm oil.	Edem (2002); USDA Foreign Agricultural Service: http://apps.fas.usda.gov/psdonline/; USDA Nutrient Database: http://ndb.nal.usda.gov/ndb/search/list
Common fig (*Ficus carica*) Middle East and western Asia	Cultivated in many temperate and subtropical countries worldwide, particularly in the Middle East and areas with a Mediterranean climate. Major producers include: Turkey, Egypt, Iran, Morocco, Algeria, Syria, USA, Greece, Spain, Afghanistan, Brazil, Tunisia and Italy.	Propagated from seeds, but more commonly by vegetative methods, i.e., cuttings, air-layering or grafting.	Figs are consumed fresh or dried and are often processed as a paste for pastries or canned. The fruit can be fermented and distilled into alcohol. Dried figs are a rich source (> 20% of the Daily Value) of dietary fibre and the essential mineral, manganese, while vitamin K and numerous other minerals are in moderate content. Figs contain diverse phytochemicals, including polyphenols such as gallic acid, chlorogenic acid, syringic acid, (+)-catechin, (−)-epicatechin and rutin.	FAOSTAT Statistical Database: http://faostat.fao.org/; Janick (2005); USDA Nutrient Database: http://ndb.nal.usda.gov/ndb/search/list; Veberic et al. (2008)
Cocoa (*Theobroma cacao*) Southeastern Mexico to Amazon Basin	Cultivated in humid tropics. Top 10 producing countries (in 2005): Cote d'Ivoire, Ghana, Indonesia, Nigeria, Brazil, Cameroon, Ecuador, Colombia, Mexico, Papua New Guinea.	Grown both in large agroindustrial plantations and by small producers, the bulk of production coming from millions of small producers; Planted under forest shade or in monocrop plantations; propagated from seeds/seedlings. Pruning, fertiliser application, pest, disease and weed control, pod harvesting and bean processing are main cultural practices for managing cocoa plantations.	Seeds or beans contain 40-50% fat as cocoa butter used for chocolate, cocoa mass and powder; pulp used for juice, smoothies, jelly and nata; fermented pulp distilled into alcoholic beverages.	CacaoNet. (2012); FAOSTAT Statistical Database: http://faostat.fao.org/; Figueira et al. (1993)

3.3 The Influence of Forest Landscape Configuration, Management and Use on Food Security and Nutrition

Forests and associated food production systems do not exist in isolation. They are part of broader economic, political, cultural and ecological landscapes. Such landscapes usually comprise diverse patches of different land use types, which may include forest and non-forest, different food production systems, and numerous other land uses. The following discussion considers the ways in which different land use-patches interact with each other in space and time to influence the productivity and sustainability of forests and tree-based systems.

3.3.1 Interactions between Landscape Components

Positive contributions of forests to agricultural productivity

Forests provide an array of direct and indirect contributions to agriculture at different scales (MA, 2005). At the broad scale, forests contribute to the recycling of nutrients, suppression of agricultural pests, detoxification of noxious chemicals, control of hydrological processes and genetic resources for future adaptation to climate change (Foley et al., 2005; MA, 2005; Plantegenest et al., 2007). In a study carried out in 56 countries in Africa, Asia and Central/South America it was found that a ten percent increase in deforestation would result in a 4-28 percent increase in flood frequency (Bradshaw et al., 2007), with large impact on rural and agrarian populations (FAO and CIFOR, 2005; Jonkman, 2005). Forests also contribute to climate change mitigation, having the capacity to absorb a significant fraction of global carbon emissions which could have positive impacts on food production (FAO, 2012).

At the local scale, forests and trees outside forests are essential for ecosystem services such as pollination (Ricketts, 2004; Ricketts et al., 2008), pest regulation and regulation of the microclimate (Kort, 1988), as discussed in Chapter 2. They can also preserve genetic diversity of domesticated and wild food species and enhance soil fertility and agricultural productivity (Tscharntke et al., 2005a; Bianchi et al., 2006; Ricketts et al., 2008; Boyles et al., 2011). For example, 75 percent of the most important crop species benefit from pollination services (Klein et al., 2007) accounting for 153 billion Euros annually (Gallai et al., 2009). In many African countries farmer-managed forest regeneration programmes are estimated to have doubled the agricultural yields over nearly five million hectares with significant potential for the future (World Bank, 2013). Green foliage collected from forests can also represent an important resource for compost to enhance productivity of field crops, such as areca nut plantations in India (Sinu et al., 2012).

As discussed in Chapter 2 and earlier in this chapter, forests are also a direct source of food, fuel, fodder and medicines, benefiting not only people living within forested landscapes (c.f. Colfer, 2008a; Kuhnlein et al., 2009), but those living elsewhere, including urban areas. For example, it is estimated that about 2.4 billion people, or 40 percent of the population of low- and middle-income countries, rely on woodfuel for cooking, with some 746 million people boiling their water with wood (FAO, 2014).

The provision of such forest benefits can be dependent on the spatial configuration of the landscape and proximity to forests. For example, Ickowitz et al. (2014) found that after controlling for confounding factors (such as distance to market and road density) children's *dietary diversity* increased with tree cover across 21 African states. Wild harvested meat also provides a significant source of food in many regions, including for example in Central Africa where a critical portion of protein and fat often comes from this source (Nasi et al., 2008). Forests can also contribute to nutrition by providing sources of income that can be spent to buy food in markets.

Negative effects of forests on agricultural productivity

Forests can also have negative impacts on nearby agricultural production, for example by harbouring agricultural pests and diseases that reduce agricultural yield, and others that more directly harm human health. New insect pests can be introduced into an area through the transportation of wood or nursery stock associated with forestry and horticultural activities (Cock, 2003). Forest wildlife species and arthropods (insects, ticks, etc.) can spread disease pathogens and parasites to livestock and humans, such as malaria, encephalitis, rabies, Ebola, SARS, and several others (Bengis et al., 2002; Belotto et al., 2005; Colfer, 2008b; Olson et al., 2010; Tomalak et al., 2011; Wilcox and Ellis, 2006). In light of the recent West African Ebola crisis, it has been argued that these risks create an opportunity to conserve forest animal species by emphasising the dangers involved in consuming wild meat (Williams, 2014). However, this argument has been rejected by others, who emphasise the complex relationship between people,forests and hunting practices that produce the risk of disease transmission (Pooley et al., 2015).

Forests are a critical habitat for wildlife species but can also be a source of human-wildlife conflict, particularly where agroforestry buffers between forests and farms provide suitable habitat for wild species (Naughton-Treves et al., 1998). When agricultural fields, agroforestry systems or homegardens are raided by wild animals, crop damage can result in significant economic losses on farms and during post-harvest stages of food production, and in some cases total crop devastation (Ntiamoa-Baidu, 1997; Hockings and McLennan, 2012). Around Kibale National Park (Uganda) – a large forested reserve harbouring crop raiding species such as baboons and chimpanzees – average financial losses for farmers in a six month period were estimated at USD 74 with more severe crop damage closer to the park boundary (Mackenzie and Ahabyona, 2012). In the struggle to protect crops, both humans and wildlife can be put in danger, undermining conservation efforts due to increased human-wildlife conflict and increasing farm labour costs (Hill, 2000; Pérez and Pacheco, 2006). In India, elephants kill over 400 people and destroy crops valued at two to three million USD every year (Bist, 2006; Rangarajan et al., 2010).

Impacts of other land use patches on forests

Forests can be impacted positively or negatively by other nearby or distant land uses in ways that affect their own role as food production systems, as habitat for biodiversity, or their structure and function more generally. Forests located near farming and urban areas may be more exposed to air, water and other types of pollution. Forests are vulnerable to emissions of reactive pollutants such as SO2, NOx, HNO3 and NH3 as well as elevated levels of ozone and excessive mineral salts (Fowler et al., 1999; Likens et al., 1996). These potentially phytotoxic pollutants, largely studied in the northern hemisphere, are damaging to forest health although it is difficult to identify specific pollutant effects given the high level of interactivity between pollutants, and between pollutants and climate change (Bytnerowicz et al., 2007; Paoletti et al., 2010). Atmospheric pollutants can also severely damage forests through acid rain (Likens et al., 1996).

Proximity to human settlements and roads can increase the likelihood of *invasive species* being introduced to, and perhaps damaging, forest environments (Bradley and Mustard, 2006; Bartuszevige et al., 2006). In most cases the introduction of non-native species may have little impact since they often fail to survive in a new habitat. However, those that do become established and thrive can cause severe and widespread economic and ecological losses, such as a reduction in forest and agricultural productivity, species population declines and even extinctions (Holmes et al., 2009). For example, in Canada the Asian longhorned beetle (*Anoplophora glabripennis*) threatens the hardwood and maple syrup industries, while the impacts of yellow star thistle (*Centaurea solstialis*) on cattle production have cost Californian ranchers and the state an estimated USD 17 million (Eagle et al., 2007). In French Polynesia and other Pacific islands, *Miconia calvescens* (an introduced tropical American tree), has shaded out native plant species in some areas and, due to its shallow rooting habit, increased erosion and frequency of landslides (Meyer and Malet, 1997; Environment Canada, 2004; Moore, 2005).

Scale and fragmentation issues

Many of the interactions described above are influenced by the scale and spatial configuration of different land use patches. The process of *forest fragmentation*, occurring when formerly forested lands are converted permanently to pastures, agricultural fields, or human-inhabited developed areas, can result in changes in ecosystem functions that alter the supply and distribution of ecosystem services vital for agriculture (Tscharntke et al., 2012). Reduced connectivity of forest patches affects the ability of pollinators, pest predators (Tscharntke et al., 2005b; Kremen, 2005), water and nutrients (Brauman et al., 2007; Power, 2010) to move across a landscape. However, there is growing evidence that in agricultural landscapes forest fragments continue to provide ecosystem services, including pollination and pest control services (Ricketts, 2004; Ricketts et al., 2008; Holzschuh et al., 2010), water regulation and purification services (Foley et al., 2005). Forest fragments in agricultural landscapes can also change dispersal patterns for fungi and soil organisms that affect decomposition (Plantegenest et al., 2007). In some cases, managing landscape configuration to enhance forest fragment connectivity may be a more effective tool for optimising agricultural landscapes for multiple ecosystem services rather than simply limiting further forest loss (Mitchell et al., 2014). It is however important that sufficiently large forest patches and connectivity are maintained, as high levels of forest loss can result in abrupt landscape-scale loss of native forest specialist species in the long term (Pardini et al., 2010).

In many parts of the world, traditional agricultural landscape management approaches have been developed to more closely link agricultural and forest (or woodland) management and ensure continuity in the provision of ecosystem services from forests. For example, Japan's traditional socio-ecological production landscapes, known as *satoyama* ("sato" =home village; "yama" =wooded hills and mountains), comprise integral social and ecological networks of villages and their surrounding agricultural lands, open forestlands and forests, in which forests are managed for multiple values, including biodiversity conservation and the ecosystem services that forests and woodlands provide to agriculture (Indrawan et al., 2014). Similar landscape management systems are found throughout Asia and elsewhere in forms that are adjusted to regional biophysical conditions (e.g. Agnoletti, 2006; Bélair et al., 2010; Johann et al., 2012; Kumar and Takeuchi, 2009; Ramakrishnan et al., 2012; Youn et al., 2012).

3.3.2 The Influence of Landscape Use and Management of Forests and Tree-based Systems on Nutrition

Many factors influence the actual or potential contributions of forests and tree-based systems to food security and nutrition of producers, their families and other consumers. These include the productivity of these management systems, the resilience of these systems to withstand shocks (weather and other events), the choice of food species cultivated and managed, and the extent to which the food products are utilised for household or local consumption, or marketed to earn income which may then be used to purchase other foods. The variety of forest and tree management practices that typically co-exist within rural landscapes may contribute to the broader food system in varying degrees, since a substantial portion of people's diet is often traded or purchased (Powell et al., 2015).

Two main types of studies can be used to evaluate how different landscape, forest and tree management approaches may impact nutrition. The first type involves studies that compare the diets of one or more ethnic groups at different stages of transition from one livelihood strategy to another, with the different livelihood strategies having different land use patterns. A selection of such studies and their main results are summarised in Table 3.4.

Table 3.4 Studies examining differences in diet between groups during livelihood and land use transitions.

Transition/ Location	Findings related to diet	Study
Shifting cultivation to plough-farming in the Philippines	Two Tiruray communities at opposite poles of this transition were studied. Hunting, fishing, and gathering of wild resources have virtually disappeared. Reliance on wild food resources diminished, with greatly increased dependence on market foods. The traditional communities had lower average intake of energy, protein, fat, calcium, iron, vitamin A and higher average intake of thiamine and riboflavin (B vitamins) compared to those in sedentary agriculture.	Schlegel and Guthrie (1973)
Comparison of diets of tribes with settled/ paddy-based agriculture, to those with shifting cultivation and those with hunting and gathering, in India	A comparison of tribes from northeast India shows that those that engaged in the most hunting (Padams) had highest percent energy from protein, highest iron, calcium and vitamin A intake. The tribe with least animal source foods (Noktoe) had second highest vitamin A intake, likely due to greater dependence on wild and cultivated vegetables. The tribes practising mixed shifting and paddy cultivation (Padam, Minyong and Galongs) had better diets than those without paddy cultivation (Nokte). In central and western India, a hunter-gatherer forest dwelling tribe (Marias) had lowest calcium, iron and vitamin A intake. Forest dwelling subsistence agriculture tribe (Baiga) had highest iron, vitamin A, compared to settled rice-based agricultural tribe (Gonds), despite much higher energy intake by Gonds.	Gupta (1980)
Hunter-gatherers in transition to settled agro-pastoralism; San of /ai/ai, in Botswana	Traditional (hunting and gathering): Percentage of caloric intake from: vegetables (85), meat (12), milk (1), maize (2). Mixed (diet of wild and domestic food): Percentage of caloric intake from: vegetables (65), meat (11), milk (17), maize (7). Settled (agro-pastoralism): Percentage of caloric intake from: vegetables (10), meat (10), milk (29), maize (43), sugar (9). Settled communities have much lower contribution to diet from vegetables and meat and much greater intake of milk, maize meal and sugar.	Hausman and Wilmsen (1985)
Comparing hunter-gatherers to neighbouring agricultural communities in Cameroon	Yassa: Agriculture and fish-based subsistence. Average daily per capita intake: 34g of vegetables; 199g fish; 24g meat. Mvae: Subsistence based on agriculture and hunting (in forest and on coast). Average daily per capita intake: 100g vegetables; 62g fish; 129g meat. Bakola: hunter-gatherer based subsistence. Average daily per capita intake: 54g vegetables; 22g fish; 216g meat. Much higher intake of meat and high animal source food intake in hunter-gatherer group, higher vegetable consumption in agricultural community.	Koppert et al. (1993)
Hunter-gatherer to sedentary urban/ agriculture in Borneo	Remote/traditional communities had more diverse diets with more meat, better nutritional status and physical fitness and greater contribution of forest resources to diet compared to sedentary agricultural or urban communities.	Dounias et al. (2007)
Hunter-gatherer to market-oriented rice cultivation in Borneo	People in resettled area with better access to markets, where people's livelihood strategies focus on market-oriented rice production had poorer diets compared to those in a remote area (possibly due to lower use of wild foods and less time for production of non-staples)	Colfer (2008a)
Agricultural community in forested landscape mosaic, transition after introduction of payments for ecosystem services (PES) in Mexico	Community perceived loss of food security, and greater dependence on purchased food. They perceived lower maize yields due to shorter fallows (less agricultural land/no new land available), lower meat consumption (no more hunting, all meat now has to be purchased and the money from PES cannot fully compensate for loss of hunting).	Ibarra et al. (2011)

Other studies that have compared the capacity of different forests and tree-based systems to produce nutritionally-important foods such as fruits and vegetables and animal sources of foods (usually done by modelling) offer insights as to their relative contribution to diet and nutrition. Differences in the diets of traditional hunter-gatherer communities and neighbouring agricultural ones in India seem to be very context specific (sometime better, sometimes worse). In many places more traditional subsistence groups had more meat in their diets, based on studies from India (Gupta, 1980), Cameroon (Koppert et al., 1993), Borneo (Colfer, 2008a; Dounias et al., 2007) and Botswana (Hausman and Wilmsen, 1985). Comparing primary forests with secondary or heavily modified forest systems, the latter provide a greater number and quantity of useful plant species (but not always animal species) than primary forests, based on studies from the Brazilian (Parry et al., 2009), Bolivian (Toledo and Salick, 2006) and Peruvian Amazon (Gavin, 2004) and from Panama (Smith, 2005). Considering shifting cultivation, the abandonment of this practice may be associated with less use of wild foods including wild meat and vegetables (and uptake of micronutrients such as iron and vitamin A), but the few existing studies have not demonstrated that shifting cultivation is associated with better dietary intake, based on studies in the Philippines (Schlegel and Guthrie, 1973) and India (Gupta, 1980). Complex agroforests have been found more likely to provide enough fruits and nutrients per unit of land than less diverse agroforestry systems, based on results of farm modelling studies from Central America and West Java) (Cerda et al., 2014; Marten and Abdoellah, 1988). Regarding home gardens, four separate reviews of the impacts of agricultural interventions on nutrition outcomes all concluded that there is convincing evidence for the positive impact of home garden interventions on nutrition, especially access to fruits and vegetables and intake of vitamin A (Berti et al., 2004; Girard et al., 2012; Masset et al., 2012; Powell et al., 2015; Tontisirin et al., 2002).

More research is needed into the detailed contribution of different forms of forest and tree management systems to nutrition.

Village near Corbett National Park, India. Photo © PJ Stephenson

3.4 The Socio-economic Organisation of Forests and Tree-based Systems

3.4.1 Introduction

The viability of production system options available to farmers, including forests and tree-based systems, is influenced by an array of biophysical and socio-economic factors. Understanding both the opportunities and constraints on the retention or adoption of these production options is of prime importance to all concerned with enhancing the food security and nutrition of farmers and rural communities as well as the urban and increasingly globalised populations whose food they produce.

Challenges faced by families and communities that rely on forests and tree-based systems for their food security and nutrition include heterogeneous and unpredictable environmental conditions (e.g. unpredictable weather exacerbated by climate change, fragile and/or marginal soils), forest degradation, deforestation and associated biodiversity losses. Production systems are also embedded in underlying "invisible" social, economic and political structures, and are influenced by social and cultural norms, values, beliefs, customs and traditions. Such factors determine social and *gender* relations and their interaction within production systems, and shape the cultural identities of different ethnic and social groups and communities and indigenous peoples, and their food and livelihood preferences and choices. Social, economic and political structures also embody power relations which determine access to land, trees and other productive resources, and participation by different stakeholders in forest and natural resource governance mechanisms and the resulting outcomes in terms of resource appropriation or sharing and conflict resolution.

The socio-economic organisation in the four production systems identified earlier in this chapter is highly diverse and complex, with considerable variations between and within continents and countries. Even a single landscape often comprises peoples or social groups of different ethnic or religious affiliation, class, caste, political ideology or agricultural profession (pastoralists, sedentary farmers, foresters, plantation managers, hunters and gatherers) who may have overlapping, complementary or quite distinct production systems.

This section concentrates on the three factors directly affecting the socio-economic organisation of production: land and tree tenure, gender relations, human capital and financial capital (including credit), with a focus at the community and household level. These factors and their interrelationships are constantly evolving in response to external changes that include: shocks (such as drought, disease, food price hikes), longer-term climate change trends, public action (policies, laws, administrative procedures), infrastructrure development, innovations and new technologies, improved extension services, changes in governance frameworks and institutions, popular demand voiced through protest and social movements, and new opportunities brought about by changes in markets for land, labour, agricultural and tree products, and forest sub-soil resources (such as minerals, fossil fuels). While the drivers of these changes are discussed in Chapter 4, the implications for the socio-economic organisation of production in forests and tree-based systems are addressed in this section, with particular focus on the livelihoods, food security and nutrition of the poor.

3.4.2 Land, Tree and Related Natural Resource Tenure

The four forest- and tree-based systems described earlier in this chapter (Section 3.2) are governed by a web of highly complex land tenure systems in which rights to land, trees and other natural resources such as water are commonly categorised as: private, communal, open access and state (Box 3.3). The related tenure rights can be defined through formal or statutory legal arrangements (*de jure*), which predominate in private or state land, or by customary practices (*de facto*) which are prevalent in communal and open access regimes.

Box 3.3 Land tenure categories

Representing the relationship, whether legally or customarily defined, among people, as individuals or groups, with respect to land (including land-related natural resources such as water and trees), land tenure is commonly categorised as:
Private: the assignment of rights to a private party who may be an individual, a married couple, a group of people, or a corporate body such as a commercial entity or non-profit organisation. For example, within a community, individual families may have exclusive rights to agricultural parcels and certain trees. Other members of the community can be excluded from using these resources without the consent of those who hold the rights.
Communal: a right of commons may exist within a community where each member has a right to use independently the holdings of the community. For example, members of a community may have the right to graze cattle on a common pasture.
Open access: specific rights are not assigned to anyone and no-one can be excluded.
State: property rights are assigned to some authority in the public sector. For example, in some countries, forest lands may fall under the mandate of the state, whether at a central or decentralised level of government.
Source: FAO, 2002a.
Note: The rights to subsoil resources such as minerals, natural gas and oil are almost always reserved for the state (RRI, 2012).

Forests and tree-based systems are characterised by different land right regimes (defined in Table 3.5), though there are marked context-specific variations in practice. Shifting cultivation is practised generally on land that is not privately owned while agroforestry is commonly practised on private land in South Asia, parts of North Africa, and Europe and on communal land in sub-Saharan Africa. Plantations and smaller tree crop stands grown by corporations/large farmers and smallholders respectively are usually on private land which provides the tenure security needed to protect costly, long-term investments. However, in countries where communal tenure is fairly secure, smallholder tree crops are also found on communal land (for example, cocoa trees in Ghana (Quisumbing et al., 2003), or oil palm on collectively-held customary land in Indonesia (Li, 2014)). Corporations quite commonly lease state land for tree plantations, for example, in Indonesia for oil palm (Li, 2014) and in many countries in Southeast Asia, Africa and Latin America for industrial timber concessions (c.f. Hatcher and Bailey, 2010). Finally, all four types of tenure can apply to managed forests, with the actual distribution by tenure varying by region and country.

Table 3.5. Generalised overview of types of tenure rights associated with forests and tree-based systems.

Forest/Tree-based system	Rights			
	Private	Communal	Open Access	State
Managed forest	✓	✓	✓	✓
Shifting cultivation		✓	✓	✓
Agroforestry	✓	✓		✓
Single-species tree crop systems	✓	✓		✓

Bundles of rights, incentives and food security

In practice, different tenure regimes can co-exist in the same landscape, and even within some tenure regimes two or more individuals or groups can have different rights to a specific area of land or related natural resources (such as trees), either simultaneously or in different seasons. Thus it is useful to think of "bundles of rights" that can be held by different holders of the rights (FAO, 2002a; Bomuhangi et al., 2011). A frequently-used classification, developed by Schlager and Ostrom (1992), distinguishes: access, withdrawal, management, exclusion and alienation rights. Access rights enable entry to the land, such as the right to walk in a forest. Withdrawal rights include the right to take something from the land, such as forest foods, firewood, timber. While in many countries communities have withdrawal rights for subsistence or small scale commercial activities, in some cases such as Thailand, legislation does not recognise customary rights of forest communities, rather criminalising extraction of forest products and land occupation (RRI, 2012). Management rights cover the right to use or change the land, such as to plant trees or crops or to graze animals, or to make improvements to the land, such as better water management. In many countries, traditional management systems developed by local communities and indigenous people to regulate access and withdrawal rights by community members have been replaced by government-authorised systems, subject to certain conditions. These can bring benefits, for example, in reducing deforestation and increasing community access to fuelwood and fodder and control over NTFPs, but they can also weaken a community's capacity to function flexibly and effectively to meet community needs for food and other livelihood requirements (Larson et al., 2010; RRI, 2012; Barry and Meinzen-Dick, 2014). Exclusion rights prevent others from using the land or resource, while alienation rights enable the transfer of land to others, by sale, lease or bequest.

Table 3.6 illustrates the complexity of these bundles of rights for the four forest- and tree-based systems. While not compatible with systems of shifting cultivation, private tenure permits all five rights (i.e. "full ownership") in the other three systems. Communal right regimes operate in all four systems, and are particularly extensive in Latin America and Africa. They are usually managed by (informal) community mechanisms (sometimes government-authorised under specific conditions) and enjoy some exclusion rights. Importantly, they do not have alienation rights. Open access regimes are confined to shifting cultivation and, in a few countries, some managed forests, where users only have access and withdrawal rights. Finally, in most countries the state owns the major share of managed forests and tree plantations, commonly delegating management rights to state bodies and/or formal community organisations under strict conditions, or leasing land for tree plantations to corporate bodies with all rights except alienation.

Table 3.6 Bundles of rights typically associated with different forest- and tree-based systems. *Source:* Adapted by authors from FAO, 2002b and Schlager and Ostrom, 1992. (The tenure categories are taken from FAO, 2002b, given in Box 3.3, and also used in Table 3.5).

Forest/ Tree based Systems and Tenure	Rights				
	Access	Withdrawal	Management	Exclusion	Alienation
Managed forest					
Private	✓	✓	✓	✓	✓
Communal	✓	✓	CG	CG	X
Open Access	✓	✓	X	X	X
State	✓	✓	✓	SB / CG (CO)	✓
Shifting cultivation					
Communal	✓	✓	CG	CG	X
Open Access	✓	✓	X	X	X
State	✓	✓	✓	CG (CO)	✓
Agroforestry					
Private	✓	✓	✓	✓	✓
Communal	✓	✓	CG	CG	X
State	✓	✓	✓	SB (CO)	✓
Single-species Tree Crop systems					
Private	✓	✓	✓	✓	✓
Communal	✓	✓	CG	CG	X
State	SB / CB	SB / CB	SB / CB	SB / CB	SB

(CG) Traditional Community Groups; (CB) Corporate Bodies; (SB) State Body; (CO) Community Organisation with formal/legal rights and obligations. X = Not permitted

More recently, the Schlager and Ostrom (1992) classification has been expanded (RRI, 2012; Stevens et al., 2014) to include the dimensions of duration and extinguishability.

Duration considers whether the rights are held in perpetuity or for a limited time period. Permanent rights are vital to safeguard the sovereignty and autonomy of indigenous peoples (RRI, 2012) and because "indigenous people's right to food is inseparable from their right to land, territories and resources, culture and self-determination" (Damman et al., 2013). Often, in customary systems the duration of rights is determined by evidence of continuous use (e.g. in Meghalaya, India (Kumar and Nongkynrih, 2005); and in Gambia (Dey, 1981)). Long-term rights provide security and incentives to invest and maintain sustainable forest and tree management practices (RRI, 2012). In Viet Nam, for example, long term (50 years or more) use rights to forest lands have been secured through Land Use Certificates, with a total of 1.8 million certificates having been issued by December 2010 (FAO, 2014).

The right of extinguishability ensures "due process and compensation" when governments exercise their universal right of "eminent domain" to expropriate lands for the "public good". While private land owners as well as communities and indigenous peoples with *de jure* use rights to state or communal forest land generally have legal entitlements to due process and compensation, communities with *de facto* rights are vulnerable to losing their land and their livelihoods (RRI, 2012). For example, herders in Mongolia protested at government issuance of gold mining rights to national and foreign companies, as they lost pastures and forests and their water was polluted by the mines (New Zealand Nature Institute, 2006). Logging concessions as well as illegal logging on indigenous peoples' land in Indonesia and Peru, have displaced thousands of people from forests on which they depend for their food and livelihoods (United Nations, 2009). Even with official *de jure* rights, in many instances weak government protection may make it difficult for communities to assert their rights. For example, Peru and Colombia have ratified various international conventions and covenants regarding indigenous peoples and the right to adequate food for all, and have demarcated and titled a large part of indigenous and community land, yet they have authorised hydrocarbon and mining companies to operate on this land, without consultation or consent by the indigenous peoples and communities concerned.

Multiple rights to a specific parcel of land or to specific natural resources on it can be held simultaneously or successsively by several people or groups (Bruce, 1999; Fuys and Dohrn, 2010). These complex rights mean that even a single landscape that might contain forests, agroforestry with trees, crops, pastures and animals, and lakes/rivers, would be subject to a web of different property rights regimes or, as conceptualised by Bruce (1999), "tenure niches". For food security and livelihoods, it is important to recognise that these "bundles of rights" can be further broken down, with different individuals, families, kinship and other groups (cross-cut by gender, class and agricultural specialisation) accessing different "rights" to the same resources. The exercise of these rights can be complementary, for example, where some people (especially men) may have ownership or usufruct rights to trees, and others (especially women) to certain products from these trees such as fruit and small branches for fuel (Rocheleau and Edmunds, 1997). In Zimbabwe for example, in communal tenure systems among the Baganda, only men use fig trees (*Ficus natalensis*) to produce bark cloth, hang beehives and create boundaries while only women use figs for soil improvement and as shade for other crops. In northern Thailand upland residents have rights to collect bamboo on individually-owned lowland farms (Fuys and Dohrn, 2010).

Rights to trees may be different from rights to the land on which they grow, particularly in the case of customary tenure systems (Howard and Nabanoga, 2007). However, even under private tenure, they may be different, for example, in Morocco the state owns argan trees even if they are grown on private land (Biermayr-Jenzano et al., 2014). Under customary tenure, an individual's rights to trees may depend on his/her rights to the land on which they are grown, while planting trees can also establish rights to land. However, bundles of rights to trees and their products can also be held by different individuals (with or without the land ownership or use rights), simultaneously or at different times, for different purposes (Fortmann and Bruce, 1988). These rights are often nested and layered in space as well as among rights holders, creating differential entitlements to benefits that are also related to the broader social structures (Howard and Nabanoga, 2007), and the social and religious/spiritual norms, values and practices of the concerned communities.

The exercise of multiple rights can cause conflicts despite the existence of mediation mechanisms (Bruce, 1999). For example, in the state-owned argan forest areas in southwestern Morocco, tensions are rife between nomadic camel and goat herders with grazing rights and local residents with rights to exploit the argan fruit (Biermayr-Jenzano et al., 2014). In Senegal, disputes between Wolof farmers and Peul herders over the use of branches from the baobab trees for fodder undermined the Peuls' food security and livelihoods. These disputes were exacerbated by a government decree protecting the baobab tree (Rose, 1996).

As Schlager and Ostrom observe (1992) "Different bundles of property rights, whether they are *de facto* or *de jure*, affect the incentives individuals face, the types of actions they take, and the outcomes they achieve". These rights are ultimately critical for ensuring food security and nutrition.

3.4.3 Gender, Rights to Land and Trees, and Food Security

Reviewing country-level statistics and a large number of field studies, Lastarria-Cornhiel et al. (2014) conclude that most land tenure systems are gender-biased, allocating primary rights to land to male members of the community and family. Gender differences in ownership or use rights to trees are particularly complex and vary by culture. In many countries, trees on state, community or open access land belong to the state. Women in matrilineal systems often have stronger rights, though sometimes these are controlled by their brothers or maternal uncles. Gender differences in the way land is accessed also contribute to differences in tenure security. In sub-Saharan Africa, men often acquire use and management rights to land through inheritance or allocation by their clan or lineage, while women more commonly acquire temporary use rights (and occasionally permanent rights) through marriage and to a considerably lesser extent through fathers and brothers (Rocheleau and Edmunds, 1997; Howard and Nabanoga, 2007; Kiptot and Franzel, 2012; Meinzen-Dick et al., 2014; Lastarria-Cornhiel et al., 2014). In such customary systems, women frequently lose their land use rights if their marriages are dissolved (through separation, divorce or death of their spouse), particularly if they do not have sons. In Latin America, women are more likely to acquire land through inheritance (so their rights are not affected if their marriages dissolve) and men through purchases in land markets (Doss et al., 2008). Paradoxically, the emergence of land rental markets in customary systems, particularly in sub-Saharan Africa, can facilitate women's access to land as male owners are more ready to rent to women because they are prohibited from acquiring permanent land rights (Giovarelli, 2006; Lastarria-Cornhiel et al., 2014).

Rural men and women often acquire different types of assets (Meinzen-Dick et al., 2014). Men are more likely to own large livestock such as cattle and buffaloes and women small livestock such as poultry and goats (Kristjanson et al., 2014). In rural Philippines women tend to have higher educational levels (and thus better access to non-farm work) while their brothers are more likely to inherit family land (Quisumbing et al., 2004). In Asia, women are more likely to own jewellery, and men are more likely to own land and assets such as farm equipment and vehicles (Agarwal, 1994b; Antonpoulos and Floro, 2005, cited in Meinzen-Dick et al., 2014).

Where the state owns trees, the use rights are either vested in the community, which exercises management responsibilities or in the male leaders of the lineage or households (Rocheleau and Edmunds, 1997). Often the effectiveness of women's rights depends on their voice in local institutions that are commonly male-dominated (Agarwal, 2010; Lastarria-Cornhiel et al., 2014). In the case of community-owned land and state land managed by communities, women often have secondary rights legitimised through their relationship to men. Howard and Nabanoga (2007) found highly complex gender-differentiated rights to trees and their products among the Baganda in Uganda that varied according to their location in homesteads, croplands, common lands or state forests. While only men owned trees on private land, women's customary rights to plant resources in gendered spaces on common or state land were as strong as men's. Rocheleau and Edmunds' (1997) review of studies in Africa also found that women's rights are substantial, particularly in customary systems where they have rights to fuelwood, medicinal plants and wild foods in the "bush" or forests, in "in-between" spaces not valued by men, such as bush along roadsides, fences, and boundaries between men's trees and crops, as well as home gardens near their houses, and also to certain tree products (e.g. fruit, fuelwood, leaves, fodder) growing on men's land. Agarwal (1994b) found that in Sri Lanka women sometimes received coconut trees as dowry and their brothers would periodically send them a share of the harvest.

However, these cases cannot be generalised, even in customary systems. For example, in Ghana, women have been able to acquire their own trees, through acquisition of private land through the market and sale of cash crops such as cocoa (Berry, 1989, 1993 cited in Rocheleau and Edmond, 1997; Lastarria-Cornhiel, 1997) or as gifts of cocoa trees from their husbands in compensation for their labour on the men's cocoa trees (Quisumbing et al., 2003). In the Colombian Pacific region, Afro-Colombians have highly complex tenure systems that permit both men and women to own trees that they have planted or inherited, and their products such as fruit and tree snails (Asher, 2009).

The nature and security of women's rights to land, trees and their products are of central importance to ensuring household food security. Gender differences in the types and relative sizes of productive assets and control of income are critical for food security as a large body of evidence shows that women are more likely to spend their income (from their own production or wage labour) on food, healthcare and education of their children (Haddad et al., 1997; Agarwal, 1997; Njuki et al., 2011; FAO, 2011; Kennedy and Peters, 1992; Duflo and Udry, 2004; Meinzen-Dick et al., 2014).

Forest and agriculture mosaic landscape, Cat Ba, Vietnam. Photo © Terry Sunderland

The interrelationships between women's rights to trees and their products and household food security and nutrition raise two major issues. The first is the need for women's security of tenure. This is clearly demonstrated by Fortmann et al. (1997), who found in their study of two Zimbabwe villages in the communal areas that women were much less likely than men to plant fruit and other trees within the homestead or on household woodlots because the trees and their produce belonged to their husbands (as household head), and they lost their use rights to the produce if he died or they divorced (even if they still lived nearby). However, both men and women were equally likely to plant trees on community woodlots where the duration of their rights to the trees was secure as long as they remained village residents. Furthermore, while richer men planted considerably more trees than poor men, indicating a greater ability to engage in commercial production, this was not the case for richer women who planted a few trees for subsistence and had less risky ways of earning, such as producing annual crops for sale, beer brewing and handicraft sales.

The second issue is the complementarity between men's and women's access to different products from the same trees, sometimes in different seasons, and from different tenure systems. For example, in Uganda, jackfruits located in different areas are used differently by men and women. Women reported 60 percent of uses in homegardens, which were mainly for subsistence especially during periods of food shortage (they use leaves for fodder and medicine) while men reported over 80 percent of uses on croplands that were for sale and subsistence, as well as fuel. Jackfruits on common land and in state forests were only used for subsistence fuel (Howard and Nabanoga, 2007).

Land ownership or use rights may not be sufficient to exercise control over the use, management and the products of trees on their land (Agarwal, 1994a; Rocheleau and Edmunds, 1997; Deere et al., 2013). Even where women have land ownership rights, research in the Gender Asset Gap Project in Ecuador, Ghana and the state of Karnataka in India found that land did not automatically translate into decision-making on what to grow, how much of the crop to sell, and over the use of the income generated from crop sales (Deere et al., 2013).

3.4.4 Human Capital, Control and Decision-making in Forests and Tree-based Systems

Rights to forests and trees and their products are embedded in the broader social systems that also determine access to human and financial capital, decision-making processes and control of the products or income from their sale, thus affecting the way in which these property rights are used. Since social systems are not static, these rights can be negotiated or changed over time (Meinzen-Dick et al., 1997; Rocheleau and Edmunds, 1997).

In many customary and open access tenure systems, the notion that individuals own their labour power and the products of their labour is widespread. Rights to forest land and trees are commonly established by the act of clearing primary forest. For example, the Lauje in Sulawesi, Indonesia, considered that the person who invested labour in clearing land or planting trees owned the land and the trees, and could alienate these through gift, sale or exchange (Li, 1998). Similarly, in sub-Saharan Africa, rights to land are derived from the labour expended to clear or cultivate the land. Land is commonly held under lineage-based systems, in which a male lineage member is entitled to land to support his family, and can use this as long as it is being cultivated. His heirs would normally be given the land that was cultivated at the time of his death (Platteau, 1992). Women are sometimes prevented by men from clearing land, for example, in The Gambia, as this would make the land "women's property" and their husbands or other male relatives would have no control over it if their husbands died or they divorced (Dey, 1981).

In open access and communal forest systems (including local and indigenous communities' formal or informal use of state land), the availability of human capital (commonly proxied as labour and education (Meinzen-Dick et al., 2014), though also covering traditional knowledge and skills and health that are less easily quantified) is one of the main factors affecting the ability of an individual, household or community to clear, maintain, and use forests and tree products. While labour is a key factor, specialised knowledge and skills that are often gender- and age-specific are also critical. For example, women often specialise in forest medicinal plants and fuelwood, and men in hunting wild animals for food, while either may have rich knowledge of other foods and fodder, depending on their cultures.

Often very poor families with few resources except their labour are highly dependent on forest products for their food security and livelihoods (Jodha,1986; Fisher, 2004; Adhikari, 2005; Narain et al., 2008). However, the literature indicates that while resource dependence (defined by Narain et al., 2008, as the share of resource income in overall income) tends to decline with overall income, the relationships are complex and there is no consistent trend. For example, Fisher (2004) and Narain et al. (2008) found that forest income declined with the household head's level of education in Malawi and Madhya Pradesh (India); similarly, Adhikari (2005) found that in Nepal, forest income declined with the household's average level of education. Both Adhikari and Narain et al. found that forest income increased with household livestock holdings as such households required more fodder. The results were also affected by the availability and type of labour, and by education/skills.

More remote villages may have higher dependence on forest resources as they have fewer opportunities for off-village labour, and are likely to have higher costs for purchasing resources and food (Narain et al., 2008). Duchelle et al. (2014) found that in the more remote communities in Pando (Bolivia) forest income made up 64 percent of total household income compared with only 12 percent in the region of Acre in Brazil, just across the border, which is better connected to markets and towns, and off-farm work opportunities.

Agroforestry systems (on private or communal land), woodlots and small tree stands are becoming an increasingly important smallholder livelihood strategy in many countries for a variety of reasons (see Section 3.2.4) of which a critical one is labour. Trees demand less labour than most field crops and are attractive where labour is scarce, expensive or difficult to manage. Households with sufficient income from non-farm sources, which therefore may not need to cultivate their land intensively, may also plant trees to provide food and other products, or to retain surplus land as an alternative to renting out or selling the land (Arnold and Dewees, 1998).

Shortages of labour (especially male labour) as well as land are leading to shorter fallows and longer cultivation periods in many shifting cultivation systems (Hunt, 1984; AIPP and IWGIA, 2014). Land shortages, for example, in the uplands of Southeast Asia, are the result of increasing population densities from endogenous growth and in-migration by large numbers of lowlanders, as well as loss of access to land taken over by the governments (Cairns and Garrity, 1999). Analyses of studies from across

Southeast Asia have shown that increasing returns to labour is usually much more important than increasing yields per unit of land area (Cairns and Garrity, 1999).

The intrahousehold division of labour and control of the product, by gender and age, is highly complex across and within forests and tree-based systems, regions, countries and cultures. In many cases women provide substantial labour and management of particular forest/tree products but men control the disposal or marketing of these products and the distribution/use of the benefits (World Bank et al,. 2009; Rocheleau and Edmunds, 1997). Case studies in seven Asian countries showed that indigenous women perform about 70 percent of the work in shifting cultivation. Men identify suitable land and do the hard physical work in land preparation. Women also help in clearing the land, selecting seeds and weeding, while both men and women harvest and conduct the rituals during the cultivation cycle together (AIPP and IWGIA, 2014). In some parts of Africa, women are involved in small retailing of forest products and men in wholesale trade (Kiptot and Franzel, 2012). This may affect incentives to increase production and sustainable resource management, with negative implications for improving food security and livelihoods. Based on her field work in Africa, Whitehead (1985) distinguishes between sex-sequential labour processes on a single product and sex-segregated labour processes on similar or different products. She considered women's claim on the product of their labour to be weaker in the first case, as their contributions were submerged in the conjugal role. In contrast, in Southeast Asia, Li (1998) found that the key issue was not the division of labour itself but the extent to which labour investment is directly connected to the creation of the property.

Women are often disadvantaged in access to and control of agricultural labour (Dey Abbas, 1997; FAO, 2011; Hill and Vigneri, 2014). Kumar and Quisumbing (2012) found that in Ethiopia, female-headed households tended to be smaller than male-headed households, and have a larger proportion of female members which disadvantaged them as many agricultural operations are male-intensive. This is particularly the case for ploughing, a task which cultural norms proscribe for women. Similar constraints were reported for Botswana (Fortmann, 1983; Peters, 1986) and Zambia (Feldstein and Poats, 1990). In many sub-Saharan African countries, women are also obliged by custom to provide labour, food and sometimes cash crops for male-controlled households. These obligations often take precedence over women's rights to work on their personal fields, trees or other income-generating activities (Dey Abbas, 1997; van Koppen, 1990; Hill and Vigneri, 2014). Women also have heavy domestic demands on their labour, which limits the time they can spare for their agricultural work (Quisumbing and Pandofelli, 2009).

Interestingly, despite women's labour and cash/credit constraints, female-managed cocoa farms in Ghana were as productive as male-managed farms (Hill and Vigneri, 2014). Women were able to compensate by using labour exchange groups and relying more on labour-intensive production methods rather than the use of purchased modern inputs. This balancing of labour and non-labour inputs confirms the review of evidence in FAO (2011) that women are as productive as men, if they have the same level of inputs.

3.4.5 Financial Capital and Credit: Using and Investing in Forests and Trees

Financial capital includes savings/debt (including in banks, credit unions, cooperatives, informal savings clubs or tontines), gold/jewellery income, credit, insurance, state transfers and remittances (Carloni, 2005; IFPRI, 2013). Savings are often in the form of livestock assets, for example, as is the case in Acre (Brazil) (Duchelle et al., 2014).

It is frequently argued that poor households (especially those headed by women) are more dependent on forest resources for food and income than richer households although the evidence is mixed (Adhikari, 2005). A growing body of evidence suggests that the role of capital and/or credit is critical in enabling households or individuals to exploit forest resources. For example, a study by Adhikari (2005) in Nepal found that households with land and livestock assets gained more from community forests because they were able to make greater use of intermediate forest products such as leaf litter, fodder and grass products. Female-headed households benefitted less than male-headed households, as they had fewer livestock and had minimal involvement as office bearers in the forest user groups. These findings are consistent with those of Velded (2000) who found that the benefits from common grazing land among the Fulani in Mali were exclusively related to capital, technology and skill levels, and those of Narain et al. (2008) in relation to complementarity of asset ownership in Jhabua (India).

For the majority of smallholders in local or indigenous communities, forest income is often insufficient to support investment in forest and tree resources. A number of countries have introduced small grants and microcredit schemes for smallholders, sometimes through the mechanisms of producer cooperatives or, particularly in Latin America, by facilitating relations beween banks and small forestry producers (FAO, 2014). In Viet Nam, through its 2007 Decision 147 on the promotion of forests for productive purposes, the government encouraged households to engage in the plan to establish 250,000 ha of new plantations per year till 2015 by providing low credit rates for smallholders (FAO, 2014).

These schemes seem to neglect earlier evidence (Arnold and Dewees, 1998) which showed that tree planting only requires low inputs of capital and that subsidies can lead to adoption of inappropriate tree species or lead to distortions in land use. Arnold and Dewees (1998) also refer to widespread evidence that seedling distribution, fertiliser and cash subsidies tend to be captured by larger farmers, who are not food insecure.

The adaptation of shifting cultivation systems to "dual economies" among many indigenous communities in Asia reflects also the importance of improved market access as well as greater opportunities to access credit or wage labour to invest earnings in farming and improve food security and livelihoods (AIPP and IWGIA, 2014). The report by AIPP and IWGIA (2014) provides examples of resulting innovative combinations of shifting cultivation with agroforestry (e.g. fruit and cashew orchards in Cambodia, rubber gardens in Indonesia), growing high value cash crops in shifting cultivation fields (e.g. vegetables, herbs, ginger, turmeric in India and Bangladesh), establishing separate, permanent fields for cash crops (e.g. tobacco, maize, flowers, pineapple, vegetables in Thailand, India, Bangladesh) and improving fallow management by planting specific trees in India.

Numerous studies cite evidence that women generally have less access to capital than men. They are often prevented by social norms or their heavy domestic and caring work from engaging in paid work outside the home or community (where wages are generally lower than in more distant, urban, jobs) and have less capacity to establish or buy tree gardens (Li, 1998). Women's lack of financial capital is often cited as a reason for their greater dependence on common property resources, as in Ethiopia (Howard and Smith, 2006).

Women selling mangoes in a roadside market in Guinea. Photo © Terry Sunderland

3.5 Conclusions

Forests and tree-based systems have historically played a major role in supporting livelihoods as well as meeting the food security and nutritional needs of people worldwide. These systems, including natural forests that are managed to optimise yields of wild foods and fodder, shifting cultivation, a wide variety of agroforestry systems and single-species tree crops, are still dominant components of rural landscapes in many parts of the world, and remain critical to food security and nutrition of hundreds of millions of people worldwide.

They offer a number of advantages over permanent (crop) agriculture given their adaptability to a broader range of environmental conditions (e.g. soils, topography and climate) and changing socio-economic conditions and the diversity of food products derived from them.

Most forests and tree-based systems we see in the world today – particularly managed forests, shifting cultivation and agroforestry systems – are underpinned by the accumulated traditional knowledge of local and indigenous communities. This knowledge has been crucial to the development and modification of these systems over generations under diverse and variable environmental conditions and to meet changing socio-economic needs.

Only rarely and relatively recently have agricultural and forest scientists, extension agents and development organisations begun to understand the importance and relevance of many of these systems, and begun to work with farmers to combine the best of traditional and formal scientific knowledge to enhance their productivity and direct (food security and nutrition) and indirect (income) benefits to their practitioners.

Despite their widespread use, particularly in regions of the world where food security and nutrition are of particular concern, the data needed for decision-makers to make informed choices is quite limited, especially at the global and national level. Further research is needed on: the actual extent of most of these systems, the numbers of people who rely on one or more such systems to meet their household food and/or income needs, and the relative value of different forests and tree-based systems on the diets and health of those who manage them. Such information is of great importance to policymakers, planners and development agencies seeking to improve the lives of food-insecure populations.

Differences in diets and nutrition associated with different subsistence strategies/ different forms of land use (e.g. managed forests, shifting agriculture, agroforests, and single-species tree crop systems) are not widely documented. Studies comparing hunter-gatherers and low-population-density forest communities to more sedentary and urbanised groups have generally shown that the former consumed more meat but their diets were not necessarily better. The few existing studies suggest that the impact of transitions from one form of subsistence and land use to another is context-specific and influenced by social, cultural and economic factors.

A number of studies have shown a link between tree cover and dietary diversity and consumption of nutritious foods. Although we do not yet understand the pathways of this relationship, it suggests that maintenance of tree cover around rural homes and communities may lead to more nutritious diets.

Forests and tree-based systems are part of broader economic, political, cultural and ecological landscapes that typically include a mosaic of different food production systems and other land uses. How these different land use patches interact with each other in space and time can profoundly influence the productivity and sustainability of forests and tree-based systems as well as their food security and nutrition outcomes.

Tenure regimes in all four forest and tree-based systems are highly complex, and rights to trees may be different from rights to the land on which they are grown.

Different bundles of rights are nested and overlap in these different systems, varying by geographical, social, cultural, economic and political factors, and affecting the access of different population groups to the trees and their products for food, income and other livelihood needs.

Most tenure systems are gender-biased, allocating primary rights to men. Since women represent 43 percent of the global agricultural labour force, and there is evidence of feminisation of agriculture in numerous developing countries, women's weak and often insecure rights of access to land, forests and trees is undermining their engagement in innovation in forests and agroforestry systems with huge costs for the food security and nutrition of their families.

Rights to land, forests and trees in customary systems are commonly based on labour expended in clearing land or planting trees. Richer households with more assets (including livestock) are able to claim or make greater use of forest common property resources. However, poorer households often have a higher dependence, as a proportion of their total income, on forest resources for food security and livelihoods.

Tree planting and management requires low inputs of capital, mainly for labour, fertilisers and pesticides, and subsidies can lead to adoption of inappropriate trees or lead to distortions in land use. Such subsidies are often captured by larger farmers, who are not food insecure. Thus policies and incentives that improve demand and market prospects for trees rather than subsidising the establishment phase are more effective in promoting food security and improved livelihoods for the poor.

References

Adhikari, B., 2005. Poverty, property rights and collective action: Understanding the distributive aspects of common property resource management. *Environment and Development Economics* 10: 7-31. http://dx.doi.org/10.1017/s1355770x04001755

Afari-Sefa,V., Gockowski, J., Agyeman, N.F and Dziwornu, A.K., 2010. *Economic Cost-benefit Analysis of Certified Sustainable Cocoa Production in Ghana.* Poster presented at the Joint 3rd African Association of Agricultural Economists (AAAE) and 48th Agricultural Economists Association of South Africa (AEASA) Conference, Cape Town, South Africa, September 19-23, 2010. http://ageconsearch.umn.edu/bitstream/97085/2/33. Cost benefit of cocoa in Ghana. pdf

Agarwal, B., 1994a. Gender and command over property: A critical gap in economic analysis and policy in South Asia. *World Development* 22: 1455-1478. http://dx.doi.org/10.1016/0305-750x(94)90031-0

Agarwal, B., 1994b. *A Field of One's Own: Gender and Land Rights in South Asia.* Cambridge: Cambridge University Press.

Agarwal, B., 1997. "Bargaining" and gender relations: Within and beyond the household. *Feminist Economics* 3(1): 1-51. http://dx.doi.org/10.1080/135457097338799

Agarwal, B., 2010. *Gender and Green Governance: The Political Economy of Women's Presence Within and Beyond Community Forestry*. Oxford: Oxford University Press.

Agnoletti, M. (ed.), 2006. *The Conservation of Cultural Landscapes.* Wallingford, UK: CAB International.

Agrawal, A., Cashore, B., Hardin, R., Shepherd, G., Benson, C. and Miller, D., 2013. *Economic Contributions of Forests*. Background Paper to UNFF tenth Session, Istanbul, 8-19 April 2013. http://www.un.org/esa/forests/pdf/session_documents/unff10/EcoContrForests.pdf

AIPP and IWGIA, 2014. *Shifting Cultivation, Livelihood and Food Security. New and Old Challenges for Indigenous Peoples in Asia.* Chiang Mai: Asia Indigenous Peoples' Pact (AIPP) and the International Work Group for Indigenous Affairs (IWGIA). http://www.iwgia.org/iwgia_files_publications_files/0694_AIPPShifting_cultivation_livelihoodfood_security.pdf

Alcorn, J.B., 1981. Huastec noncrop resource management: Implications for prehistoric rain forest management. *Human Ecology* 9(4): 395-417. http://dx.doi.org/10.1007/bf01418729

Alexiades, M.N., 2009. *Mobility and Migration in Indigenous Amazonia: Contemporary Ethnoecological Perspectives*. Oxford: Berghahn.

Altieri, M.A., 2002. Agroecology: The science of natural resource management for poor farmers in marginal environments. *Agriculture, Ecosystems and Environment* 93: 1-24. http://dx.doi.org/10.1016/s0167-8809(02)00085-3

Altieri, M.A., 2004. Linking ecologists and traditional farmers in the search for sustainable agriculture. *Frontiers in Ecology and the Environment* 2(1): 35-42. http://dx.doi.org/10.2307/3868293

Anderson, A.B. and Posey, D.A., 1989. Management of a tropical scrub savanna by the Gorotire Kayapo. *Advances in Economic Botany* 7: 159-173.

Anderson, K., 2006. *Tending the Wild: Native American Knowledge and the Management of California's Natural Resources*. Berkeley, CA, USA: University of California Press. http://permaculteur.free.fr/ecoanarchisme/tending_the_wild.pdf

Andrade, G.I. and Rubio-Torgler, H., 1994. Sustainable use of the tropical rain forest: Evidence from the avifauna in a shifting-cultivation habitat mosaic in the Colombian Amazon. *Conservation Biology* 8(2): 545-554. http://dx.doi.org/10.1007/978-1-4612-4018-1_27

Angelsen, A., 2008. *Moving Ahead with REDD: Issues, Options and Implications*. Bogor, Indonesia: Center for International Forestry Research. http://dx.doi.org/10.17528/cifor/002601

Arnold, M. and Dewees, P. 1998. *Rethinking Approaches to Tree Management by Farmers*. Overseas Development Institute (ODI) Natural Resource Perspectives, No. 26. http://www.odi.org/sites/odi.org.uk/files/odi-assets/publications-opinion-files/2414.pdf

Asher, K., 2009. *Black and Green: Afro-Colombians, Development, and Nature in the Pacific Lowlands*. Durham, NC: Duke University Press.

Balée, W., 2006. The research program of historical ecology. *Annual Review of Anthropology* 35: 75-98. http://dx.doi.org/10.1146/annurev.anthro.35.081705.123231

Bannister, M.E. and Nair, P.K.R., 2003. Agroforestry adoption in Haiti: The importance of household and farm characteristics. *Agroforest Systems* 57: 149-157. http://dx.doi.org/10.1023/a:1023973623247

Barros, H.R., Ferreira, T.A. and Genovese, M.I., 2012. Antioxidant capacity and mineral content of pulp and peel from commercial cultivars of citrus from Brazil. *Food Chemistry* 134(4): 1892-8. PMID 23442635. http://dx.doi.org/10.1016/j.foodchem.2012.03.090

Barry, D. and Meinzen-Dick, R., 2014. The invisible map: Community tenure rights. In: *The Social Lives of Forests: Past, Present and Future of Woodland Resurgence*, edited by S. Hecht, K. Morrison and C. Padoch. Chicago and London: University of Chicago Press. http://dx.doi.org/10.7208/chicago/9780226024134.001.0001

Bartuszevige, A.M., Gorchov, D.L. and Raab, L., 2006. The relative importance of landscape and community features in the invasion of an exotic shrub in a fragmented landscape. *Ecography* 29: 213-222. http://dx.doi.org/10.1111/j.2006.0906-7590.04359.x

Bélair, C., Ichikawa, K., Wong, B.Y.L. and Mulongoy K.J. (eds.), 2010. *Sustainable Use of Biological Diversity in Socio-ecological Production Landscapes*, Technical Series No. 52. Montreal: Secretariat of the Convention on Biological Diversity. https://www.cbd.int/doc/publications/cbd-ts-52-en.pdf

Belotto, A., Leanes, L.F., Schneider, M.C., Tamayo, H. and Correa, E., 2005. Overview of rabies in the Americas. *Virus Research* 111: 5-12. http://dx.doi.org/10.1016/j.virusres.2005.03.006

Bengis, R.G., Kock, R.A. and Fischer, J., 2002. Infectious animal diseases: The wildlife/livestock interface. *Revue Scientifique et Technique* (International Office of Epizootics) 21: 53-65. PMID: 11974630

Berkes, F., Colding, J. and Folke, C., 2000. Rediscovery of traditional ecological knowledge as adaptive management. *Ecological Applications* 10: 1251-1262. http://dx.doi.org/10.1890/1051-0761(2000)010[1251:roteka]2.0.co;2

Berti, P.R., Krasevec, J. and Fitzgerald, S., 2004. A review of the effectiveness of agriculture interventions in improving nutrition outcomes. *Public Health Nutrition* 7(5): 599-609. http://dx.doi.org/10.1079/phn2003595

Bianchi, F.J.J., Booij, C.J. and Tscharntke, T., 2006. Sustainable pest regulation in agricultural landscapes: A review on landscape composition, biodiversity and natural pest control. *Proceedings of the Royal Society B: Biological Sciences* 273: 1715-1727. http://dx.doi.org/10.1098/rspb.2006.3530

Biermayr-Jenzano, P., Kassam S.N. and Aw-Hassan, A., 2014. *Understanding Gender and Poverty Dimensions of High Value Agricultural Commodity Chains in the Souss-Masaa-Draa Region of South-western Morocco*. ICARDA working paper, Mimeo. Amman, Jordan.

Bist, S.S., 2006. Elephant conservation in India—an overview. *Gajah* 25: 27-35. http://www.asesg.org/PDFfiles/Gajah/25-27-Bist.pdf

Blate, G.M., 2005. Modest trade-offs between timber management and fire susceptibility of a Bolivian semi-deciduous forest. *Ecological Applications* 15: 1649-1663. http://dx.doi.org/10.1890/04-0385

Boerboom, J.H.A. and Wiersum, K.F., 1983. Human impact on tropical moist forest. In: *Man's Impact on Vegetation*, edited by W. Holzner, M.J.A. Werger and I. Ikusima. The Hague: W. Junk.

Boffa, J.-M., 1999. *Agroforestry Parklands in Sub-saharan Africa.* FAO Conserv. Guide 34. Rome: Food and Agriculture Organization of the United Nations. http://www.fao.org/docrep/005/x3940e/x3940e00.htm

Bomuhangi, A., Doss, C. and Meinzen-Dick, R., 2011. Who owns the land? Perspectives from rural Ugandans and implications for land acquisitions. *IFPRI Discussion Paper 01136.* Washington DC: IFPRI. http://dx.doi.org/10.1080/13545701.2013.855320

Borggaard, O.K., Gafur, A. and Petersen, L., 2003. Sustainability appraisal of shifting cultivation in the Chittagong Hill Tracts of Bangladesh. *Ambio* 32(2): 118-123. http://dx.doi.org/10.1579/0044-7447-32.2.118

Boyer, J. and Liu, R.H., 2004. Apple phytochemicals and their health benefits. *Nutrition Journal* 3(1): 5. http://dx.doi.org/10.1186/1475-2891-3-5

Boyles, J.G., Cryan, P.M., McCracken, G.F. and Kunz, T.H., 2011. Economic importance of bats in agriculture. *Science* 332: 41-42. http://dx.doi.org/10.1126/science.1201366

Bradley, B.A. and Mustard, J.F., 2006. Characterizing the landscape dynamics of an invasive plant and risk of invasion using remote sensing. *Ecological Applications* 16: 1132-1147. http://dx.doi.org/10.1890/1051-0761(2006)016[1132:ctldoa]2.0.co;2

Bradshaw, C.J.A., Sodhi, N.S., Peh, K.S.-H. and Brook, B.W., 2007. Global evidence that deforestation amplifies flood risk and severity in the developing world. *Global Change Biology* 13: 2379-2395. http://dx.doi.org/10.1111/j.1365-2486.2007.01446.x

Brauman, K.A., Daily, G.C., Duarte, T.K. and Mooney, H.A., 2007. The nature and value of ecosystem services: An overview highlighting hydrologic services. *Annual Review of Environment and Resources* 32: 1-32. http://dx.doi.org/10.1146/annurev.energy.32.031306.102758

Brondizio, E.S., 2008. *The Amazonian Caboclo and the Açaí Palm: Forest Farmers in the Global Market.* New York: New York Botanical Garden Press.

Brondizio, E.S., Safar, C.A.M. and Siqueira, A.D., 2002. The urban market of açaí fruit (*Euterpe oleracea* Mart.) and rural land use change: Ethnographic insights into the role of price and land tenure constraining agricultural choices in the Amazon estuary. *Urban Ecosystems* 6(1-2): 67-97. http://dx.doi.org/10.1023/a:1025966613562

Brown, H.C.P., Smit, B., Sonwa, D.J., Somorin, O.A. and Nkem, J., 2011. Institutional perceptions of opportunities and challenges of REDD+ in the Congo Basin. *Journal of Environment & Development* 20(4): 381-404. http://dx.doi.org/10.1177/1070496511426480

Bruce, J., 1999. Legal bases for the management of forest resources as common property. *Forests, Trees and People Community Forestry Note 14.* Rome: FAO. http://www.fao.org/docrep/012/x2581e/x2581e00.pdf

Bruun, T.B., de Neergaard, A., Lawrence, D. and Ziegler, A., 2009. Environmental consequences of the demise in swidden agriculture in Southeast Asia: Carbon storage and soil quality. *Human Ecology* 37: 375-388. http://dx.doi.org/10.1007/s10745-009-9257-y

Bytnerowicz, A., Omasa, K. and Paoletti, E., 2007. Integrated effects of air pollution and climate change on forests: A northern hemisphere perspective. *Environmental Pollution* 147: 438-445. http://dx.doi.org/10.1016/j.envpol.2006.08.028

CacaoNet, 2012. *A Global Strategy for the Conservation and Use of Cacao Genetic Resources, as the Foundation for a Sustainable Cocoa Economy* (B. Laliberté, compiler). Montpellier, France: Bioversity International. http://www.bioversityinternational.org/uploads/tx_news/A_global_strategy_for_the_conservation_and_use_of_cacao_genetic_resources__as_the_foundation_for_a_sustainable_cocoa_economy_1588.pdf

Cairns, M.F. (ed.), 2007. *Voices from the Forest: Integrating Indigenous Knowledge into Sustainable upland Farming*. Washington, DC: Resources for the Future.

Cairns, M.F. (ed.), 2015. *Shifting Cultivation and Environmental Change: Indigenous People, Agriculture and Forest Conservation*. London: Earthscan Publications (Routledge).

Cairns, M. and Garrity, D.P., 1999. Improving shifting cultivation in Southeast Asia by building on indigenous fallow management strategies. *Agroforestry Systems* 47: 37-48. http://dx.doi.org/10.1023/a:1006248104991

Carloni, A. 2005. *Rapid Guide for Missions. Analysing Local Institutions and Livelihoods*. Institutions for Rural Development 1. Rome: FAO. http://www.fao.org/3/a-a0273e.pdf

Carney, J and Elias, M., 2014. Gendered knowledge and the African shea-nut tree. In: *The Social Lives of Forests: Past, Present and Future of Woodland Resurgence*, edited by S. Hecht, K. Morrison and C. Padoch. Chicago and London: University of Chicago Press. http://dx.doi.org/10.7208/chicago/9780226024134.001.0001

Castella, J.-C., Lestrelin, G., Hett, C., Bourgoin, J., Fitriana, Y.R., Heinimann, A. and Pfund, J.-L., 2013. Effects of landscape segregation on livelihood vulnerability: Moving from extensive shifting cultivation to rotational agriculture and natural forests in Northern Laos. *Human Ecology* 41.1 (Feb. 2013): 63-76. http://dx.doi.org/10.1007/s10745-012-9538-8

Cerda, R., Deheuvels, O., Calvache, D., Niehaus, L., Saenz, Y., Kent, J., Vilchez, S., Villota, A., Martinez, C. and Somarriba, E., 2014. Contribution of cocoa agroforestry systems to family income and domestic consumption: Looking toward intensification. *Agroforestry Systems* 88(6): 1-25. http://dx.doi.org/10.1007/s10457-014-9691-8

Chao, S., 2012. *Forest Peoples: Numbers Across the World*. Moreton-in-Marsh, UK: Forest Peoples Programme. http://www.forestpeoples.org/sites/fpp/files/publication/2012/05/forest-peoples-numbers-across-world-final_0.pdf

Chazdon, R.L. 2014. *Second Growth: The Promise of Tropical Forest Regeneration in an Age of Deforestation*. Chicago: University of Chicago Press.

Chen, H., Morrell, P.L., Ashworth, V.E.T.M., De La Cruz, M. and Clegg, M.T., 2008. Tracing the geographic origins of major avocado cultivars. *Journal of Heredity* 100(1): 56-65. PMID 18779226. http://dx.doi.org/10.1093/jhered/esn068

Cissé, M.I., 1995. *Les parcs agroforestiers au Mali. Etat des connaissances et perspectives pour leur amélioration*. AFRENA Rep. 93. Nairobi: ICRAF.

Clough, Y., Abrahamczyk, S., Adams, M.-O., Anshary, A., Ariyanti, N., et al., 2010. Biodiversity patterns and trophic interactions in human-dominated tropical landscapes in Sulawesi (Indonesia): Plants, arthropods and vertebrates. In: *Tropical Rainforests and Agroforests Under Global Change*, edited by T. Tscharntke, C. Leuschner, E. Veldkamp, H. Faust, E. Guhardja and A. Bidin. Environmental Science and Engineering Series. Berlin: Springer Verlag. http://dx.doi.org/10.1007/978-3-642-00493-3

Cochrane, M.A., 2003. Fire science for rainforests. *Nature* 421: 913-919. http://dx.doi.org/10.1038/nature01437

Cock, M.J.W., 2003. *Biosecurity and Forests: An Introduction—with Particular Emphasis on Forest Pests.* FAO Forest Health and Biosecurity Working Paper FBS/2E. Rome: FAO. ftp://ftp.fao.org/docrep/fao/006/J1467E/J1467E.pdf

Colfer, C.J.P., 2008a. *The Longhouse of the Tarsier: Changing Landscapes, Gender and Well Being in Borneo.* Phillips, Maine: Borneo Research Council, in cooperation with CIFOR and UNESCO.

Colfer, C.J.P., 2008b. *Human Health and Forests: A Global Overview of Issues, Practice and Policy.* London: Earthscan.

Colfer, C.J.P., Colchester, M., Joshi, L., Puri, R.K., Nygren, A., Lopez, C., 2005. Traditional knowledge and human well-being in the 21st century. In: *Forests in the Global Balance—Changing Paradigms,* IUFRO World Series No. 17, edited by G. Mery, R. Alfaro, M. Kanninen. and M. Lovobikov. 2005. Helsinki: International Union of Forest Research Organizations. http://www.iufro.org/science/special/wfse/forests-global-balance

Colfer, C.J.P., Minarchek, R.D., Cairns, M., Aier, A., Doolittle, A., Mashman, V., Odame, H.H., Roberts, M., Robinson, K. and Van Esterik, O., 2015. Gender analysis and indigenous fallow management. In: *Shifting Cultivation and Environmental Change: Indigenous People, Agriculture and Forest Conservation,* edited by M. Cairns. London: Earthscan.

Colfer, C.J.P., Peluso, N.L. and Chin, S.C., 1997. *Beyond Slash and Burn: Building on Indigenous Management of Borneo's Tropical Rain Forests.* Bronx, NY: New York Botanical Garden.

Collings, N., 2009. Environment. In: *The State of the World's Indigenous Peoples.* United Nations. Department of Economic and Social Affairs, Division for Social Policy and Development, Secretariat of the Permanent Forum on Indigenous Issues Report No. ST/ESA/328. http://www.un.org/esa/socdev/unpfii/documents/SOWIP/en/SOWIP_web.pdf

Condominas, G., 1977. *We Have Eaten the Forest: The Story of a Montagnard Village in the Central Highlands of Vietnam.* New York: Hill and Wang.

Conklin, H. C., 1957. *Hanunoo Agriculture: A Report on an Integral System of Shifting Cultivation in the Philippines.* Rome: FAO.

Cramb, R.A., 1993. Shifting cultivation and sustainable agriculture in East Malaysia: A longitudinal case study. *Agricultural Systems* 42: 209-226. http://dx.doi.org/10.1016/0308-521x(93)90055-7

CTA, 2012. Climate change: Concerns for cocoa. *SPORE* No. 159: 9.

Damman, S., Kuhnlein, H.V. and Erasmus, B., 2013. Human rights implications of Indigenous Peoples' food systems and policy recommendations. In: *Indigenous Peoples' Food Systems and Well-being. Interventions and Policies for Healthy Communities,* edited by H.V. Kuhnlein, B. Erasmus, D. Spigelski and B. Burlingame. Rome: FAO and CINE (Centre for Indigenous Peoples' Nutrition and Environment). http://www.fao.org/3/a-i3144e.pdf

De Foresta, H. and Michon, G., 1997. The agroforest alternative to *Imperata* grasslands: When smallholder agriculture and forestry reach sustainability. In: *Agroforestry Innovations for Imperata Grassland Rehabilitation,* edited by D.P. Garrity. Dordrecht, Netherlands: Kluwer Academic Publishers and Nairobi: International Centre for Research in Agroforestry. http://dx.doi.org/10.1007/bf00142877

DeFries, R.S., Rudel. T., Uriarte, M. and Hansen, M., 2010. Deforestation driven by urban population growth and agricultural trade in the twenty-first century. *Nature Geoscience* 3: 178-181. http://dx.doi.org/10.1038/ngeo756

Deere, C.D., Boakye-Yiadom, L., Doss, C., Oduro, A.D., Swaminathan, H., Twyman, J. and Suchitra, J.Y. 2013. *Women's Land Ownership and Participation in Agricultural Decision-making: Evidence from Ecuador, Ghana and Karnataka, India.* The Gender Asset Gap Project Research Brief Series No. 2. Bangalore: Indian Institute of Management. http://genderassetgap.org/sites/default/files/ResearchBrief2.pdf

Denevan, W.M., 1992. The pristine myth: The landscape of the Americas in 1492. *Annals of the Association of American Geographers* 82(3): 369-385. http://dx.doi.org/10.1111/j.1467-8306.1992.tb01965.x

Denevan, W.M. and Padoch, C., 1987. *Swidden-fallow Agroforestry in the Peruvian Amazon*. New York: New York Botanical Garden.

Denevan, W.M., Treacy, J.M., Alcorn, J.B., Padoch, C., Denslow, J. and Paitan, S.F., 1984. Indigenous agroforestry in the Peruvian Amazon—Bora Indian management of swidden fallows. *Interciencia* 9: 346-357. http://pdf.usaid.gov/pdf_docs/PNAAT727.pdf

Dey, J. 1981. Gambian women: Unequal partners in rice development projects? *Journal of Development Studies* 17(3): 109-122. http://dx.doi.org/10.1080/00220388108421801

Dey Abbas, J., 1997. Gender asymmetries in intrahousehold resource allocation in Sub-Saharan Africa: Some policy implications for land and labor productivity. In: *Intrahousehold Resource Allocation in Developing Countries: Methods, Models and Policy*, edited by L. Haddad, J. Hoddinott and H. Alderman. Baltimore: John Hopkins University Press, for IFPRI. https://www.pep-net.org/sites/pep-net.org/files/typo3doc/pdf/intrahhres1.pdf

Doss, C., Grown, C. and Deere, C.D., 2008. *Gender and Asset Ownership: A Guide to Collecting Individual Level Data*. Policy research working paper 4704. Washington DC: World Bank. http://dx.doi.org/10.1596/1813-9450-4704

Dounias, E., Selzner, A., Koizumi, M. and Levang, P. 2007. From sago to rice, from forest to town: The consequences of sedentarization for the nutritional ecology of Punan former hunter-gatherers of Borneo. *Food and Nutrition Bulletin* 28(2, suppl.): 294S-302S(9). http://www.ncbi.nlm.nih.gov/pubmed/17658075

Dove, M., 1983., Theories of swidden agriculture, and the political economy of ignorance. *Agroforestry Systems* 1: 85-99. http://dx.doi.org/10.1007/bf00596351

Dove, M.R., Smith, D.S., Campos, M.T., Mathews, A. S., Rademacher, A., Rhee, S. and Yoder, L.M., 2013. Globalisation and the construction of Western and non-Western knowledge. In: *Local Science vs Global Science: Approaches to Indigenous Knowledge in International Development*, edited by P. Sillitoe. Oxford and New York: Berghan Books.

Dreher, M.L. and Davenport, A.J., 2013. Hass avocado composition and potential health effects. *Critical Reviews in Food Science and Nutritio* 53 (7): 738-50. PMID 23638933. http://dx.doi.org/10.1080/10408398.2011.556759

Duchelle, A., Almeyda Zambrano, A.M., Wunder, S., Börner, J. and Kainer, K. 2014. Smallholder Specialization Strategies along the Forest Transition Curve in Southwestern Amazonia. *World Development*. http://dx.doi.org/10.1016/j.worlddev.2014.03.001

Duflo, E. and Udry, C., 2004. *Intrahousehold Resource Allocation in Côte d'Ivoire: Social Norms, Separate Accounts and Consumption Choices*. NBER working paper No. 10498. Cambridge MA: National Bureau of Economic Research. http://dx.doi.org/10.3386/w10498

Eagle, A.J., Eiswerth, M.E., Johnson, W.S., Schoenig, S.E. and Cornelis van Kooten, G., 2007. Costs and losses imposed on California ranchers by yellow starthistle. *Rangeland Ecology & Management* 60: 369-377. http://dx.doi.org/10.2111/1551-5028(2007)60[369:calioc]2.0.co;2

Edem, D.O., 2002. Palm oil: Biochemical, physiological, nutritional, hematological and toxicological aspects: A review. *Plant Foods for Human Nutrition* 57(3): 319-341. http://dx.doi.org/10.1023/a:1021828132707

Environment Canada, 2004. *An Invasive Alien Species Strategy for Canada*. Ottawa: Environment Canada. http://publications.gc.ca/collections/collection_2014/ec/CW66-394-2004-eng.pdf

FAO, 1982. *Fruit-bearing Forest Trees: Technical Notes*. FAO Forestry Paper 34. Rome: Food and Agriculture Organization of the United Nations. http://www.fao.org/docrep/015/t0006e/t0006e00.pdf

FAO, 2002a. *Land Tenure and Rural Development*. FAO Land Tenure Studies 3. Rome: FAO. ftp://ftp.fao.org/docrep/fao/005/y4307E/y4307E00.pdf

FAO, 2002b. *Gender and Access to Land*. FAO Land Tenure Studies 4. Rome: FAO. ftp://ftp.fao.org/docrep/fao/005/y4308e/y4308e00.pdf

FAO, 2010. *Global Forest Resources Assessment 2010: Main Report*. FAO Forestry Paper 163. Rome: Food and Agriculture Organization of the United Nations. http://www.fao.org/docrep/013/i1757e/i1757e.pdf

FAO, 2011. *State of Food and Agriculture. Women in Agriculture: Closing the Gender Gap for Development*. Rome: FAO. http://www.fao.org/docrep/013/i2050e/i2050e.pdf

FAO, 2012. *Roles of Forests in Climate Change*. http://www.fao.org/forestry/climatechange/53459/en/

FAO, 2014. *State of the World's Forests. Enhancing the Socioeconomic Benefits from Forests*. Rome: FAO. http://www.fao.org/3/a-i3710e.pdf

FAO and CIFOR, 2005. *Forests and Floods: Drowning in Fiction or Thriving on Facts?* Bangkok: Centre for International Forestry Research, Bogor and Food and Agriculture Organisation of the United Nations. http://www.fao.org/docrep/008/ae929e/ae929e00.htm

FAOSTAT Statistical Database, 2010. Rome: Food and Agriculture Organization of the United Nations. http://faostat.fao.org/

Feary, S.A., Eastburn, D., Sam, N. and Kennedy, J., 2012. Western Pacific. In: *Traditional Forest-related Knowledge: Sustaining Communities, Ecosystems and Biocultural Diversity*, edited by J.A. Parrotta and R.L. Trosper. Dordrecht, the Netherlands: Springer. http://dx.doi.org/10.1007/978-94-007-2144-9_11

Feldstein, H.S. and Poats, S.V., 1990. *Working Together. Gender Analysis in Agriculture*. West Hartford, CT: Kumarian.

Figueira, A., Janick, J. and BeMiller, J.N., 1993. New products from *Theobroma cacao*: Seed pulp and pod gum. In: *New Crops*, edited by J. Janick and J.E. Simon. New York: Wiley.

Finegan, B. and Nasi, R., 2004. The biodiversity and conservation potential of swidden agricultural landscapes. In: *Agroforestry and Biological Conservation in Tropical Landscapes*, edited by G. Schroth, G.A.B. da Fonseca, C.A. Harvey, C. Gascon, H.L. Vasconcelos and A.-M.N. Izac, Washington, DC: Island Press. http://library.uniteddiversity.coop/Permaculture/Agroforestry/Agroforestry_and_Biodiversity_Conservation_in_Tropical_Landscapes.pdf

Fisher, M., 2004. Household welfare and forest dependence in southern Malawi. *Environment and Development Economics* 9(2): 135-154. http://dx.doi.org/10.1017/s1355770x03001219

Foley, J.A., DeFries, R., Asner, G.P., Barford, C., Bonan, G., Carpenter, S.R., Chapin, F.S., Coe, M.T., Daily, G.C., Gibbs, H.K., Helkowski, J.H., Holloway, T., Howard, E.A., Kucharik, C.J., Monfreda, C., Patz, J.A., Prentice, I.C., Ramankutty, N. and Snyder, P.K., 2005. Global Consequences of Land Use. *Science* 309: 570-574. http://dx.doi.org/10.1126/science.1111772

Foli, S., Reed, J., Clendenning, J., Petrokofsky, G., Padoch, C. and Sunderland, T., 2014. To what extent does the presence of forests and trees contribute to food production in humid and dry forest landscapes?: A systematic review protocol. *Environmental Evidence* 3: 15. http://www.environmentalevidencejournal.org/content/3/1/15 http://dx.doi.org/10.17528/cifor/005476

Fortmann, L., 1983. Who plows? The effect of economic status on women's participation in agriculture in agriculture in Botswana. Mimeo.

Fortmann, L., Antinori, C. and Nabanne, N., 1997. Fruits of their labors: Gender, property rights, and tree planting in two Zimbabwe villages. *Rural Sociology* 62(3): 295-314. http://dx.doi.org/10.1111/j.1549-0831.1997.tb00653.x

Fortmann, L. and Bruce, J. W. (eds.), 1988. *Whose Trees? Proprietary Dimensions of Forestry*. Boulder and London: Westview Press.

Fowler, D., Cape, J.N., Coyle, M., Flechard, C., Kuylenstierna, J., Hicks, K., Derwent, D., Johnson, D. and Stevenson, D., 1999. The global exposure of forests to air pollutants. In: *Forest Growth Responses to the Pollution Climate of the 21st Century*, edited by L.J. Sheppard and J.N. Cape. Dordrecht, Netherlands: Springer. http://dx.doi.org/10.1007/978-94-017-1578-2_1

Fox, J., Truong, D.M., Rambo, A.T., Tuyen, N.P., Cuc, L.T. and Leisz, S., 2000. Shifting cultivation: A new old paradigm for managing tropical forests. *BioScience* 50(6): 521-528. http://dx.doi.org/10.1641/0006-3568(2000)050[0521:scanop]2.0.co;2

Fox, J., Fujita, Y., Ngidang, D., Peluso, N., Potter, L., Sakuntaladewi, N., Sturgeon, J. and Thomas, D., 2009. Policies, Political-Economy, and Swidden in Southeast Asia. *Human Ecology* 37(3): 305-322. http://dx.doi.org/10.1007/s10745-009-9240-7

Franzel, S., 1999. Socioeconomic factors affecting the adoption potential of improved tree fallows in Africa. *Agroforestry Systems* 47: 305-321. http://dx.doi.org/10.1023/a:1006292119954

Fuys, A. and Dohrn, S. 2010. Common property regimes: Taking a closer look at resource access. In: *Beyond the Biophysical. Knowledge, Culture and Power in Agriculture and Natural Resource Management*, edited by L. German, J. Ramisch and R. Verma. Dordrecht, Heidelberg, London, New York: Springer. http://dx.doi.org/10.1007/978-90-481-8826-0_9

Gallai, N., Salles, J.-M., Settele, J. and Vaissiere, B.E., 2009. Economic valuation of the vulnerability of world agriculture confronted with pollinator decline. *Ecological Economics* 68: 810-821. http://dx.doi.org/10.1016/j.ecolecon.2008.06.014

Galloway-McLean, K., 2010. *Advance Guard: Climate Change Impacts, Adaptation, Mitigation and Indigenous Peoples—A Compendium of Case Studies*. Darwin, Australia: United Nations University-Traditional Knowledge Initiative. http://www.unutki.org/news.php?doc_id=101&news_id=92

García Latorre, J. and García Latorre, J., 2012. Globalization, local communities, and traditional forest-related knowledge. In: *Traditional Forest-related Knowledge: Sustaining Communities, Ecosystems and Biocultural Diversity*, edited by J.A. Parrotta and R.L. Trosper. Dordrecht, the Netherlands: Springer. http://dx.doi.org/10.1007/978-94-007-2144-9_12

Gavin, M.C., 2004. Changes in forest use value through ecological succession and their implications for land management in the Peruvian Amazon. *Conservation Biology* 18(6): 1562-70. http://dx.doi.org/10.1111/j.1523-1739.2004.00241.x

Giovarelli, R., 2006. Overcoming gender biases in established and transitional property rights systems. In: *Land Law Reform: Achieving Development Policy Objectives*. Law, Justice, and Development Series, edited by J.W. Bruce, R. Giovarelli, L. Rolfes, D. Bledsoe and R. Mitchell. Washington, DC: World Bank Publications. http://dx.doi.org/10.1596/978-0-8213-6468-0

Girard, A.W., Self, J.L., McAuliffe, C. and Olude, O., 2012. The effects of household food production strategies on the health and nutrition outcomes of women and young children: A systematic review. *Paediatric and Perinatal Epidemiology* 26: 205-22. http://dx.doi.org/10.1111/j.1365-3016.2012.01282.x

Gockowski, J. and Sonwa, D., 2011. Cocoa intensification scenarios and their predicted impact on CO2 emissions, biodiversity conservation, and rural livelihoods in the Guinea rain forest of West Africa. *Environmental Management* 48: 307-321. http://dx.doi.org/10.1007/s00267-010-9602-3

Goldammer, J.G., 1988. Rural land-use and wildland fires in the tropics. *Agroforestry Systems* 6: 235-252. http://dx.doi.org/10.1007/bf02220124

Gupta, P.NS., 1980. Food consumption and nutrition of regional tribes of India. *Ecology of Food and Nutrition* 9(2): 93-108. http://dx.doi.org/10.1080/03670244.1980.9990587

Haddad, L. Hoddinott, J. and Alderman, H. (eds.), 1997. *Intrahousehold Resource Allocation in Developing Countries: Methods, Models and Policy.* Baltimore: John Hopkins University Press, for IFPRI. http://dx.doi.org/10.2307/1244597

Hall, J.B., Aebischer, D.P., Tomlinson, H.F., Osei-Amaning, E. and Hindle, J.R., 1996. *Vitellaria Paradoxa: A Monograph.* Bangor, UK: University of Wales.

Hammond D.S., ter Steege, H. and van der Borg, K., 2007. Upland soil charcoal in the wet tropical forests of central Guyana. *Biotropica* 39: 153-160. http://dx.doi.org/10.1111/j.1744-7429.2006.00257.x

Hatcher, J. and Bailey, L. 2010. *Tropical Forest Tenure Assessment: Trends, Challenges and Opportunities.* Washington DC and Yokohama: RRI and ITTO.

Hausman, A.J. and Wilmsen, E.N., 1985. Economic change and secular trends in the growth of San children. *Human Biology* 57(4): 563-571.

Hecht, S.B., 2009. Kayapó savanna management: Fire, soils, and forest islands in a threatened biome. In: *Amazonian Dark Earths: Wim Sombroek's Vision,* edited by W.I. Woods, W.G. Teixeira, J. Lehmann, C. Steiner, A.M.G.A. WinklerPrins and L. Rebellato. Heidelberg: Springer. http://dx.doi.org/10.1007/978-1-4020-9031-8_7

Hecht, S.B., Morrison, K.D. and Padoch, C. (eds.), 2014. *The Social Lives of Forests: Past, Present, and Future of Woodland Resurgence.* Chicago: University of Chicago Press. http://dx.doi.org/10.5860/choice.52-0834

Hett, C., Castella, J.-C., Heinimann, A., Messerli, P. and Pfund, J.-L., 2012. A landscape mosaics approach for characterizing swidden systems from a REDD+ perspective. *Applied Geography* 32: 608-618. http://dx.doi.org/10.1016/j.apgeog.2011.07.011

Hill, C.M., 2000. Conflict of interest between people and baboons: Crop raiding in Uganda. *International Journal of Primatology* 21: 299-315. http://sanrem.cals.vt.edu/1048/Conflict of Interest Between People and Baboons.pdf

Hill, R.V. and Vigneri, M. 2014. Mainstreaming gender sensitivity in cash crop market supply Chains. In: *Gender in Agriculture. Closing the Knowledge Gap,* edited by A. Quisumbing, R. Meinzen-Dick, T. Raney, A. Croppenstedt, J. Behrman and A. Peterman. Rome and Dordrecht: FAO and Springer. http://dx.doi.org/10.1007/978-94-017-8616-4_13

Hiraoka, M., 1994. Mudanças nos padrões econômicos de uma população ribeirinha do estuário do Amazonas. In: *Povos das Águas: Realidade e perspectivas na Amazônia,* edited by L. Furtado, A.F. Mello and W. Leitão. Belém, Para, Brazil: MPEG/Universidade Federal do Pará.

Hiraoka, M., 1995. Aquatic and land fauna management among the floodplain riberenos of the Peruvian Amazon. In: *The fragile tropics of Latin America: Sustainable Management of Changing Environments,* edited by T. Nishizawa and J.I. Uitto. Tokyo: United Nations University.

Hladik, C.M., Linares, O.F., Hladik, A., Pagezy, H. and Semple, A., 1993. Tropical forests, people and food: An overview. In: Tropical Forests, People and Food. *Biocultural Interactions and Applications to Development,* Man and Biosphere Series No. 13, edited by C.M. Hladik, A. Hladik, O.F. Linares, H. Pagezy, A. Semple and M. Hadley. Paris: UNESCO and New York: Parthenon.

Hockings, K.J. and McLennan, M.R., 2012. From forest to farm: Systematic review of cultivar feeding by chimpanzees—management implications for wildlife in anthropogenic landscapes. *PLoS ONE* 7, e33391. http://dx.doi.org/10.1371/journal.pone.0033391

Holmes, T.P., Aukema, J.E., Von Holle, B., Liebhold, A. and Sills, E., 2009. Economic impacts of invasive species in forests. *Annals of the New York Academy of Sciences* 1162: 18-38. http://dx.doi.org/10.1111/j.1749-6632.2009.04446.x

Holzschuh, A., Steffan-Dewenter, I. and Tscharntke, T., 2010. How do landscape composition and configuration, organic farming and fallow strips affect the diversity of bees, wasps and their parasitoids? *Journal of Animal Ecology* 79: 491-500. http://dx.doi.org/10.1111/j.1365-2656.2009.01642.x

Howard, P.L. and Smith, E., 2006. *Leaving Two Thirds out of Development: Female Headed Households and Common Property Resources in the Highlands of Tigray, Ethiopia*. Livelihood Support Programme (LSP) Working Paper 40. Rome: FAO. http://www.fao.org/3/a-ah624e.pdf

Howard, P. L. and Nabanoga, G., 2007. Are there customary rights to plants? An inquiry among the Baganda (Uganda), with special attention to gender. *World Development* 35(9): 1542-1563. http://dx.doi.org/10.1016/j.worlddev.2006.05.021

Hunt, D., 1984. *The Labour Aspects of Shifting Cultivation in African Agriculture*. Rome: FAO.

Hurni, K., Hett, C., Heinimann, A., Messerli, P. and Wiesmann, U., 2013. Dynamics of shifting cultivation landscapes in northern Lao PDR between 2000 and 2009 based on an analysis of MODIS time series and Landsat images. *Human Ecology* 41(1): 21-36. http://dx.doi.org/10.1007/s10745-012-9551-y

Ibarra, J.T., Barreau, A., Del Campo, C., Camacho, C.I., Martin, G.J. and McCandless S.R., 2011. When formal and market-based conservation mechanisms disrupt food sovereignty: Impacts of community conservation and payments for environmental services on an indigenous community of Oaxaca, Mexico. *International Forestry Review* 13(3): 318-337. http://dx.doi.org/10.1505/146554811798293935

Ickowitz, A., Powell, B., Salim, M.A. and Sunderland, T.C.H., 2014. Dietary quality and tree cover in Africa. *Global Environmental Change* 24: 287-294. http://dx.doi.org/10.1016/j.gloenvcha.2013.12.001

Indrawan, M., Yabe, M., Nomura, H. and Harrison, R., 2014. Deconstructing satoyama—The socio-ecological landscape in Japan. *Ecological Engineering* 64: 77-84. http://dx.doi.org/10.1016/j.ecoleng.2013.12.038

IFPRI, 2013. *Reducing the Gender Asset Gap through Agricultural Development. A Technical Resource Guide*. Washington DC: IFPRI. http://dx.doi.org/10.1007/978-94-017-8616-4_5

ITTO, 2002. *ITTO Guidelines for the Restoration, Management and Rehabilitation of Degraded and Secondary Tropical Forests*. ITTO Policy Development Series No. 13. Yokohama, Japan: International Tropical Timber Organization. http://www.cifor.org/library/1175/itto-guidelines-for-the-restoration-management-and-rehabilitation-of-degraded-and-secondary-tropical-forests/

Janick, J., 2005. The origins of fruits, fruit growing, and fruit breeding. *Plant Breeding Review* 25: 255-320. http://dx.doi.org/10.1002/9780470650301.ch8

Jarosz, L., 1993. Defining and explaining tropical deforestation: Shifting cultivation and population growth in colonial Madagascar (1896-1940). *Economic Geography* 69(4): 366-379. http://dx.doi.org/10.2307/143595

Jodha, N.S., 1986. Common Property Resources and Rural Poor in Dry Regions of India. *Economic and Political Weekly* 21: 1169-81. http://www.jstor.org/stable/4375858

Johann, E., Agnoletti M., Bölöni, J., Erol, S.Y., Holl, K., Kusmin, J., García Latorre, J., García Latorre, J., Molnár, Z., Rochel, X., Rotherham, I.D., Saratsi, E., Smith, M., Tarang, L., van Benthem, M. and van Laar, J., 2012. Europe. In: *Traditional Forest-related Knowledge: Sustaining Communities, Ecosystems and Biocultural Diversity*, edited by J.A. Parrotta and R.L. Trosper. Dordrecht, the Netherlands: Springer. https://dx.doi.org/10.1007/978-94-007-2144-9_6

Johns, T., 1996. *The Origins of Human Diet and Medicine: Chemical Ecology*. Tucson AZ, USA: University of Arizona Press.

Johnson, D., 1973. The botany, origin, and spread of the cashew *Anacardium occidentale* L. *Journal of Plantation Crops* 1(1-2): 1-7.

Jonkman, S.N., 2005. Global perspectives on loss of human life caused by floods. *Natural Hazards* 34(2): 151-175. http://dx.doi.org/10.1007/s11069-004-8891-3

Kennedy, E. and Peters, P. 1992. Household food security and child nutrition: The interaction of income and gender of household head. *World Development* 20(8): 1077-1085. http://dx.doi.org/10.1016/0305-750x(92)90001-c

Kerr, W.E. and Posey D.A., 1984. Notas sobre a agricultura dos índios Kayapó. *Interciência* 9(6): 392-400.

Kiptot, E. and Franzel, S., 2012. Gender and agroforestry in Africa: A review of women's participation. *Agroforestry Systems* 84: 35-58. http://dx.doi.org/10.1007/s10457-011-9419-y

Kislev, M.E., Hartmann, A. and Bar-Yosef, O., 2006. Early domesticated fig in the Jordan Valley. *Science* 312(5778): 1372-1374 (2 June 2006). http://dx.doi.org/10.1126/science.1125910

Kiyingi, I. and Gwali, S., 2013. Productivity and profitability of robusta coffee agroforestry systems in central Uganda. *Uganda Journal of Agricultural Sciences* 13(1): 85-93. http://www.researchgate.net/publication/236901244_Productivity_and_profitability_of_Robusta_coffee_agroforestry_systems_in_central_Uganda

Klein, A.-M., Vaissière, B.E., Cane, J.H., Steffan-Dewenter, I., Cunningham, S.A., Kremen, C. and Tscharntke, T., 2007. Importance of pollinators in changing landscapes for world crops. *Proceedings of the Royal Society B: Biological Sciences* 274: 303-313. http://dx.doi.org/10.1098/rspb.2006.3721

Kleinman, P.J.A., Pimentel, D. and Bryant, R.B., 1996. Assessing ecological sustainability of slash-and-burn agriculture through soil fertility indicators. *Agronomic Journal* 88: 122-127. http://dx.doi.org/10.2134/agronj1996.00021962008800020002x

Kort, J., 1988. Benefits of windbreaks to field and forage crops. *Agriculture, Ecosystems and Environment* 22-23: 165-190. http://dx.doi.org/10.1016/b978-0-444-43019-9.50018-3

Koppert, G.J.A., Dounias, E., Froment, A. and Pasquet, P., 1993. Food consumption in three forest populations of the southern coastal areas of Cameroon: Yassa—Mvae—Bakola. In: *Tropical Forests, People and Food. Biocultural Interactions and Applications to Development.* Man and Biosphere Series No. 13. Paris: UNESCO/The Parthenon Publishing Group. http://www.cifor.org/publications/pdf_files/research/forests_health/17.pdf

Kremen, C., 2005. Managing ecosystem services: What do we need to know about their ecology? *Ecology Letters* 8: 468-479. http://dx.doi.org/10.1111/j.1461-0248.2005.00751.x

Kristjanson, P., Waters-Buyer, A., Johnson, N., Tipilda, A., Njuki, J., Baltenweck, I., Grace, D., MacMillan, S., 2014. Livestock and women's livelihoods. In: *Gender in Agriculture. Closing the Knowledge Gap*, edited by A. Quisumbing, R. Meinzen-Dick, T. Raney, A. Croppenstedt, J. Behrman and A. Peterman. Rome and Dordrecht: FAO and Springer. http://dx.doi.org/10.1007/978-94-017-8616-4_9

Kuhnlein, H.V., Erasmus, B. and Spigelski, D., 2009. *Indigenous Peoples' Food Systems*. Rome: Food and Agriculture Organization of the United Nations. ftp://ftp.fao.org/docrep/fao/012/i0370e/i0370e00.pdf

Kumar, B.M. and Takeuchi, K., 2009. Agroforestry in the Western Ghats of peninsular India and the satoyama landscapes of Japan: A comparison of two sustainable land use systems. *Sustainability Science* 4(2): 215-232. http://dx.doi.org/10.1007/s11625-009-0086-0

Kumar, C. and Nongkynrih, K. 2005. Customary tenurial forest practices and the poor in Khasi—Jaintia Society of Meghalaya. Case study submitted for the joint study *Rural Common Property in a Perspective of Development and Modernization*. Delhi: CIFOR and North Eastern Hill University.

Kumar, N. and Quisumbing, A., 2012. *Policy Reform Toward Gender Equality in Ethiopia: Little by Little the Egg Begins to Walk.* IFPRI Discussion Paper 1126. Washington DC: IFPRI. http://dx.doi.org/10.2139/ssrn.2184985

Landscan, 2010. *Landscan Global Population Database.* http://www.eastview.com/online/landscan

Lambin, E.F., Turner, B.L., Geist, H.J., Agbola, S.B., Angelsen, A., Bruce, J.W., Coomes, O.T., Dirzo, R., Fischer, G., Folke, C., George, P.S., Homewood, K., Imbernon, J., Leemans, R., Li, X., Moran, E.F., Mortimore, M., Ramakrishnan, P.S., Richards, J.F, Skånes, H., Steffen, W., Stone, G.D., Svedin, U., Veldkamp, T.A., Vogel, C. and Xu, J., 2001. The causes of land-use and land-cover change: Moving beyond the myths. *Global Environmental Change* 11: 261-269. http://dx.doi.org/10.1016/s0959-3780(01)00007-3

Larson, A.M., Barry, D. and Dahal G.R., 2010. New rights for forest-based communities? Understanding processes of forest tenure reform. *International Forestry Review* 12(1): 78-96. http://dx.doi.org/10.1505/ifor.12.1.78

Lastarria-Cornhiel, S., 1997. Impact of privatization on gender and property rights in Africa. *World Development* 25(8): 1317-1333. http://dx.doi.org/10.1016/s0305-750x(97)00030-2

Lastarria-Cornhiel, S., Behrman, J., Meinzen-Dick, R. and Quisumbing, A., 2014. Gender equity ann land: Toward secure and effective access for rural women. In: *Gender in Agriculture. Closing the Knowledge Gap*, edited by A. Quisumbing, R. Meinzen-Dick, T. Raney, A. Croppenstedt, J. Behrman and A. Peterman. Rome and Dordrecht: FAO and Springer. http://dx.doi.org/10.1007/978-94-017-8616-4_6

Lauri, P.-É., Maguylo, K. and Trottier, C., 2006. Architecture and size relations: An essay on the apple (*Malus* x *domestica*, Rosaceae) tree. *American Journal of Botany* 93(93): 357-368. http://dx.doi.org/10.3732/ajb.93.3.357

Li, T.M., 1998. Working separately but eating together: Personhood, property and power in conjugal relations. *American Ethnologist* 25(4): 675-694. http://dx.doi.org/10.1525/ae.1998.25.4.675

Li, T.M., 2014. *Social Impacts of Oil Palm in Indonesia. A Gendered Perspective from West Kalimantan.* Toronto: University of Toronto, for CIFOR. http://dx.doi.org/10.17528/cifor/005579

Likens, G.E., Driscoll, C.T. and Buso, D.C., 1996. Long-term effects of acid rain: Response and recovery of a forest ecosystem. *Science* 272: 244-246. http://dx.doi.org/10.1126/science.272.5259.244

Lim, H.F., Liang, L., Camacho, L.D., Combalicer, E.A. and Singh, S.K.K., 2012. Southeast Asia. In: *Traditional Forest-related Knowledge: Sustaining Communities, Ecosystems and Biocultural Diversity*, edited by J.A. Parrotta and R.L. Trosper. Dordrecht, the Netherlands: Springer. http://dx.doi.org/10.1007/978-94-007-2144-9_10

Linares, O.F., 1976. "Garden hunting" in the American tropics. *Human Ecology* 4(4): 331-349. http://dx.doi.org/10.1007/bf01557917

MA (Millennium Ecosystem Assessment), 2005. *Ecosystems and Human Well-being: Synthesis.* Washington DC: Island Press. http://www.millenniumassessment.org/documents/document.356.aspx.pdf

Mackenzie, C.A. and Ahabyona, P., 2012. Elephants in the garden: Financial and social costs of crop raiding. *Ecological Economics* 75: 72-82. http://dx.doi.org/10.1016/j.ecolecon.2011.12.018

Maffi L., 2005. Linguistic, cultural, and biological diversity. *Annual Review of Anthropology* 29: 599-617. http://dx.doi.org/10.1146/annurev.anthro.34.081804.120437

Marten, G.G. and Abdoellah, O.S., 1988. Crop diversity and nutrition in west java. *Ecology of Food and Nutrition* 21(1): 17-43. http://dx.doi.org/10.1080/03670244.1988.9991016

Martin, P.A., Newton, A.C. and Bullock, J.M., 2013. Carbon pools recover more quickly than plant biodiversity in tropical secondary forests. *Proceedings of the Royal Society B* 280: 20132236. http://dx.doi.org/10.1098/rspb.2013.2236

Masset, E., Haddad, L., Cornelius, A. and Isaza-Castro, J., 2012. Effectiveness of agricultural interventions that aim to improve nutritional status of children: Systematic review. *BMJ (British Medical Journal)* 344: d8222. http://dx.doi.org/10.1136/bmj.d8222

Masters, E.T., Yidana, J.A. and Lovett, P.N. [n.d.]. *Reinforcing Sound Management through Trade: Shea Tree Products in Africa*. Rome: Food and Agriculture Organization. http://www.fao.org/docrep/008/y5918e/y5918e11.htm

McDaniel, J., Kennard, D. and Fuentes, A., 2005. Smokey the tapir: Traditional fire knowledge and fire prevention campaigns in lowland Bolivia. *Society and Natural Resources* 18: 921-931. http://dx.doi.org/10.1080/08941920500248921

Meinzen-Dick, R., Brown, L., Feldstein, H. and Quisumbing, A., 1997. Gender, property rights, and natural resources. *World Development* 25(8): 1303-1315. http://dx.doi.org/10.1016/s0305-750x(97)00027-2

Meinzen-Dick, R., Johnson, N., Quisumbing, A., Njuki, J., Behrman, J., Rubin, D., Peterman, A. and Waithanji, E., 2014. The gender asset gap and its implications for agricultiral and rural development. In: *Gender in Agriculture. Closing the Knowledge Gap*, edited by A. Quisumbing, R. Meinzen-Dick, T. Raney, A. Croppenstedt, J. Behrman, and A. Peterman. Rome and Dordrecht: FAO and Springer. http://dx.doi.org/10.1007/978-94-017-8616-4_5

Mercer, D.L., 2004. Adoption of agroforestry innovations in the tropics: A review. *Agroforestry Systems* 61-62(1-3): 311-328. http://dx.doi.org/10.1007/978-94-017-2424-1_22

Mertz, O., 2009. Trends in shifting cultivation and the REDD mechanism. *Current Opinion in Environmental Sustainability* 1(2): 156-160. http://dx.doi.org/10.1016/j.cosust.2009.10.002

Mertz, O., Leisz, S., Heinimann, A., Rerkasem, K., Thiha, Dressler,W., Cu, P.V., Vu, K.C., Schmidt-Vogt, D., Colfer, C.J.P., Epprecht, M., Padoch, C. and Potter, L., 2009a. Who counts? The demography of swidden cultivators. *Human Ecology* 37: 281-289. http://dx.doi.org/10.1007/s10745-009-9249-y

Mertz, O., Padoch, C., Fox, J., Cramb, R.A., Leisz, S.J., Lam, N.T. and Vien, T.D., 2009b. Swidden change in Southeast Asia: Understanding causes and consequences. *Human Ecology* 37(3): 259-264. http://dx.doi.org/10.1007/s10745-009-9245-2

Mertz, O., Wadley, R.L., Nielsen, U., Bruun, T.B., Colfer, C.J.P., de Neergaard, A., Jepsen, M.R., Martinussen, T., Zhao, Q., Noweg, G.T. and Magid, J., 2008. A fresh look at shifting cultivation: Fallow length an uncertain indicator of productivity. *Agricultural Systems* 96: 75-84. http://dx.doi.org/10.1016/j.agsy.2007.06.002

Meyer, J.-Y. and Malet, J.-P., 1997. Study and management of the alien invasive tree *Miconia calvescens* DC. (Melastomataceae) in the Islands of Raiatea and Tahaa (Society Islands, French Polynesia): 1992-1996. Report, Cooperative National Park Resources Studies Unit. Honolulu: University of Hawaii at Manoa, Department of Botany. http://hdl.handle.net/10125/7368

Mitchell, M.G.E., Bennett, E.M. and Gonzalez, A., 2014. Forest fragments modulate the provision of multiple ecosystem services. *Journal of Applied Ecology* 51: 909-918. http://dx.doi.org/10.1111/1365-2664.12241

Molnar, T.J., Zaurov, D.E., Capik, J.M., Eisenman, S.W., Ford, T., Nikolyi, L.V. and Funk, C.R, 2011. Persian Walnuts (Juglans regia L.) in Central Asia. In: *Northern Nut Growers Association (NNGA) 101st Annual Report*. http://www.ippfbe.org

Moore, B.A., 2005. *Alien Invasive Species: Impacts on Forests and Forestry*. Rome: Food and Agriculture Organisation of the United Nations. http://www.fao.org/docrep/008/j6854e/j6854e00.htm

Mukherjee, S.K., 1972. Origin of mango (*Mangifera indica*). *Economic Botany* 26(3): 260-264. http://dx.doi.org/10.1007/bf02861039

Muleta, D., 2007. *Microbial Inputs in Coffee (Coffea arabica L.) Production Systems, Southwestern Ethiopia: Implications for Promotion of Biofertilizers and Biocontrol Agents*. PhD thesis, Swedish University of Agricultural Sciences/Uppsala. http://pub.epsilon.slu.se/1657/1/Am_Thesis_template_LC_PUBL..pdf

Nair, P.K.N., 1993. *An Introduction to Agroforestry*. Dordrecht, Netherlands: Kluwer Academic Publishers. http://dx.doi.org/10.1007/978-94-011-1608-4

Nair, P.K.R. and Fernandes, E.,1984. *Agroforestry as an Alternative to Shifting Cultivation*. FAO Soils Bulletin *No. 53*: 169-182.

Nair, P.K.R., Kumar, B.M. and Nair, V.D., 2009. Agroforestry as a strategy for carbon sequestration. *Journal of Plant Nutrition and Soil Science* 172(1): 10-23. http://dx.doi.org/10.1002/jpln.200800030

Narain, U., Gupta, S. and van 't Veld, K., 2008. Poverty and the environment: Exploring the relationship between household incomes, private assets and natural assets. *Land Economics* 84(1): 148-167. http://dx.doi.org/10.3368/le.84.1.148

Nasi, R., Brown, D., Wilkie, D., Bennett, E., Tutin, C., Van Tol, G. and Christophersen, T., 2008. *Conservation and Use of Wildlife-based Resources: The Bushmeat Crisis*. In: CBD Technical Series No. 33, Montreal: Secretariat of the Convention on Biological Diversity. https://www.cbd.int/doc/publications/cbd-ts-33-en.pdf

Naughton-Treves, L., Treves, A., Chapman, C. and Wrangham, R., 1998. Temporal patterns of crop-raiding by primates: Linking food availability in croplands and adjacent forest. *Journal of Applied Ecology* 35: 596-606. http://dx.doi.org/10.1046/j.1365-2664.1998.3540596.x

Nepstad, D., Carvalho, G., Barros, A.C., Alencar, A., Capobianco, J.P., Bishop, J., Moutinho, P., Lefebvre, P., Silva, U.L. and Prins, E., 2001. Road paving, fire regime feedbacks, and the future of Amazon forests. *Forest Ecology and Management* 154: 395-407. http://dx.doi.org/10.1016/s0378-1127(01)00511-4

New Zealand Nature Institute, 2006. *Rural Livelihoods and Access to Forest Resources in Mongolia*. LSP Working Paper 32. Rome: FAO, Livelihood Support Programme (LSP). http://www.fao.org/3/a-ah253e.pdf

Njuki, J, Kaaria, S., Chamunorwa, A. and Chiuri, W., 2011. Linking smallholder farmers to markets, gender, and intrahousehold dynamics: Does the choice of commodity matter? *Eur European Journal of Development Research* 23(3): 426-443. http://dx.doi.org/10.1057/ejdr.2011.8

Ntiamoa-Baidu, Y., 1997. *Wildlife and Food Security in Africa*. Rome: FAO. http://www.fao.org/docrep/w7540e/w7540e00.htm

Obiri, D.B., Bright, G.A., McDonald, M.A., Anglaaere, L.C.N. and Cobbina, J., 2007. Financial analysis of shaded cocoa in Ghana. *Agroforestry Systems* 71(2): 139-149. http://dx.doi.org/10.1007/s10457-007-9058-5

Obiri, D.B., Depinto, A. and Tetteh, F., 2011. *Cost-benefit Analysis of Agricultural Climate Change Mitigation Options: The Case of Shaded Cocoa in Ghana*. Research report prepared for IFPRI, Washington. 56pp.

Ohler, J.G., 1979. Cashew processing. *Tropical Abstracts* 21(9): 1792-2007.

Olson, S.H., Gangnon, R., Silveira, G.A. and Patz, J.A., 2010. Deforestation and malaria in Mâncio Lima County, Brazil. *Emerging Infectious Diseases* 16(7): 1108-1115. http://dx.doi.org/10.3201/eid1607.091785

Opeke, L.K., 1982. *Tropical Tree Crops*. Chichester, UK: John Wiley & Sons.

Oteng-Yeboah, A., Mutta, D., Byarugaba, D. and Mala, W.A., 2012. Africa. In: *Traditional Forest-related Knowledge: Sustaining Communities, Ecosystems and Biocultural Diversity*, edited by J.A. Parrotta and R.L. Trosper. Dordrecht, the Netherlands: Springer. http://dx.doi.org/10.1007/978-94-007-2144-9_2

Padoch, C., Coffey, K., Mertz, O., Leisz, S., Fox, J., Wadley, R.L., 2007. The demise of swidden in Southeast Asia? Local realities and regional ambiguities. *Geografisk Tidsskrift-Danish Journal of Geography* 107: 29-41. http://dx.doi.org/10.1080/00167223.2007.10801373

Padoch, C., Brondizio, E., Costa, S., Pinedo-Vasquez, M., Sears, R.R. and Siqueira, A., 2008. Urban forest and rural cities: Multi-sited households, consumption patterns, and forest resources in Amazonia. *Ecology and Society* 13(2): 2. http://www.ecologyandsociety.org/vol13/iss2/art2/

Padoch, C. and de Jong, W., 1992. Diversity, variation, and change in ribereño agriculture. In: *Conservation of Neotropical Forests: Working from Traditional Resource Use*, edited by K. Redford and C. Padoch. Columbia University Press: New York.

Padoch, C. and Peluso, N., 1996. *Borneo in Transition: People, Forests, Conservation and Development*. Kuala Lumpur: Oxford University Press.

Padoch, C. and Peters, C., 1993. Managed forest gardens in West Kalimantan, Indonesia. In: *Perspectives on Biodiversity: Case Studies of Genetic Resource Conservation and Development*, edited by C.S. Potter, J.I. Cohen and D. Janczewski. Washington, DC: AAAS.

Palm, C.A., Sanchez, P.A., Ericksen, P.J. and Vosti, S.A. (Eds.), 2005. *Slash-and-burn Agriculture: The Search for Alternatives*. New York: Columbia University Press.

Paoletti, E., Schaub, M., Matyssek, R., Wieser, G., Augustaitis, A., Bastrup-Birk, A.M., Bytnerowiczg, A., Günthardt-Goergb, M.S., Müller-Starckc, G. and Serengilh, A., 2010. Advances of air pollution science: From forest decline to multiple-stress effects on forest ecosystem services. *Environmental Pollution* 158: 1986-1989. http://dx.doi.org/10.1016/j.envpol.2009.11.023

Pardini, R., Bueno, A.A., Gardner, T.A., Prado, P.I. and Metzger, J.P., 2010. Beyond the fragmentation threshold hypothesis: Regime shifts in biodiversity across fragmented landscapes. *PLoS ONE* 5: e13666. http://dx.doi.org/10.1371/journal.pone.0013666

Parrotta, J.A., 1993. *Cocos nucifera* L. Coconut, palma de coco. *Res. Note SO-ITF-SM-57*. New Orleans, LA: USDA Forest Service, Southern Forest Experiment Station. http://www.fs.fed.us/global/iitf/pubs/sm_iitf057 (7).pdf

Parrotta, J.A. and Agnoletti, M., 2012. Traditional forest-related knowledge and climate change. In: *Traditional Forest-related Knowledge: Sustaining Communities, Ecosystems and Biocultural Diversity*, edited by J.A. Parrotta and R.L. Trosper. Dordrecht, the Netherlands: Springer. http://dx.doi.org/10.1007/978-94-007-2144-9_13

Parrotta, J.A. and Trosper, R.L. (eds.), 2012. *Traditional Forest-related Knowledge: Sustaining Communities, Ecosystems and Biocultural Diversity*. Dordrecht, Netherlands: Springer. http://dx.doi.org/10.1007/978-94-007-2144-9

Parry, L., Barlow, J. and Peres, C.A., 2009. Hunting for sustainability in tropical secondary forests. *Conservation Biology* 23(5): 1270-1280. http://dx.doi.org/10.1111/j.1523-1739.2009.01224.x

Pérez, E. and Pacheco, L.F., 2006. Damage by large mammals to subsistence crops within a protected area in a montane forest of Bolivia. *Crop Protection* 25: 933-939. http://dx.doi.org/10.1016/j.cropro.2005.12.005

Perfecto, I., Vandermeer, J., Mas, A. and Soto Pinto, L., 2005. Biodiversity, yield, and shade coffee certification. *Ecological Economics* 54: 435-446. http://dx.doi.org/10.1016/j.ecolecon.2004.10.009

Peters, P.E. 1986. Household management in Botswana: Cattle, crops and wage labor. In: *Understanding Africa's Rural Households and Farming Systems*, edited by J.L. Moock. Boulder and London:Westview.

Pinedo-Vasquez, M., Hecht, S. and Padoch, C., 2012. Amazonia. In: *Traditional Forest-related Knowledge: Sustaining Communities, Ecosystems and Biocultural Diversity*, edited by J.A. Parrotta and R.L. Trosper. Dordrecht, the Netherlands: Springer. http://dx.doi.org/10.1007/978-94-007-2144-9_4

Plantegenest, M., Le May, C. and Fabre, F., 2007. Landscape epidemiology of plant diseases. *Journal of the Royal Society Interface* 4: 963-972. http://dx.doi.org/10.1098/rsif.2007.1114

Platteau, J-P. 1992. *Land Reform and Structural Adjustment in Sub-Saharan Africa: Controversies and Guidelines*. FAO Economic and Social Development Paper 107. Rome: FAO.

Pooley, S., Fa, J.E. and Nasi, R., 2015. No conservation silver lining to Ebola. *Conservation Biology* 29(3): 965-967. http://dx.doi.org/10.1111/cobi.12454

Posey, D.A. (ed.), 1999. *Cultural and Spiritual Values of Biodiversity*. London: Intermediate Technology Publications, UNEP. http://dx.doi.org/10.3362/9781780445434

Powell, B., Thilsted Haraksingh, S., Ickowitz, A., Termote, C., Sunderland, T. and Herforth, A., 2015. Improving diets with wild and cultivated biodiversity from across the landscape. *Food Security*, in press. http://dx.doi.org/10.1007/s12571-015-0466-5

Power, A.G., 2010. Ecosystem services and agriculture: Tradeoffs and synergies. *Philosophical Transactions of the Royal Society B: Biological Sciences* 365: 2959-2971. http://dx.doi.org/10.1098/rstb.2010.0143

Quisumbing, A.R., Payongayong, E., Aidoo, J.B. and Otsuka, K., 2003. Women's land rights in the transition to individualized ownership: Implications for the management of tree resources in western Ghana. In : *Household Decisions, Gender, and Development. A Synthesis of Recent Research*, edited by A. Quisumbing. Washington DC: IFPRI.

Quisumbing, A.R., Estudillo, J.P. and Otsuka, K., 2004. *Land and Schooling: Transferring Wealth Across Generations*. Baltimore: John Hopkins University Press, for IFPRI.

Quisumbing, A.R. and Pandofelli, L,. 2009. *Promising Approaches to Address the Needs of Poor Female Farmers: Resources, Constraints, and Interventions*. IFPRI Discussion Paper 00882. Washington DC: IFPRI. http://citeseerx.ist.psu.edu/viewdoc/download?doi=10.1.1.227.3028&rep=rep1&type=pdf

Raintree, J. and Warner, K., 1986. Agroforestry pathways for the intensification of shifting cultivation. *Agroforestry Systems* 4(1): 39-54. http://dx.doi.org/10.1007/bf01834701

Ramakrishnan, P.S., 1992. *Shifting Agriculture and Sustainable Development: An Interdisciplinary Study from North-eastern India*. Man and Biosphere Book Series No. 10. Paris: UNESCO and Caernforth, Lancaster, UK: Parthenon Publishing.

Ramakrishnan, P.S., Rao, K.S., Chandrashekara, U.M., Chhetri, N., Gupta, H.K., Patnaik, S., Saxena, K.G. and Sharma, E., 2012. South Asia. In: *Traditional Forest-related Knowledge: Sustaining Communities, Ecosystems and Biocultural Diversity*, edited by J.A. Parrotta and R.L. Trosper. Dordrecht, the Netherlands: Springer. http://dx.doi.org/10.1007/978-94-007-2144-9_9

Rangarajan, M., Desai, A., Sukumar, R., Easa, P.S., Menon, V., Vincent, S., Ganguly, S., Talukdar, B.K., Singh, B., Mudappa, D., Chowdhary, S. and Prasad, A.N., 2010. *Securing the Future for Elephants in India*. The Report of the Elephant Task Force. New Delhi: Ministry of Environment and Forests. http://www.moef.nic.in/downloads/public-information/ETF_REPORT_FINAL.pdf

Reij, C., 2014. Re-greening the Sahel: Linking adaptation to climate change, poverty reduction, and sustainable development in drylands. In: *The Social Lives of Forests: Past, Present and Future of Woodland Resurgence*, edited by S. Hecht, K. Morrison and C. Padoch. Chicago and London: University of Chicago Press.

Rerkasem, K., Lawrence, C., Padoch, D. Schmidt-Voght, D., Zeigler, A.D. and Bruun, T.B., 2009. Consequences of swidden transitions for crop and fallow biodiversity in Southeast Asia. *Human Ecology* 37: 347-360. http://dx.doi.org/10.1007/s10745-009-9250-5

Rice, R.A. and Greenberg, R., 2000. Cacao cultivation and the conservation of biological diversity. *AMBIO: A Journal of the Human Environment* 29(3): 81-87, 167-173. http://dx.doi.org/10.1639/0044-7447(2000)029[0167:ccatco]2.0.co;2

Ricketts, T.H., 2004. Tropical forest fragments enhance pollinator activity in nearby coffee crops. *Conservation Biology* 18: 1262-1271. http://dx.doi.org/10.1111/j.1523-1739.2004.00227.x

Ricketts, T.H., Regetz, J., Steffan-Dewenter, I., Cunningham, S.A., Kremen, C., Bogdanski, A., Gemmill-Herren, B., Greenleaf, S.S., Klein, A.M., Mayfield, M.M., Morandin, L.A., Ochieng, A. and Viana B.F., 2008. Landscape effects on crop pollination services: Are there general patterns? *Ecology Letters* 11: 499-515. http://dx.doi.org/10.1111/j.1461-0248.2008.01157.x

Rival, L.M., 2002. *Trekking Through History: The Huaorani of Amazonian Ecuador*. New York: Columbia University Press.

Rocheleau, D. and Edmunds, D., 1997. Women, men and trees: Gender, power and property in forest and agrarian landscapes. *World Development* 25(8): 1351-1371. http://dx.doi.org/10.1016/s0305-750x(97)00036-3

RRI, 2012. *What Rights? A Comparative Analysis of Developing Countries' National Legislation on Community and Indigenous Peoples' Forest Tenure Rights*. Washington DC: Rights and Resources Initiative. http://www.rightsandresources.org/

Rose, L., 1996. *Disputes in Common Property Regimes*. Land Tenure Center Paper 154. Madison: Land Tenure Center, University of Wisconsin-Madison.

Rudel, T.K., DeFries, R., Asner, G.P. and Laurance, W.F., 2009. Changing Drivers of Deforestation and New Opportunities for Conservation. *Conservation Biology* 23(6): 1396-1405. http://dx.doi.org/10.1111/j.1523-1739.2009.01332.x

Ruf, F. and Schroth, G., 2004. Chocolate forests and monocultures: A historical review of cocoa growing and its conflicting role in tropical deforestation and forest conservation. In: *Agroforestry and Biodiversity Conservation in Tropical Landscapes*, edited by G. Schroth, G.A.B. Da Fonseca, C.A. Harvey, C. Gascon, H.L. Lasconcelos and A.N. Izac. Washington, DC: Island Press. http://library.uniteddiversity.coop/Permaculture/Agroforestry/Agroforestry_and_Biodiversity_Conservation_in_Tropical_Landscapes.pdf

Russell, W.M S., 1988. Population, swidden farming and the tropical environment. *Population and Environment* 10: 77-94. http://dx.doi.org/10.1007/bf01359134

Sanchez P.A., 1995. Science in agroforestry. *Agroforest Systems* 30: 5-55. http://dx.doi.org/10.1007/bf00708912

Sanchez, P.A., Palm, C.A., Vosti, S.A., Tomich, T. and Kasyoki, J., 2005. Alternatives to slash and burn: Challenges and approaches of an international consortium. In: *Slash-and-burn Agriculture: The Search for Alternatives*, edited by C.A. Palm, S.A. Vosti, P.A. Sanchez and P.J. Ericksen. New York: Columbia University Press.

Saroj, P.L. and Rupa, T.R., 2014. Cashew research in India: Achievements and strategies. *Progressive Horticulture* 46(1): 1-17.

Sauer, C.O., 1969. *Agricultural Origins and Dispersals*, 2nd edn. Cambridge and London: MIT Press.

Scherr, S.J., 1995. Economic factors in farmer adoption: Patterns observed in Western Kenya. *World Development* 23: 787-804. http://dx.doi.org/10.1016/0305-750x(95)00005-w

Schlager, E. and Ostrom, E., 1992. Property rights and natural resources: A conceptual analysis. *Land Economics* 68(3): 249-62. http://dx.doi.org/10.2307/3146375

Schlegel, S.A. and Guthrie, H.A., 1973. Diet and the tiruray shift from swidden to plow farming. *Ecology of Food and Nutrition* 2(3): 181-191. http://dx.doi.org/10.1080/03670244.1973.9990335

Schmidt-Vogt, D., Leisz, S.J., Mertz, O., Heinimann, A., Thiha, T. Messerli, P., Epprecht, M., Cu, P.V., Chi, V.K., Hardiono, M. and Dao, T.M., 2009. An assessment of trends in the extent of swidden in Southeast Asia. *Human Ecology* 37: 269-280. http://dx.doi.org/10.1007/s10745-009-9239-0

Schroth, G. and Harvey, C., 2007. Biodiversity conservation in cocoa production landscapes: An overview. *Conservation and Biology* 16(8): 2237-2244. http://dx.doi.org/10.1007/s10531-007-9195-1

Scott, J., 1999. *Seeing Like a State: How Certain Schemes to Improve the Human Condition have Failed*. New Haven, CT: Yale University Press.

Sinu, P.A., Kent, S.M. and Chandrashekara, K., 2012. Forest resource use and perception of farmers on conservation of a usufruct forest (Soppinabetta) of Western Ghats, India. *Land Use Policy* 29: 702-709. http://dx.doi.org/10.1016/j.landusepol.2011.11.006

Smith, D.A., 2005. Garden game: Shifting cultivation, indigenous hunting and wildlife ecology in western Panama. *Human Ecology* 33(4): 505-37. http://dx.doi.org/10.1007/s10745-005-5157-y

Smith, N.J.H., Williams, J.T., Plucknett, D.L. and Talbot, J.P., 1992. *Tropical Forests and their Crops*. Ithaca and London: Cornell University Press.

Spencer, J.E., 1966. *Shifting Cultivation in Southeastern Asia*. Berkeley and Los Angeles: University of California Press.

Stevens, C., Winterbottom, R., Springer, J. and Reytar, K., 2014. *Securing Rights, Combating Climate Change: How Strengthening Community Forest Rights Mitigates Climate Change*. Washington DC: World Resources Institute. http://www.wri.org/securingrights

Swift, M.J., Vandermeer, J., Ramakrishnan, P.S., Anderson, J.M., Ong, C.K. and Hawkins, B., 1996. Biodiversity and agroecosystem function. In: *Functional Roles of Biodiversity: A Global Perspective*, SCOPE Series, edited by H.A. Mooney, J.H. Cushman, E. Medina, O.E. Sala and E.-D. Schulze. Chichester, UK: Wiley.

Tejeda-Cruz, C., Silva-Rivera, E., Barton, J.R. and Sutherland, W.J., 2010. Why shade coffee does not guarantee biodiversity conservation. *Ecology and Society* 15: 13.

Thompson, I.D., Ferreira, J., Gardner, T., Guariguata, M., Koh, L.P., Okabe, K., Pan, Y., Schmitt, C.B., Tylianakis, J., Barlow, Kapos, V., Kurz, W.A., Spalding, M. and van Vliet, N., 2012. Forest biodiversity, carbon and other ecosystem services: Relationships and impacts of deforestation and forest degradation. In: *Understanding Relationships Between Biodiversity, Carbon, Forests and People: The Key to Achieving REDD+ Objectives*, IUFRO World Series No. 31, edited by J.A. Parrotta, C. Wildburger and S. Mansourian. Vienna: International Union of Forest Research Organizations.

Thrupp, L.A., Hecht, S.B. and Browder, J.O., 1997. *The Diversity and Dynamics of Shifting Cultivation: Myths, Realities and Policy Implications*. Washington, DC: World Resources Institute. http://pdf.wri.org/diversitydynamicscultivation_bw.pdf

Toledo, M. and Salick, J. 2006. Secondary succession and indigenous management in semideciduous forest fallows of the Amazon Basin. *Biotropica* 38(2): 161-170. http://dx.doi.org/10.1111/j.1744-7429.2006.00120.x

Tomalak, M., Rossi, E., Ferrini, F. and Moro, P.A., 2011. Negative aspects and hazardous effects of forest environments on human health. In: *Forests, Trees and Human Health*, edited by K. Nilsson, M. Sangster, C. Gallis, T. Hartig, S. de Vries, K. Seeland and J. Schipperijn. New York: Springer. http://dx.doi.org/10.1007/978-90-481-9806-1_4

Tontisirin, K., Nantel, G. and Bhattacharjee, L., 2002. Food-based strategies to meet the challenges of micronutrient malnutrition in the developing world. *Proceedings of the Nutrition Society* 61(2): 243-250. http://dx.doi.org/10.1079/pns2002155

Tscharntke, T., Klein, A.M., Kruess, A., Steffan-Dewenter, I. and Thies, C., 2005a. Landscape perspectives on agricultural intensification and biodiversity—ecosystem service management. *Ecology Letters* 8: 857-874. http://dx.doi.org/10.1111/j.1461-0248.2005.00782.x

Tscharntke, T., Rand, T.A. and Bianchi, F.J.J.A., 2005b. The landscape context of trophic interactions:insect spillover across the crop-noncrop interface. *Annual Zoology Fennici*: 421-432. http://www.annzool.net/PDF/anzf42/anzf42-421.pdf

Tscharntke, T., Tylianakis, J.M., Rand, T.A., Didham, R.K., Fahrig, L. and Batáry, P., Bengtsson, J., Clough, Y., Crist, T.O., Dormann, C.F., Ewers, R.M., Fründ, J., Holt, R.D., Holzschuh, A., Klein, A.M., Kleijn, D., Kremen, C., Landis, D.A., Laurance, W., Lindenmayer, D., Scherber, C., Sodhi, N., Steffan-Dewenter, I., Thies, C., van der Putten, W.H. and Westphal, C., 2012. Landscape moderation of biodiversity patterns and processes—eight hypotheses. *Biological Reviews*: 87: 661-685. http://dx.doi.org/10.1111/j.1469-185x.2011.00216.x

Turner, N.J., Łuczaj, Ł.J., Migliorini, P., Pieroni, A., Dreon, A.L., Sacchetti, L.E. and Paoletti, M.G., 2011. Edible and tended wild plants, traditional ecological knowledge and agroecology, *Critical Reviews in Plant Sciences* 30: 1-2, 198-225. http://dx.doi.org/10.1080/07352689.2011.554492

United Nations, 2009. *The State of the World's Indigenous People.* New York: UN Department of Economic and Social Affairs, Permanent Forum on Indigenous Issues. http://www.un.org/esa/socdev/unpfii/documents/SOWIP/en/SOWIP_web.pdf

Uriarte, M., Pinedo-Vasquez, M., DeFries, R.S., Fernandes, K., Gutierrez-Velez, V., Baethgen, W. E. and Padoch, C., 2012. Depopulation of rural landscapes exacerbates fire activity in the western Amazon. *Proceedings of the National Academy of Sciences* 109(52): 21546-21550. http://dx.doi.org/10.1073/pnas.1215567110

Van Koppen, B., 1990. *Women and the Design of Irrigation Schemes: Experiences from Two Cases in Burkina Faso.* Paper presented at the International Workshop on Design for Sustainable Farmer-Managed Irrigation Schemes in Sub-Saharan Africa. Wageningen Agricultural University, The Netherlands, 5-8 February.

van Vliet, N., Mertz, O., Heinimann, A., Langanke, T., Pascual, U., Schmook, B., Adams, C., Schmidt-Vogt, D., Messerli, P., Leisz, S., Castella, J.-C., Jørgensen, L., Birch-Thomsen, T., Hett, C.,Bech-Bruun, T., Ickowitz, A., Vum, K.C., Yasuyuki, K., Fox, J., Padoch, C., Dressler, W. and Ziegler, A.D., 2012. Trends, drivers and impacts of changes in swidden cultivation in tropical forest-agriculture frontiers: A global assessment. *Global Environmental Change* 22: 418-429. http://dx.doi.org/10.1016/j.gloenvcha.2011.10.009

Veberic, R., Colaric, M. and Stampar, F., 2008. Phenolic acids and flavonoids of fig fruit (*Ficus carica* L.) in the northern Mediterranean region. *Food Chemistry* 106(1): 153-157. http://dx.doi.org/10.1016/j.foodchem.2007.05.061

Velded, T., 2000. Village politics: Heterogeneity, leadership and collective action. *Journal of Development Studies* 36: 105-134. http://dx.doi.org/10.1080/00220380008422648

Wadley, R. L. and Colfer, C.J.P., 2004. Sacred forest, hunting, and conservation in West Kalimantan, Indonesia. *Human Ecology* 32: 313-338. http://dx.doi.org/10.1023/b:huec.0000028084.30742.d0

Warner, K., 1991. *Shifting Cultivators: Local and Technical Knowledge and Natural Resource Management in the Humid Tropics*. Community Forestry Note 8. Rome: FAO. ftp://ftp.fao.org/docrep/fao/u4390e/u4390e00.pdf

Watson, G.A., 1990. Tree crops and farming systems development in the humid tropics. *Experimental Agriculture* 26: 143-159. http://dx.doi.org/10.1017/s0014479700018147

Whitehead, A. 1985. Effects of technological change on rural women: A review of analysis and concepts. In: *Technology and Rural Women: Conceptual and Empirical Issues*, edited by I. Ahmed. London: George Allen and Unwin.

Wiersum, K.F., 1997. Indigenous exploitation and management of tropical forest resources: An evolutionary continuum in forest-people interactions. *Agriculture, Ecosystems and Environment* 63(1997): 1-16. http://dx.doi.org/10.1016/s0167-8809(96)01124-3

Wilcox, B.A. and Ellis, B., 2006. Forests and emerging infectious diseases of humans. *Unasylva* 57: 11-18. ftp://ftp.fao.org/docrep/fao/009/a0789e/a0789e03.pdf

Williams, T., 2014. Ebola's silver lining. *New Scientist* 223: 26-27. http://dx.doi.org/10.1016/s0262-4079(14)61720-6

WOCAT, 2007. *SWC Technology: Shade-grown Coffee.* San José, Costa Rica: Ministerio de Agricultura y Ganadería.

World Bank, 2013. *Forest, Trees and Woodlands in Africa. An Action Plan for World Bank Engagement.* Washington, DC: World Bank. https://openknowledge.worldbank.org/bitstream/handle/10986/11927/730260REPLACEM0tion0Plan06014012web.pdf?sequence=1

World Bank, FAO and IFAD, 2009. *Gender and Forestry, Module 15. Gender in Agriculture Sourcebook.* Washington DC: World Bank. http://siteresources.worldbank.org/INTGENAGRLIVSOUBOOK/Resources/Module15.pdf

Youn, Y.-C., Liu, J., Daisuke, S., Kim, K., Ichikawa, M., Shin, J.-H. and Yuan, J., 2012. Northeast Asia. In: *Traditional Forest-related Knowledge: Sustaining Communities, Ecosystems and Biocultural Diversity*, edited by J.A. Parrotta and R.L. Trosper. Dordrecht, the Netherlands: Springer. http://dx.doi.org/10.1007/978-94-007-2144-9_8

Ziegler, A.D., Phelps, J., Yuen, J.Q., Webb, E.L., Lawrence, D., Fox, J.M., Bruun, T.B., Leisz, S.J., Ryan, C.M., Dressler, W., Mertz, O., Pascual, U., Padoch, C. and Koh, L.P., 2012. Carbon outcomes of major land-cover transitions in SE Asia: Great uncertainties and REDD+ policy implications. *Global Change Biology* 18(10): 3087-3099. http://dx.doi.org/10.1111/j.1365-2486.2012.02747.x

Zomer, R.J., Bossio, D.A., Trabucco, A., Yuanjie, L., Gupta, D.C. and Singh, V.P., 2007. *Trees and Water: Smallholder Agroforestry on Irrigated Lands in Northern India.* IWMI Research Reports, No. 122. Colombo, Sri Lanka: International Water Management Institute. http://indiaenvironmentportal.org.in/files/RR122.pdf

Zomer, R.J., Trabucco, A., Coe, R. and Place, F., 2009. *Trees on Farm: Analysis of Global Extent and Geographic Pattern of Agroforestry.* ICRAF Working Paper No. 89. Nairobi: World Agroforestry Centre. http://dx.doi.org/10.5716/wp16263.pdf

Zomer, R.J., Trabucco, A., Coe, R. Place, F., van Noordwijk, M. and Xu, J.C., 2014. *Trees on Farms: An Update and Reanalysis of Agroforestry's Global Extent and Socio-ecological Characteristics.* Working Paper 179. Bogor, Indonesia: World Agroforestry Centre (ICRAF) Southeast Asia Regional Program. http://www.worldagroforestry.org/downloads/Publications/PDFS/WP14064.pdf

4. Drivers of Forests and Tree-based Systems for Food Security and Nutrition

Coordinating lead author: *Daniela Kleinschmit*
Lead authors: *Bimbika Sijapati Basnett, Adrian Martin, Nitin D. Rai and Carsten Smith-Hall*
Contributing authors: *Neil M. Dawson, Gordon Hickey, Henry Neufeldt, Hemant R. Ojha and Solomon Zena Walelign*

In the context of this chapter, drivers are considered to be natural or anthropogenic developments affecting forests and tree-based systems for food security and nutrition. They can improve and contribute to food security and nutrition, but they can also lead to food insecurity and malnutrition. For analytical purposes, drivers are separated here into the following four interconnected categories: (i) environmental, (ii) social, (iii) economic and (iv) governance. When reviewing scientific findings twelve major drivers (i.e. population growth, urbanisation, governance shifts, climate change, commercialisation of agriculture, industrialisation of forest resources, gender imbalances, conflicts, formalisation of tenure rights, rising food prices and increasing per capita income) were identified within these four categories. They affect food security and nutrition through land use and management; through consumption, income and livelihood; or through both. These drivers are interrelated and can have different consequences depending on the social structure; for example, they can support food security for elite groups but can increase the vulnerability of other groups.

 http://dx.doi.org/10.11647/OBP.0085.04

4.1 Introduction

Drivers of change are the subject of a vast scholarly literature. In the context of this book, drivers are understood as natural or anthropogenic developments affecting *forests and tree-based systems*[1] for *food security* and *nutrition*. This chapter aims to provide a structured and comprehensive overview of major findings from this literature in an effort to better understand the interrelations among these drivers and how they impact on food security and nutrition. It covers drivers that improve and contribute to food security and nutrition as well as those leading to increased food insecurity and poor nutrition. Changes to improve food security and nutrition result for example in an increased availability of food and better nutrition. In contrast, changes that lead to food insecurity increase the vulnerability of both *ecosystems* and humanity. Identifying drivers and understanding their impact pathways is essential to determine options for effective interventions by enhancing positive and minimising negative effects (see Chapter 6).

Driving forces can originate at different spatial scales and can be distant or proximate. In this chapter, attention is paid in particular to those drivers constraining forests and tree-based systems for food security and nutrition in vulnerable regions, i.e. the tropics, neo-tropics and sub-Saharan Africa. Both the environmental and the human components of forests and tree-based systems for food security and nutrition are subject to changing dynamics presenting a different picture over time. These dynamics also imply that drivers and effects are strongly interrelated and can mutually affect each other. Consequently a simplified classification as driver or effect sometimes falls short in addressing this complex relationship.

The chapter builds on a framework (see Figure 4.1) to categorise drivers and trace their impact pathways. According to their content, drivers can be separated for analytical reasons into the following four interconnected categories: (i) environmental, (ii) social, (iii) economic, and (iv) *governance*. Environmental drivers refer to developments in nature (many of which have themselves anthropogenic causes) that change food security. Social drivers include the role of patterns of social differentiation, inequalities and changes in influencing forests and tree- based systems for food security and nutrition. Economic drivers relate to direct and indirect impacts from economic activities that are both economy-wide and site specific. Governance refers to those institutions setting the rules of the game, differentiating between state and non-state governance. The drivers identified in these four groups mainly impact food security and nutrition through two major pathways: changes in land use and (*forest*) *management* or changes in consumption, incomes and livelihoods (see Figure 4.1). Both pathways determine food availability, access and stability that ensure food security and nutrition.

This chapter presents findings from available scholarly literature for each category of driver and ends with a summary of major results. Literature about drivers referring

1 All terms that are defined in the glossary (Appendix 1), appear for the first time in italics in a chapter.

to the interrelation between forests and tree-based systems on the one hand, and food security and nutrition on the other, is rare. For this reason, the authors of this chapter reviewed literature on the subject of change from both scientific areas – *forests* and food security – and linked them to present a comprehensive overview of relevant drivers.

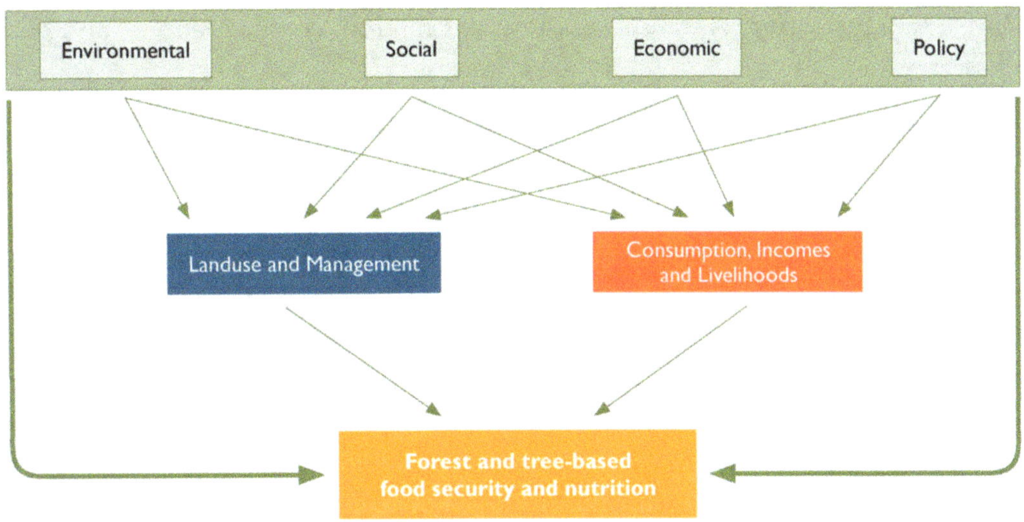

Fig. 4.1 Framework of drivers directly and indirectly impacting on forests and tree-based systems for food security and nutrition

4.2 Environmental Drivers

Before reviewing different environmental drivers it should be highlighted once again that environmental and anthropogenic developments are marked by a complex, mutual relationship. Environmental drivers are themselves consequences of human action, policies and societal processes. Consequently the underlying interactions between social, political, economic and ecological processes are difficult to isolate from each other. The effect of human activities on ecosystems has been large enough to warrant the call to rename the current geological era as the Anthropocene (Crutzen, 2006). Many critical thresholds of the earth's biophysical systems have already been crossed as a result of human activity (Rockström et al., 2009). These processes have uneven impacts on different sections of humanity (Mohai et al., 2009). People living directly off production from the earth's ecosystems are particularly affected by these changes. Factors such as increasing temperatures, variable precipitation, *fragmentation, deforestation, invasive species* and loss of biological diversity affect not only the extent of forest but also the structure and species composition within forests (and therefore, forest products) thus impacting on the availability of food and nutrition.

Three significant larger scale environmental drivers that impact directly on the forests-food nexus will be discussed in this section: *climate change*, deforestation and forest transitions, and invasive species. The identification of larger scale drivers of forest change and food security is important given the conventional understanding that forest *degradation* is often the result of local processes such as conversion to agriculture, grazing and harvest of forest produce. While these local practices do have impacts they are generally small-scale, reversible and are often regulated by local bodies. The larger drivers listed below however call for action at national and global scales.

Climate change

Climate change is affecting global and local ecological processes in many ways. Though the consequences are complex, there is enough evidence that ongoing and future changes are going to be drastic. The Intergovernmental Panel on Climate Change (IPCC) notes that by the end of this century rates of climate change as a result of medium to high emission levels "pose high risk of abrupt and irreversible regional-scale change in the composition, structure, and function of terrestrial ecosystems". There is widespread evidence that the poorest regions in the world, such as sub-Saharan Africa, will be affected the hardest by climate change. The IPCC report notes that "increased tree mortality and associated forest dieback is projected to occur in many regions over the 21st century, due to increased temperatures and drought" (IPCC, 2014). The effects of climate change, combined with land cover change such as reduced forest cover and fragmentation, exacerbate impacts (Afreen et al., 2010). These climate-induced changes affecting forest cover imply both direct and indirect consequences for food security and nutrition: direct consequences result from changes in the availability and quality of food and nutrition, while indirect consequences result from changes in income and livelihoods related to forest products. The consequences of climate change for forests and tree-based systems for food security and nutrition, however, are not well understood although comprehensive reviews of climate and agricultural *food systems* have been published (Vermeulen et al., 2012).

The IPCC more specifically forecasts the following changes concerning forests: a decrease in tree densities in parts of North Africa, range shifts of several southern plants and animals, changes in plant phenology and growth in many parts of Asia (earlier greening), distribution shifts of many plant and animal species upwards in elevation and an increase in tree mortality and forest fire in the Amazon. Climate-induced effects will interact with ongoing *landscape* changes to produce a range of synergistic outcomes with significant effect on plant and forest health (Pautasso et al., 2010). Studies have demonstrated that climatic impacts interact with other landscape level drivers of change to affect biological assemblages and ecosystems. For instance, some landscapes might hinder the dispersal of species and thus prevent species from shifting range (adapting) as climate regimes change (Garcia et al., 2014). Tropical tree species are going to be the most affected by climate change as they are already close to their thermal tolerances (IPCC, 2014). The inability of species to adapt to changing climates combined with phenological changes such as earlier flowering (and thus reduced fruit yields and production) could result in direct impacts on the amount of forest resources available for harvest and use by local communities, particularly impacting those communities that are most dependent on forests.

There is a shared understanding in the literature that climate change will affect the most vulnerable groups, especially women (Brody et al., 2008). An indirect effect of climate change is expected from increasing world food prices with harsh consequences for the poorest, including women. Literature dealing with gender imbalances of climate change impacts mainly refers to food security and agriculture. However, findings could be transferred to forests and tree-based food systems and food security and nutrition. In particular, the literature identifies a number of reasons for the gender-differentiated vulnerability of climate change impacts. Amongst these, are different coping and adaptive abilities of men and women (UNDP, 2012) depending on the inequalities in access to assets as well as legal socio-cultural barriers preventing women from effectively responding to climate change (UNDP, 2012).

Deforestation and forest transition

Deforestation and forest transitions interact with food security and nutrition in many ways, directly impacting on the extent of forest available for the harvest of fruit and other forest- and tree-based diets. In particular, deforestation and forest degradation affect biodiversity and the variety of food available through habitat loss and forest transformation.

The process of deforestation is complex and goes beyond the simple removal of trees; there is a continuum of forest structures that complicates what is understood as forests, and is accompanied often by rapid regrowth. The relationship between deforestation and forest dependence is neither inverse nor linear. It has been demonstrated that forests with intermediate levels of diversity are as viable for *livelihoods* as diverse forests, and *secondary forests* have been shown to provide more forest products than highly diverse forests (Saw et al., 1991). There is increasing evidence that in some instances areas that were deforested are now indistinguishable from *primary forests* (Willis et al., 2004). The conventional understanding that current forested landscapes are remnants of past deforestation and degradation has been revisited by studies that have shown that these forests might have been raised by people through active management and customary practice (Fairhead and Leach, 1996; Virah-sawmy, 2009).

While during the past decade deforestation rates have decreased globally, some countries are showing increasing rates of reforestation (Meyfroidt and Lambin, 2011). Reforestation is occurring due to a host of factors such as flows of labour, capital, conservation policies, and the valuation of and markets for *ecosystem services* (Hecht et al., 2006). The valuation of ecosystem services and reducing emissions from deforestation and forest degradation, enhancing forest carbon stocks, sustainable management and conservation of forests (REDD+) has implications for the governance and local use of forests (Phelps et al., 2010) (see Section 4.5). Consequently, policies that encourage reforestation risk having equally adverse impacts on local communities – by preventing access to forest resources – as those that encourage deforestation. Equally, policies that are aimed at reducing deforestation and degradation by local communities often lead to deprivation and livelihood insecurity (DeFries et al., 2010).

Studies have shown that there is a direct relationship between tree cover, tree species diversity and food security especially of vulnerable groups (Ickowitz et al., 2014; van Noordwijk et al., 2014). Changes in the extent and type of forest have implications for the provisioning of food, and for food security and nutrition of local and distant human populations.

Global rates of deforestation have been high for the last few centuries and have been driven by such factors as agriculture (commercial and subsistence), mining, urbanisation and infrastructure expansion (Hosonuma et al., 2012; Williams, 2003) (see also Chapter 3). Globalisation and urbanisation trends starting in the 1980s have changed the agents of deforestation from local population use to capital-intensive commercial farming that supplies distant markets (Rudel et al., 2009). In a review of the history of forest clearing, Williams (2008) concludes, on the basis of four estimates, that the total area of forest that has been lost is between 19 and 36 percent which, while still a large area, is not as devastating as commonly perceived. Recent trends show that agriculture is the biggest driver of deforestation accounting for 73 percent of deforestation worldwide, while mining accounts for seven percent, infrastructure for 10 percent and urban expansion for 10 percent (Hosonuma et al., 2012). Agri-businesses such as cattle ranching, soybean farming and oil palm plantations are now the most important drivers of forest loss globally (Boucher et al., 2011; Rudel et al., 2009). There are regional variations in the significance of drivers of deforestation, with urban expansion for example, being the most important in Asia. Commercial agriculture accounts for 68 percent of forest loss in Latin America and 35 percent in Africa and Asia (Hosonuma et al., 2012). DeFries et al. (2010) show that forest loss is strongly correlated with urban population growth and the export of agricultural products. Furthermore, the interrelation between forests and water has been highlighted in many studies (e.g. Malmer et al., 2010). Rainfall does not provide sufficient water supply in many countries so households depend on sources of groundwater that are often found in or near forested land. Deforestation affects water supply in different ways depending on local conditions.

Invasive species

Another ecological driver of local forest change is invasive species which are often a result of altered management. Plants and animals have been constantly moved to new areas for a range of purposes and have been agents of positive as well as adverse change (Kull and Rangan, 2008; Robbins, 2004). Managing landscapes for the control of invasive species has implications for food security through the increase in resources such as fodder, game and tree species. The change in composition of forests or the dominance of certain species has ecological determinants but as has been often recorded these follow from changes in management regimes and policy contexts (Dove, 1986; Robbins, 2004). While there has been some effort in defining and identifying invasive species, what is equally important is to identify the reasons that species were introduced and what social, economic and ecological contexts enabled their spread. Often the introduction of species has had positive impacts on food systems such as with many tropical *agroforestry* systems (Ewel, 1999), while at other times it has had negative impacts, such as with forestry tree species for fuelwood and timber that have had adverse impacts on food species and water availability (Richardson and van Wilgen, 2004).

The incursion of non-native species into terrestrial ecosystems has a long history through the exchange of plants and animals as a result of human movement. Some of these species become invasive and lead to structural and species changes in the forest as well as altering ecological processes, ultimately affecting food availability. Recent research shows that invasive species are resulting in *biodiversity* loss and low regeneration rates of other native species (Ticktin et al., 2012). The ecological consequences of invasive species are high, as are socio-economic outcomes (Pysek and Richardson, 2010). For example, in South Africa, the value of native fynbos ecosystems has been reduced by over USD11.75 billion because of invasive species (van Wilgen et al., 2001).

Ginger or Kahili lily (*Hedychium gardnerianum*) is a plant native to the Himalayas that has become an invasive species in the Azores where it was introduced as an ornamental plant in the 19th century. Photo © PJ Stephenson

Furthermore, the changes induced by climate change encourage certain species to move into forest habitats. The increase in energy availability in forests as a result of the shift in ranges of native species enables invasive species to fill the available capacity in these areas (Chown et al., 2013). In a study of the impacts of an invasive species on a *non-timber forest product* species in India, Ticktin et al. (2012) have demonstrated that *Lantana camara* suppresses the regeneration of seedlings of the forest tree species and leads to drastic changes in population growth of this species.

A strong relationship exists between governance regimes and ecological outcomes. Forest areas that have seen recent incursions of invasive species are also areas in which customary practice has been suspended as a result of territorial governance regimes. Management regimes that have banned fires have seen significant increases in densities of invasive species (Debuse and Lewis, 2014) leading Robbins (2004) to assert that "it is not species, but sociobiological networks that are invasive". Zero burn policies that are practised in India and Brazil have resulted in a range of adverse outcomes due to policies not being socially contextualised (Carmenta et al., 2011). The continued narrative of the degradation caused by local practices such as fire has been shown to be unfounded in many forest types. For example, Welch et al. (2013) have shown how indigenous burning in the Cerrado savannah of Brazil has assisted vegetation recovery. Ensuring that the management of landscapes for forest regeneration or to control invasive species is consistent with historical practices promotes both local culture and food security.

The effects of invasive species on livelihood and food security are not uniform within or across communities. As Shackleton et al. (2006) have shown in South Africa, invasive species that have adverse impacts on some sections of the local population are being used to considerable advantage by other sections of the community. While ecological studies have highlighted the impact of invasive species on biological diversity and provisioning of resources, it should be noted that humans have historically relied on such species for food and other requirements. Equally, as forests are being transformed by species introductions and by changed management regimes, people are evolving coping strategies to maintain their livelihood systems. Many communities have optimised the use of introduced species to their benefit and enhanced their livelihood options through the use of such species whether directly for food (e.g. fruits of *Opuntia* sp. in South Africa) or through the sale of products derived from these species (e.g. charcoal from *Prosopis juliflora* in India). The outcomes of environmental changes on food security are complex and require context-specific responses and strategies.

4.3 Social Drivers

This section discusses conflict, relative poverty and inequality, and demographic changes as social drivers that influence forests and tree-based systems for food security and nutrition. It highlights the role of deeply-rooted patterns of social differentiation and inequalities in influencing forests and food systems, both in terms of land use and management as well as income and livelihood. Conflict is considered since forests are

Donkey cart crivers, Senegal.
Photo © Terry Sunderland

often at the centre of conflicting interests, whereas the sheer movement of people from rural, urban and transnational spaces are some of the defining characteristics of the contemporary era with considerable effects on tropical forests and forest-dependent communities.

Conflicts in and about forests

About 243 million hectares of the world's closed forests are located in areas affected by conflicts since 1990 (De Koning et al., 2008). A substantial body of scholarly and grey literature has been devoted to conflicts that emerge from competing claims and interests – commercial, subsistence and cultural – over resources in forested landscapes. This section focuses on how conflicts, spill over into forested landscapes as well as on conflicts that are endemic to forests themselves. The impact of such conflicts on forests and food security can be understood in terms of direct access to foods sourced from forests and indirect effects on food security, for example, via wood for fuel which is essential for cooking in many countries of the world.

During the past 20 years, armed conflicts have struck forest areas in more than 30 countries in the tropics. The prominent examples include Cambodia, Democratic Republic of the Congo, Liberia, Myanmar and Sierra Leone where rebel warfare largely played out in remote cross-border forest areas (De Koning et al., 2008). Africa is home to most of the forest at risk whereas Asia has the highest number of forest dwellers at risk (De Koning et al., 2008). An estimated three-quarters of Asian forests, two-thirds of African forests and one-third of Latin American forests have been affected by violent conflict (de Jong et al., 2007). The mere overlap between forest and conflict areas does not necessarily mean that the forest or forest rights have any role to play in motivating or perpetuating conflicts. However, because of the risks involved due to instability and insecurity, it can be assumed that these areas only serve in a limited way as a source of food.

Studies on the correlation between countries' forest cover and the emergence and duration of civil conflict show contradictory results (e.g. Collier and Hoeffler, 2001; Lujala, 2003; Rod and Rustad, 2006). Nevertheless, different studies identified that forests can facilitate or prolong conflicts, for example through flows of finances to competing parties, use of forests for patronage, transport of weapons by loggers, agriculture and hunting pressures, and social and economic buffers. For instance, forests and forest products have been exploited by armed groups (e.g. military and rebels) to strengthen their fighting capacities (see for example, Baral and Heinmen, 2006, for Nepal; Dudley et al., 2002; de Merode et al., 2007, for the Democratic Republic of the Congo). General implications of such conflict on food security of forest dependent communities are difficult to predict. Armed conflict can weaken pre-existing institutions governing access to forest food but it can also offer new and extra-legal channels. Effects on food security and nutrition depend on the larger political economy in which the conflict is situated and the interaction with the formal and informal institutions that govern the forests. For instance, de Merode et al. (2007) note that there was a fivefold increase in illegal trade of bushmeat for local consumption and sale during the civil war in the Democratic Republic of the Congo. In contrast Baral and Heinmen (2006) highlight that the Maoist movement and ensuing civil conflict in Nepal between 1996 and 2006 largely undermined both conservation efforts and the livelihoods of local people by hampering their ability to derive income from forests and limiting households' access to food and nutrition.

Conflicts can also be more endemic to forested landscapes (de Jong et al., 2007). These tend to be localised and non-violent, though some may escalate to violent armed conflicts. Through an analysis of forest-related conflicts in five Asian countries, De Koning et al. (2008) classify such conflicts as emerging from the following, interrelated factors: (a) contested statutory and customary *tenure*, (b) exclusionary conservation and economic development policies, and (c) poor coordination between land use planning agencies. For example, conflict between local communities and oil palm plantation corporations in Indonesia due to overlapping claims over land and weak protection for customary land rights, illustrates the first type of conflict (Colchester and Chao, 2013; Li, 2014; Sheil et al., 2009). The implications of such conflicts over oil palm expansion (as is the case of large-scale land acquisition for other agricultural commodities such as soy) are ambiguous from a food security perspective. On the one hand, the rapid expansion of oil palm is driven, to a large part, by demand for cooking oil among poor and middle class households in Indonesia domestically (26 percent) and internationally (73 percent) (Obidzinski et al., 2012). On the other hand, such expansion is displacing local people and undermining their source of food and income through loss of direct access to landscapes that were previously used for food provisioning and thus changing incomes and livelihoods. Similar conflicts can be observed in other countries where industrial use of forestry and weak forest tenure interplay, affecting in particular indigenous peoples. These issues are not only prevalent in the tropics, such as in South America where forests have been replaced with forest plantations by global forest enterprises (Kröger, 2012), but also in temperate regions such as in the north of Scandinavia, where there are reindeer herding conflicts (Raitio, 2008).

From a gender perspective, conflicts over forest products are often covert and confined to the intra-household level due to different preferences for forest and tree products and unequal access to them (Agarwal, 2010; Rocheleau and Edmunds, 1997; Sarin, 2001; Schroeder, 1999). In their seminal research in sub-Saharan Africa, Rocheleau and Edmunds (1997) find that although property rights are gender exclusive and women lack formal titles to individual or communal land, women still enjoy *de facto* rights to fuelwood, certain plants and animals. However, in areas that have undergone commercialisation of forest products, a remapping of the boundaries often occurs. Men as strategic actors bypass women's micro-rights and maintain their privilege in the landscape. Similarly, Elias and Carney's (2007) study of the shea tree (*Vitellaria paradoxa*) in Burkina Faso shows that rural women have historically collected, marketed and transformed shea nuts into multi-purpose butter for consumption and sale. The growing global trade in shea butter supplied to food and cosmetic industries represents an opportunity to further poor women's incomes, although such international sales have also led to a re-configuration of rights and claims over shea tree with many women losing access as a consequence. Both these studies suggest that women contest their loss of access at the household level but this does not amount to substantive changes in tenure regimes.

Relative poverty and inequality

This sub-section mainly focuses on social and gender balance questions concerning relative poverty and inequality. The relationship between poverty and food security from the perspective of per capita income will be described in Section 4.4 on economic drivers.

A wide range of studies note that rural poverty and remaining natural forests tend to share overlapping spaces. A significant proportion of people suffering from extreme poverty live in forest-based ecosystems (Mehta and Shah, 2003 for India; Sunderlin and Huynh, 2005 for Viet Nam; World Bank, 2003; Zhou and Veeck, 1999 for China). Sunderlin et al. (2005) posit that these are likely to be a product of some of the following interrelated factors: (a) most forests and extremely poor people are located in remote areas and out of the reach of the market economy and technological processes; (b) forests are often a refuge for poor and powerless peoples; (c) forests have "pro-poor characteristics" because they are open access or have low barriers to access. Nevertheless, communities who either live in forested landscapes and/or who rely on forests are neither homogeneous nor uniformly-dependent on forests. Existing distribution of power and the structure of incentives mediate who can access, use and control forest products for consumption, income and livelihoods (see also Section 4.5 in this chapter).

Livestock feeding, Labe, Guinea. Photo © Terry Sunderland

Research on the equity dimensions of community forestry in Nepal demonstrates that poor and rich households do not have symmetric opportunities to benefit from forest resources. Adhikari et al. (2004) present an econometric analysis of the impact of the private endowments of forest user group member households on forest access for consumption purposes. Using data from the Middle Hills they find that poorer households face more restrictions in accessing forest products than less poor or relatively better off households (see also Chapter 3). In the Terai region of Nepal where societal inequalities are even more pronounced and forest products are of higher value, Iversen et al. (2006) found that households that belong to the richer echelons of user groups have a vested interest in maintaining the widely observed practice of charging a subsidised member price for high-value products such as timber. By being required to pay in advance, poorer households are excluded from accessing high-valued products. Richer households on the other hand, derive considerable income by ensuring that there is a high margin between member price and market price when re-selling high-valued products in the local market, and thereby, siphoning off disproportionate benefits from communal resources. Similar findings of "elite capture" and the spill over of pre-existing societal divisions in the allocation of forest products for consumption, income and livelihoods have also been observed by other analysts in the context of sub-Saharan Africa, such as by Coulibaly-Lingani et al. (2009) for Burkina Faso, and Jumbe and Angelsen (2006) in Malawi.

Most analysts agree that increasing women's active participation in the institutions established to govern access and command over forests would support both women's empowerment and household food provisioning (UN Women, 2014). This is particularly relevant since both unpaid care work (cooking, taking care of children and the elderly etc.) and collection of food, firewood and fodder from forests are acknowledged as highly feminised tasks across the world (UN Women, 2014). Research findings show that women's participation in forest governance is lagging behind in many different contexts from South and East Asia to Latin America and sub-Saharan Africa (Agarwal, 1997; Agarwal, 2001; Mai et al., 2011; Mairena et al., 2012; Mukasa et al., 2012; Nightingale, 2002; and Sarin, 2001). Agarwal (2001) attributes women's limited voice and influence in forest governance regimes to gender inequalities in men and women's personal and household endowments. These inequalities manifest themselves in terms of women's low bargaining power vis-à-vis men in negotiating for their interests in forests at the household and community levels. Coleman and Mwangi's (2013) cross-country study in Bolivia, Mexico, Kenya and Nicaragua has identified two main determinants affecting women's participation in forest governance: education of household heads and institutional exclusion, which in turn supports Agarwal's analysis from South Asia (for more information about governance and gender inequalities see Chapter 3 and Section 4.5 in this chapter).

Demographic change: migration, urbanisation and agrarian transformation

In 2013, the world population totalled 7.2 billion and it is projected to reach 9.6 billion by 2050, with most growth in developing regions, especially Africa (UN, 2013). Consequently the demand for food, feed and fibre will increase and the land area per capita to feed all the people will decline. Some analysts such as Vanhanen et al. (2010) conclude that without improved agricultural productivity, rising food demands will result in increasing deforestation and forest degradation to make way for agriculture. But others point out that trade-offs between agricultural intensification and food production are also equally possible. Through a review of historical and cross-country studies, Angelsen and Kaimowitz (2001) concluded that the impact of intensification, is dependent on technology type (labour-intensive or capital-intensive); farmer characteristics (income and asset level, resource constraints); and context (policy incentives, market conditions etc.) (Angelsen and Kaimowitz, 2001). While of high relevance, population growth is not the only demographic driver pressuring on the forest-food system; interrelated drivers include changes in consumption patterns (see Section 4.4 in this chapter), migration, urbanisation and agrarian transformation.

Although migration of people is by no means a new phenomenon, the sheer number of people moving between rural and urban areas and transnationally is unprecedented. The UN's Economic and Social Council (ECOSOC) (2013) estimated that in 2010 alone, the number of international migrants was approximately 214 million, while internal migrants totalled 700 million. These were merely documented figures and are likely to be far surpassed. While international migration has become one of the defining features of globalisation, the world's population is also increasingly becoming urban. To date 54 percent of the world's population resides in urban areas with an expected increase of 11 percent by 2050 (UN, 2014). North America, Latin America and the Caribbean, and Europe are considered the world's most urbanised regions. Although Africa and Asia remain mostly rural, urbanisation is expected to be faster in these regions than in the others (UN, 2014). Small cities and towns in Asia, Africa and Latin America that lie in or near tropical forest areas are likely to experience the greatest magnitude of urbanisation (UN, 2011).

From a food security perspective, these trends have important implications for availability of, access to and relative dependence on forest products for food and income. However, research on the nexus between migration, urbanisation and forests remains very limited, let alone from a food security perspective. Forest governance involves territorialisation and the bounding of people to specific geographies. Hecht (forthcoming) calls for going beyond the conventional wisdom that sees migration as either disruptive to forest systems or as a livelihood failure. Instead, rural communities are increasingly multi-sited and dispersed, continuing rural production even as they depend also on other sources of off-farm income.

Major mechanisms through which migration and urbanisation affect rural communities and forests include: land abandonment, remittances, changes in rural labour availability, variations in the gender composition of households, and shifting demands of urban consumers on agricultural land and rural resources (Brondizio et al., 2014; Padoch et al., 2008; Parry et al., 2010). Much of this research is focused on forest cover and income and has not yet been concerned with implications for food security and nutrition more directly, although potential implications can be inferred. For instance, research in the state of Amazonas in Brazil is showing that the persistent marginalisation of remote, forest-based communities (due to a combination of long distances from markets, persistent under-investment in infrastructure and educational facilities) is compelling these communities to migrate to peri-urban areas.

While such an exodus might present opportunities from a conservation perspective, it is likely to be changing their use of forests for food and fuel, and rendering them more dependent on market vagaries for food provisioning (Parry et al., 2010). Findings from scholarly literature suggest that migration and associated remittances are reducing relative dependence on forests for consumption, income and livelihoods. This is in turn, leading to a decline in land change from forest to agricultural land (Eloy et al., 2014 for Brazilian Amazon; Hecht and Saatchi, 2007 for El Salvador; and Schmook and Radel, 2008 for Southern Yucatan, Mexico) as remittances are being used to buy food rather than to produce and source food from forests.

Urbanisation can have contradictory implications on forests and tree-based systems for food security and nutrition. On the one hand, urbanisation can lead to a reduction in forest food consumption patterns, with more emphasis on processed products and food safety (see Section 4.5). On the other hand, research also shows that urban populations can maintain their rural consumption patterns with considerable effects on land use and management. In sub-Saharan Africa, for instance, the rate of urbanisation is level with rising demand for fuelwood consumption. In other words, urbanisation has not accompanied a decline in fuelwood consumption patterns as previously expected (e.g. Zulu, 2010 for Malawi). These findings are also supported by research in Amazonia showing that rural-urban migrants keep their forest product consumption patterns in cities and continue to play a role in rural land use decisions (Padoch et al., 2008 for the Amazon; Tritsch et al., 2014 for French Guiana). Arnold et al. (2006) undertook a global analysis of woodfuel demand and supply which showed that there is no need for large-scale forestry interventions devoted to the provision of fuelwood for urban consumers as was hypothesised in the 1970s due to steady supplies from rural areas. But the growing urban demand for charcoal is likely to impact on tropical forests and poor, rural users in Africa in particular as they compete with urban consumers.

Migration and urbanisation have led to profound changes in socio-economic systems and have contributed to the feminisation of rural landscapes in many contexts (Deere, 2005; DeSchutter, 2013). Agarwal (2012) is careful not to insinuate that migration is causing feminisation, rather that the agrarian transition or the shift of workers to industry and services, and from rural to urban areas in developing countries, has been gendered. The proportion of women workers in agriculture increased across developing countries, in particular in South America and Oceania. According to Agarwal (2012), women farmers lack access and command over credit, land, production inputs, technology and markets. Hence, she argues that effects of volatile food prices and projected effects of climate change will very likely have a disproportionate impact on women as farmers and providers of household food. Scholarly literature focusing on the nexus of gender, migration and forest governance yields contradictory results in terms of whether male out-migration can empower women to play a bigger role in forest decision-making and enjoy greater access to forest products for themselves and their families. In Nepal, for example, transnational migration and remittances are emerging as major sources of employment. Migrants in Nepal are exclusively men due to a combination of intra-household constraints and governmental restrictions on women migrating. Giri and Darnhofer (2010) understand male out-migration as an opportunity for increasing women's access to forest resources and power over forest governance. An ethnographic study by Basnett (2013) indicates that this opportunity very much depends on interlocking gender and social differentiation.

While many countries are experiencing a "disappearing of peasantries" with declining contribution of agriculture to national economies and labour allocation to agriculture, others are witnessing a "repeasantrisation" as is evident in tropical forested landscapes (Rigg and Vandergeest, 2012). The latter trends are particularly evident in Southeast Asia and the Amazon in the face of commodity booms and large-scale, agro-industrial plantations of oil palm, rubber, pulp etc. (Kaimowitz and Smith, 2001; Wunder, 2001) and the resultant absorption of labour back to rural areas. For instance, Li (forthcoming) points out that rapid expansion of oil palm plantations in West Kalimantan (Indonesia) has accompanied significant deforestation, the dispossession of indigenous peoples' access to rubber and rice smallholdings and a casualisation of employment of plantation workers. Migrant labourers are compelled to bear all risks associated with migrating and being apart from their families; they have limited control over their work environment and scant means to negotiate for change. Indigenous Dayaks in comparison, lose access to forests and trees on which they had relied for direct food provisioning, income and livelihood. Food security amidst declining mosaic landscapes is therefore a challenge for both migrants and indigenous people alike.

4.4 Economic Drivers

This section provides an overview of the main economic drivers affecting the relationships between on the one hand, forests and tree-based systems, and on the other, food security and nutrition, documenting and illustrating the main points using materials featuring a range of products (e.g. bushmeat, fruit, nuts etc.) at different scales (global, regional, national, local). It does not include related general topics, such as the identification of economic drivers that contribute to increased urbanisation which affects general food consumption patterns. We distinguish two types of driver impacts: (i) economy-wide derived impacts, such as the impact of a new national food safety policy on bushmeat trade and consumption; and (ii) site specific indirect impacts from economic activities that only influence food security through other mechanisms, such as the construction of roads into forest areas supplying forest foods.

As is the case for food security more broadly, there are no generally accepted indicators to measure the diverse and contextually variable forest and food security relationships (Carletto et al., 2013; Coates, 2013). Economic drivers may hence impact differently across locations, actors and time. This is illustrated and exemplified in the forest-food security reliance continuum in Figure 4.2. Forest foods contribute to food security in two main ways: (i) directly through the provision of nutrients, and (ii) indirectly through generation of income, typically through cash sales (see also Chapter 2).

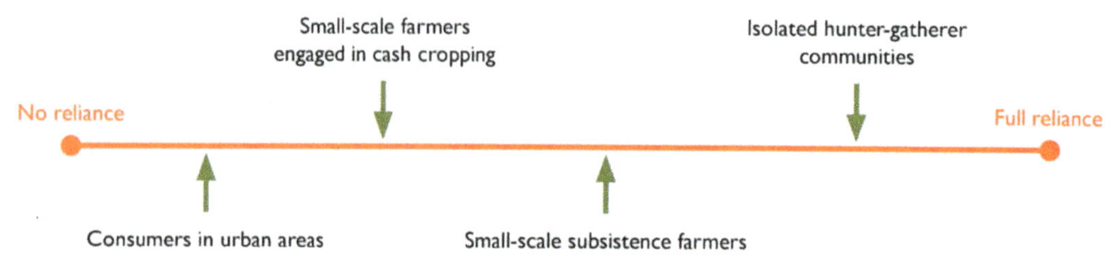

Fig. 4.2 The forest food security reliance continuum, with examples (based on Smith-Hall et al., 2012)

While it has long been recognised that forest foods can be important for food security (e.g. Bharucha and Pretty, 2010; FAO, 1989; Pimentel et al., 1997) there is scant quantitative information on the economic importance of forest foods in most locations and at all scales, including household and nationally. Angelsen et al. (2014), in a study of around 8,000 households in 24 developing countries found that, in terms of household incomes, food products constituted the second most important group of forest products and the most important from non-forest environments. Forests have also been found to provide famine foods in response to multiple adverse events (covariate shocks) and income in response to isolated adverse events (idiosyncratic shocks) (Dewees, 2013; Ngaga et al. 2006; Wunder et al., 2014). Forest foods are mostly traded locally or sub-nationally in non-transparent markets marred by problems including inefficiencies leading to high losses, seasonal supplies, lack of credit facilities and rudimentary knowledge of consumer preferences (e.g. Grote, 2014; Jamnadass et al., 2011; Vinceti et al., 2013).

There appears to be a continuum of forest food product commercialisation (see also Chapter 2): from products that are occasionally bartered in villages, to small-scale trade in regional markets along informal chains, to national and international trade along formal chains. The private sector appears to play a prominent role everywhere. Furthermore, products move along the continuum in response to changes in demand and supply; for example, cashew and shea nuts have moved from wild collection to domestication and cultivation in West Africa as has grasscutter farming. Homma (1992) provides an overview of domestication processes. Products and actors at different points on the continuum are impacted differently by economic drivers. For example, the impact of sustainable harvest certification initiatives will have a larger impact on internationally-traded spices than on locally-bartered fruits. In addition, there will be variations in data availability; for instance, there is usually no national data available on products that are harvested and traded in informal markets, such as forest foods in West Africa (Bertrand et al., 2013) and spices from the Himalayas (Olsen and Helles, 2009). The nature of appropriate public policy responses will also vary along the continuum. In the following sub-sections, we review state-of-knowledge on four key economic drivers.

Income per capita

Global per capita gross domestic product (GDP) is steadily increasing, except for a short downturn in 2008-09 due to the global financial crisis, and has been termed a global mega-trend driving per capita demand for food (Cassman, 2012). As income increases, households' demand for food increases less than proportionally (Engel's Law, see Cirera and Masset, 2010) and there is generally a dietary shift with decreasing importance of starchy staples (e.g. rice, wheat) and increased consumption of meat, fish, fruits and vegetables (Cassman, 2012; MEA, 2005). It has been noted, however, that higher incomes may not lead to improved food security if the additional income is spent on other items such as clothes, cell phones or tobacco (Dewees, 2013). In their above-mentioned survey, Angelsen et al. (2014) found forest food income to range from USD 49 (in purchasing power parity) in Africa (five percent of total household income) to 717 (15 percent) in Latin America, with a global average of 128 (six percent). There are few examples of how price changes impact forest food resources and their management; in an excellent study of timur (*Zanthoxylum armatum*, a small tree yielding fruits used for spices and condiments), Hertog and Wiersum (2000) show how increasing market prices drive intensification of forest management including a shift of production from public to private lands.

There are four main issues to note in relation to rising incomes and forest foods. First, many forest foods are likely, in economic terms, to be inferior goods (demand decreases with rising incomes and increases with declining incomes) and rising incomes would thus mean less forest food production extraction and reliance. Delang (2006) notes, however, that forest food gathering is important in many rural communities with low economic growth, and likely to remain so. It is also noteworthy that forest food consumption is increasing in some high income countries, e.g. in northern Europe apparently in response to perceptions that food should be locally grown, organic and aesthetic, indicating that we need to understand the dynamics of forest food consumption better. Second, rising per capita income is one of the factors driving the expansion of supermarkets in much of the global south (e.g. Humphrey, 2007; Reardon and Hopkins, 2007). This is likely to have long term impacts on traditional markets and outlets for forest foods, such as fresh fruit, including through a shift away from spot purchasing and introduction of grades and standards; all changes indicating an ongoing fundamental restructuring of many forest food markets. It may also potentially lead to impacts on nutrition. Third, there is large variation geographically and along the commercialisation continuum. For instance, per capita GDP in Africa grew from USD 1,400 in 1991 to USD 1,700 in 2008 (Groningen Growth and Development Centre: Total Economy Database) and the level of per capita income appears to be too low for households to make the transition to consumption of industrially-processed products, meaning that forest food products still play a key role in calorie intake and dietary composition in poorer households (Okojie and Shimeles, 2006). Fourthly, with rising incomes and urbanisation, people tend to eat more meat and milk products, for example, increasing the demand for crops as feed (Grote, 2014). This may also impact forest foods, notably demand for bushmeat. Urbanisation may also increase demand for semi-processed foods as the opportunity costs increase as women find employment which could increase processing of forest foods, and is also likely to lead to more focus on food safety and labelling, which could challenge small-scale producers of forest foods.

Absolute and relative food prices

Absolute food price levels directly affect household-level food consumption choices. Recent spikes in key agricultural food commodity prices in 2008-09 and 2010-11 affected the livelihoods and food security of millions of people (Akter and Basher, 2014; Brown et al., 2009; FAO, 2008) and led to riots (Berazneva and Lee, 2013). Studies indicate that country and regional agricultural food prices behave differently from international food prices (Brown et al., 2012; Minot, 2014) due to low integration of local and regional markets into international markets. Data on forest food prices appear very limited; it is difficult to collect, due to the high number of products and the frequency of informal trading, and thus not systematically monitored. Such data deficiencies impede our ability to analyse and understand forest food price dynamics.

There is also a price volatility difference within regions or countries. For instance, Minot (2014) observed higher agricultural food price volatility in main cities than in secondary cities, indicating different effects of price changes on different segments of society. Assuming that forest foods are mainly traded outside main cities, this would indicate less price volatility for such products. Agricultural food price increases impact most severely on the poorest households and particularly so if they are female-headed (Akter and Basher, 2014; Benfica, 2014; Drimie and McLachlan, 2013); the reason may be that such households do not have the asset base required to smooth income or consumption shortfall. A recent study specifically found that asset-poor households are most likely to use forests as part of their coping strategies (Wunder et al., 2014). It has also been observed that fluctuating agricultural food prices can decrease local food production and reduce employment opportunities (Tiwari and Joshi, 2012) and that price spike patterns and associated inadequate public policy responses may follow a repeating pattern in some countries (Ellis and Manda, 2012). Balancing national food availability with affordable food prices can be challenging (Haug and Hella, 2013). In terms of household-level food security, however, it has been noted that households can respond to rising agricultural food prices in different ways, such as downgrading food quality to maintain quantity, that serve to limits the nutritionally harmful effects of higher prices (Gibson, 2013).

There are close linkages between food, energy and financial markets that may explain much of the recently observed agricultural food price spikes and volatility (Grote, 2014; Tadesse et al., 2014). In particular, rising fossil fuel prices and biofuel policies may be a key driver of high grain and oilseed prices as biofuel production becomes financially more attractive (Gorter et al., 2013; Grote, 2014; Tokgoz et al., 2012), a situation that may be exacerbated by increased speculation on agricultural commodity markets (Grote, 2014). These processes, in combination with insecure property rights (Godfray et al., 2010), could contribute to land grabbing – the process of appropriating land and resources to produce commodities and accumulate wealth (Nevins and Peluso, 2008) – in the global south, with differential geographical impacts on forest food products. For instance, economic growth in China and the associated demand for wood fibre has arguably led to state-sanctioned corporate land grabbing in Lao PDR that has negatively affected local people's access to forest food products (Barney, 2008) while the same economic growth has simultaneously improved the possibilities for high altitude Himalayan communities to sell wild harvested products on new markets (Shrestha and Bawa, 2014).

Training tree nursery workers in Zaraninge Forest Reserve, Tanzania. Photo © PJ Stephenson

Markets and policies

The linkages between forests and food security reflect a wide range of policy interests related to health, development, human rights, biodiversity conservation, forests, food, trade and agriculture. Forest foods present a highly complex challenge to public policy institutions at different scales due to the wide range of potential user groups, the diverse motivations that drive collection and the lack of reliable data and information on trade flows, nutritional values and consumer preferences (Johns and Eyzaguirre, 2006; Toledo and Burlingame, 2006; Vinceti et al., 2013). There is also a high degree of variability in the levels of product collection, processing and marketing in different forest food product value chains making monitoring and regulation difficult. According to the FAO (1997), the non-timber forest product (NTFP) sector is generally dominated by the rural poor and labour-intensive small-scale industries, making it important for policy mechanisms to carefully differentiate between subsistence and commercial forest food activities. Here, equitably managing resource access becomes a key challenge for policy due to the generally low barriers to market entry and broad participation by both women and men in forest food collection (Arnold, 2008). While the subsistence-based forest food sector tends to have less impact on forest resources than the commercial sector (Neumann and Hirsch, 2000), it is often difficult for policy frameworks to effectively separate these activities due to the dynamic nature of forest food markets, which are often highly seasonal, and where products classified as "traditional"can quickly become commercial and where "commercial" products can be replaced by substitutes (FAO, 1997). When considering the ongoing structural transformations that have been occurring in the agricultural sector (in a wide sense), other important policy issues affecting commercial forest food production and trade include the urban demand for safe, responsibly-produced and high quality foods that is driving processes of certification and labelling (Grote, 2014) and the need to simplify regulatory regimes to reduce transaction costs for producers and develop a framework supporting producer organisations (Dewees, 2013).

International bodies of particular importance to forest food markets include the Convention on International Trade in Endangered Species of Wild Fauna and Flora (CITES) which limits and regulates the trans-border trade of many wild food species also in relation to forest foods such as bushmeat (e.g. Bennett, 2011); the Millennium Development Goals (MDGs) which seek to reduce hunger and poverty while maintaining ecosystem services, *inter alia* leading to more focus on the green economy including the importance of forest food products to livelihoods (e.g. Rasul et al., 2012); the Convention on Biological Diversity (CBD) which assures the protection of genetic, species and ecosystem diversity and the World Trade Organization (WTO) which regulates the trade policies of nations and products, requiring clear and agreed standards and definitions to enable commercialisation (Precillia Ijang et al., 2011). The impacts of these global institutions are contested in scholarly literature as they might lead to perverse effects. For example, critics of the WTO argue that the liberalisation of commodity trade and reduction of farm protection resulted in food dependency of substantial areas in the global South (Lawrence and McMichael, 2012).

Recognising that forest ecosystems are likely to play their most important role in household food security through diversifying diets and providing essential sources of nutrients, a number of observers have called for greater policy integration focused around meeting the nutritional and health needs of local resource users (Arnold, 2008; Bharucha and Pretty, 2010; Johns and Eyzaguirre, 2006). Improving the sustainable utilisation of diverse forest foods to support food security and nutrition likely will involve engaging local users in research and decision-making processes, facilitating information flows, enabling access to credit and markets, developing community-based education programmes, supporting the development of user/producer organisations and improving efficiency by reducing transaction costs or encouraging technology adoption and innovation (King, 2008; Shumsky et al., 2014; Tontisirin et al., 2002).

Production system changes

Production systems refer to the general production structure in a country that influences land use patterns. The type, size, location, and dynamics of production systems are *inter alia* determined by economic incentives, for example, in response to new or collapsing markets influenced by processes of globalisation, certification or changes in market efficiency. Two examples in relation to forest food products are: (i) formerly subsistence bushmeat products being commercialised and entering informal value chains as new demands and urban bushmeat markets were created by processes of urbanisation in Benin (Bertrand et al., 2013), and (ii) the currency devaluation in Brazil in 1999, combined with an international price increase of soybeans and beef, and control of hoof and mouth disease, leading to large scale soybean and cattle production in central-west Brazil (Chomitz, 2007), replacing forest food producing savannah woodland (de Souza and Felfili, 2006). Predicting rates of change for individual production systems is difficult, as is quantifying the impact of changes on forest food production.

4.5 Governance

Governance includes traditional state-centric decision-making as well as broader-based processes at a range of different scales. These broader systems of "governance" are not just driven by states and their domestic ambitions, but also by global markets and by a range of non-state actors that include civil society, businesses and international non-governmental and governmental organisations. This section explores the role of governance as a driver of forest-related goals and policies, and the implications for food security and nutrition of different stakeholders.

In recent years, three main drivers can be identified in the shift from state to more broad-based decision-making regimes in the forest sector: globalisation, ecosystem service thinking and economic valuation. Firstly, regarding globalisation, forest governance has historically been driven by social, economic and environmental imperatives of states (Sikor et al., 2013; Vandergeest and Peluso, 1995), but the interests and influence of global and non-state actors have progressively widened and deepened due to both expanding and new frontiers of financial investments (Muradian et al., 2013; Murray Li, 2007; Sullivan, 2013). These local to global stakeholders are connected across scales by value chains and their incipient public and private producer and trade standards regimes (McDermott et al., 2012), by civil society mobilisations for forest and food justice (Martinez-Alier, 2014; Schlosberg, 2013; Sikor and Newell, 2014), and by emerging global socio-ecological narratives such as that on planetary boundaries (Rockström et al., 2009).

Secondly, regarding ecosystem services thinking, this framework has gained enormous buy-in as a means of (re)conceiving the relationship between humans and ecosystems, including the view of humans as separate to nature, and nature as a provider of services to humans. The ecosystem services framing has influenced thinking about the relationship between forests and food security (Poppy et al., 2014a; Poppy et al., 2014b) and has been successfully promoted by important science-policy platforms, including the recently formed Intergovernmental Platform on Biodiversity and Ecosystem Services (IPBES), and major conservation non-governmental organisations (NGOs) (Turnhout et al., 2013).

Thirdly, regarding valuation, there has been a revalorisation of rural landscapes, in terms of financial, political and cultural values attached to particular goals and practices (Sikor et al., 2013). For example, in the wake of food price inflation in the late 2000s, crop yield narratives received a boost and there was a re-emphasis on highlighting lands that were considered underutilised or producing only a fraction of their yield potential (e.g. in statistical databases, maps, World Bank reports). It has been argued that this shift in how lands were valued globally has contributed to governments supporting policies that facilitated the global land rush (Li, 2014). Similarly, use of the ecosystem services framework has generated financial valuation of forest hydrological and carbon storage services. Such new forms of valuation provide legitimacy to particular forms of governance such as state regulation to protect downstream and global citizens, or public-private partnerships to market forest carbon offsets. However, the incorporation of new, global values as drivers of forest governance also pose risks, with some stakeholders under threat of losing control over previous ways of valuing and governing forests (Hunsberger et al., 2014; Martin et al., 2013b; Pascual et al., 2014).

State-focused governance

Although the rising influence of global discourses, institutions and markets has created significant shifts in governance regimes over tropical forest-agrarian landscapes, there are numerous instances in which the influence of national states and sub-national actors has been retained and even reinforced. Some states still exercise considerable control over the way land is allocated to different uses (Sunderlin et al., 2008; White and Martin, 2002), and the way in which property rights and tenure are regulated, including public versus private and commercial use, the establishment of protected areas and the exploitation of land for agriculture (Sikor et al., 2013).

There is considerable variation amongst countries which continue to adopt a state-centric approach to governance. The dominant discourse and scope of state interests differs dramatically. Instruments may range from top-down implementation of policies to delineate the landscape into categories with associated rules, to participatory land use planning exercises or even the designation of indigenous lands for decentralised governance. Hence the impacts of such approaches on the food security of forest-adjacent populations are varied.

State-focused approaches to governance of forests and surrounding landscapes endure particularly in certain circumstances: where states seek to maintain political control over economic activity and development; in circumstances in which tenure and land use are considered by central governments to be related to issues of internal security; where the state seeks to reconcile the interests of different ethnic and minority groups; and where land management is part of the process of defining citizenship itself (Beswick, 2011; Lestrelin et al., 2012; Li, 2010). Under these conditions, rapid modernisation of agriculture is commonly promoted and traditional practices such as *shifting cultivation* and inter-cropping are disincentivised or even discriminated against (Fox et al., 2009; Padoch and Pinedo-Vasquez, 2010). For example, in Lao PDR participatory land use planning has been employed by the national government with the explicit purpose of ending shifting agricultural practices and stabilising cultivation among ethnic minorities in mountainous outlying regions (Lestrelin et al., 2012). Some, or even many, people may benefit from formalisation of tenure and modernisation of farming methods.

However, rapid, state-driven agrarian change can also have detrimental impacts on food security and nutrition among poor or minority local actors, including indigenous communities whose livelihoods and culture have been particularly tied to forest habitats, and who are least able and willing to adapt (Baird and Shoemaker, 2005; Dounias and Froment, 2011). Even where policies appear to decentralise forest governance and grant additional local powers, women or minority groups may be further excluded from decision-making processes or suffer from restricted access to food from forests (Sikor and Ngoc Thanh, 2007). Similarly, formalisation of property rights does not always equate to maintained or improved access to resources because negotiation processes – both formal and informal – involve many actors, and the effects on access within local food systems are uncertain (Andersson, 2004). Where ultimate control of tenure is exercised by the state, smallholder tenure over farmland and forests (and associated subsistence needs) may also be undermined by decisions made by powerful non-local actors, such as private corporations granted government concessions for industry, infrastructure or energy projects (Agrawal et al., 2008). This can occur through not only large scale land grabs but also "control grabs" which may involve the imposition of state-influenced contract farming arrangements to the exclusion of poor local actors and a reduced ability to grow or to buy food (Huggins, 2014).

State-focused governance regimes often create zones for different land uses which tend to partition the landscape (de Groot et al., 2010). This division of land is often mirrored in states' institutional structures, with separation of responsibilities across different government departments. The separation of forestry from agricultural decision-making is generally detrimental to integrated landscape management. However, national scale approaches are not without their merits and may protect local populations from adverse effects of global market forces. Without state intervention, global markets and agendas can drive increased inequality and dispossession, through which local perspectives can easily be given lower priority than global goals such as carbon sequestration or biodiversity conservation (Arts and Buizer, 2009). Bolivia's approach represents a good example where the movement for indigenous rights has supported the granting of substantial autonomy over indigenous land rights as an "ethno-environmental fix" (Anthias and Radcliffe, 2013; Cronkleton et al., 2011). However, in many developing countries there is poor capacity to effectively decentralise environmental management (Tacconi, 2007).

State-focused governance can include renegotiation or even rejection of internationally-designed conservation instruments, to better fit the national context. While international discourses and influences are far from absent in these situations, they are instead transformed or negotiated to serve state interests. By such means, payments for ecosystem services (PES) and Reducing Emissions from Deforestation and Forest Degradation (REDD/REDD+) schemes have been "demarketised" into tax and subsidy arrangements. These may promote increased participation including of stakeholders with limited productive assets, for whom potential impacts on land and forest tenure, and associated food access may be averted (Bennett, 2008 and Sikor, 2013 for China; Cronkleton et al., 2011 and Uberhuaga et al., 2011 for Bolivia; Milne and Adams, 2012; Phelps et al., 2010). The development of REDD+ is also connected with fears about "recentralisation" (e.g. Phelps et al., 2010) affecting the use and dependence of local people on forests. Greater state and non-local control, for instance through the designation of protected areas directly impacts livelihoods and decreases access to food from the forest (West et al., 2006).

Governance beyond the state: markets and non-state actors

Multi-sectoral and multi-scale forms of governance do not replace state-focused governance of forests and tree-based systems for food security and nutrition, but become integrated in different ways, as noted above with reference to PES and REDD+. Systems for certifying ethical and sustainable forest management took off in the early 1990s after it became clear that a global forest governance convention was not going to emerge from the UN process (Strassburg et al., 2012). Certification is a market-based intervention typically involving standards that are established and monitored through networks of producers, NGOs and private sector partners. Some certification schemes like the Forest Stewardship Council (FSC) particularly stress their independence from governments whilst others prefer to have state government involvement. Even in the former case, governments remain influential through their control of the legal and policy levers that provide the operational context for forestry and food production (Hysing, 2009). In Tanzania, for example, the state maintains some control over price setting for commodities across the forest-farm landscape, including for tree food crops such as cashews.

Payments for ecosystem services were also originally promoted as non-state forms of governance, using market-based approaches to reducing deforestation and forest degradation (Ferraro and Kiss, 2002; Wunder, 2005). In practice, however, states have either been significant gatekeepers, determining what kind of PES is legitimate, or have actually instigated PES schemes as federal programmes for transferring resources to rural forest management (McAfee and Shapiro, 2010; Milne and Adams, 2012; Shapiro-Garza, 2013). Payments for ecosystem services exemplify the growing presence of hybrid governance approaches, operating across scales and with public, private and civil society involvement. Other key forestry sector examples include REDD+ and the EU's Forest Law Enforcement, Governance and Trade (FLEGT) initiative (Glück et al., 2010).

The past ten years have seen considerable optimism for the potential opportunities presented by these new governance configurations. In some cases, new market and network-based governance approaches are explicitly targeted at generating synergies between forest conservation and food security. For example, Brazil's Sustainable Agricultural Network is reported to be a rigorous system for ensuring that the Brazilian cattle supply chain is managed to reduce deforestation (Newton et al., 2014). This has a very high potential to achieve synergies between food security and reduced deforestation (Strassburg et al., 2014). More generally, forest certification, PES, REDD+ and FLEGT are not promoted as directly addressing relationships between forests and food security. However, it is probably fair to argue that managing this relationship has often been part of their rationale. First and foremost, these forms of governance respond to past concerns that state-based forest management has not often succeeded in linking forest conservation with local livelihood and food security (Adams et al., 2004; Ferraro, 2001; Salafsky and Wollenberg, 2000; Wunder, 2001). Secondly, PES schemes in particular respond to ecosystem services research that provides evidence of forest-food security linkage, such as the role that landscape biodiversity plays in achieving more productive and stable agricultural systems (Cardinale et al., 2012). Thirdly, concern about tropical deforestation has led global consumers to reflect more on how global food systems impact on both environmental sustainability and the livelihood needs of southern producers (Schlosberg, 2013).

Whilst optimism has been high, evaluations of the effects of certification and PES-based forms of governance present a mixed picture. There are already a number of reviews of the state of knowledge about the impacts of certification (Blackman and Rivera, 2011; Romero, 2013; Romero et al., 2013; SCR, 2012) and the impacts of PES (Miteva et al., 2012; Pattanayak et al., 2010; Samii et al., 2014; Wunder, 2013). These reviews highlight that evidence for both environmental and social outcomes remains quite weak, in part because of the difficulties and costs involved in undertaking robust impact evaluations, but also highlighting that these market-based approaches do not provide easy and readily scalable ways of improving sustainability across forest-food landscapes. The evidence base for fledgling REDD+ and FLEGT is even more limited, whilst specific evidence relating forestry policies with food security outcomes is almost absent.

Often, market or incentive-based governance interventions are ill-suited to bringing about synergies between environmental and social goals, as shown by a growing body of research. One reason stated is that the logic of market efficiency stands in opposition to the need for equity that is so fundamental to distribution of basic needs such as food security. There is a specific literature related to forest and agricultural carbon markets that identifies constraints on achieving synergies between carbon mitigation and local livelihoods. Firstly, policy-making and funding for mitigation and adaptation tend to be separate (Klein et al., 2005; Locatelli et al., 2011), meaning that livelihoods and food security are not integrated with thinking about landscape carbon. Secondly, there is uncertainty about the effects of different carbon mitigation interventions on food security, partly due to lack of adequate monitoring (Harvey et al., 2014). Thirdly, there are factors that constrain communities and individuals from taking part in mitigation-oriented carbon and agricultural projects. Such access problems can result from unsuitable financing (Siedenburg et al., 2012), problems of tenure (Robledo et al., 2012), local inequalities arising from, for example, wealth and gender constraints (Brown and Corbera, 2003; Lee et al., forthcoming), and discrimination based on ethnicity or social histories (Martin et al., 2013a).

There remains considerable disagreement about whether market and incentive-based approaches to forest governance can overcome such problems and deliver synergies with local livelihood and food security. Some scholars argue that they have the potential to bring new streams of revenue to rural communities as well as enhancing ecosystem services that support food security in the longer term (Harvey et al., 2014; Smith et al., 2013). There is also some evidence that PES and certification improve land tenure security for local people. Despite being market-based, FSC certification can contribute to more rather than less democratic governance of forests (Dare et al., 2011; Meidinger, 2011). Furthermore, a major review of the effects of certification found cases where it enhances land tenure security for local people (SCR, 2012). Improvements in land security are also noted for PES schemes (Tacconi et al., 2013) and in some REDD+ pilot projects (Hoang et al., 2013; Maraseni et al., 2014).

In contrast, there are also many studies that highlight the risks associated with market- and incentive-based approaches. Studies of FSC and Rainforest Alliance operations have found that the costs of accessing certification outweigh the benefits, meaning low uptake of FSC in developing countries (Marx and Cuypers, 2010) and among smallholders (Auer, 2012; Gullison, 2003; McDermott et al., 2015), and a bias towards large producers (McDermott et al., 2015; Pinto and McDermott, 2013). Scholars show that in order for PES and REDD+ schemes to target those who are most able to (competitively) provide services, access to schemes has often been restricted to those with appropriate assets such as land (Porras et al., 2008) or education (Zbinden and Lee, 2005), favouring larger operations and wealthier farmers (Pagiola and Platais, 2007) and reducing opportunities for women (Boyd, 2002; Lee et al., forthcoming). The fact that certification and PES schemes tend to offer small returns also means that those who sign up tend to have low entry costs, suggesting that they are already at or near to achieving the required practices with very little management change required (Arriagada et al., 2009; Blackman et al., 2014; Gómez-Zamalloa et al., 2011; Honey-Roses et al., 2009).

Local market, Chittagong, Bangladesh. Photo © Terry Sunderland

One particular concern, expressed primarily in theoretical works, is that the economic valuation of ecosystem services, and their incorporation into global commodity markets, enhances the risk of local and indigenous communities being dispossessed of land and related rights and access (Büscher et al., 2012; Li, 2014; Matulis, 2014; McAfee, 2012). This is an important concern because it suggests that recently popular approaches to governing forests could directly threaten local conditions for food security. Careful research is required and it is important to note that empirical evidence to date is limited and suggestive that risks and outcomes vary considerably according to context. Studies of some certification processes, such as on the Roundtable on Sustainable Palm Oil (Silva-Castañeda, 2012) and the Roundtable on Responsible Soy (Elgert, 2012), find a democratic deficit that leads to marginalisation of smallholder concerns for food and livelihood security. In these cases, certification legitimises new partnerships between environmentalists, private sector agro-industrialists and recent migrant populations, threatening land security for indigenous and peasant communities and weakening their pre-existing relationship with NGOs (Elgert, 2012). In similar fashion, Ibarra et al. (2011) describe how an indigenous community in Mexico withdrew from a PES scheme because of growing concerns about loss of self-determination over its own food systems.

The role of market-based mechanisms in the provision of food security and nutrition from forests and tree-based systems is complex and ambiguous. It remains impossible to generalise and as with broader governance effects, context is essential to understanding the relationships.

4.6 Conclusions

This chapter aimed to provide an overview of natural and anthropogenic drivers affecting forests and tree-based systems, to understand how they affect food security and nutrition and to identify interrelations among them. For analytical reasons, these drivers were categorised as environmental, social, economic and governance.

Following our framework introduced at the beginning of this chapter forest- and tree-based drivers can affect food security and nutrition through changes in land use and management or through changes in consumption, income and livelihood. Some drivers affect food security in both ways (Figure 4.3).

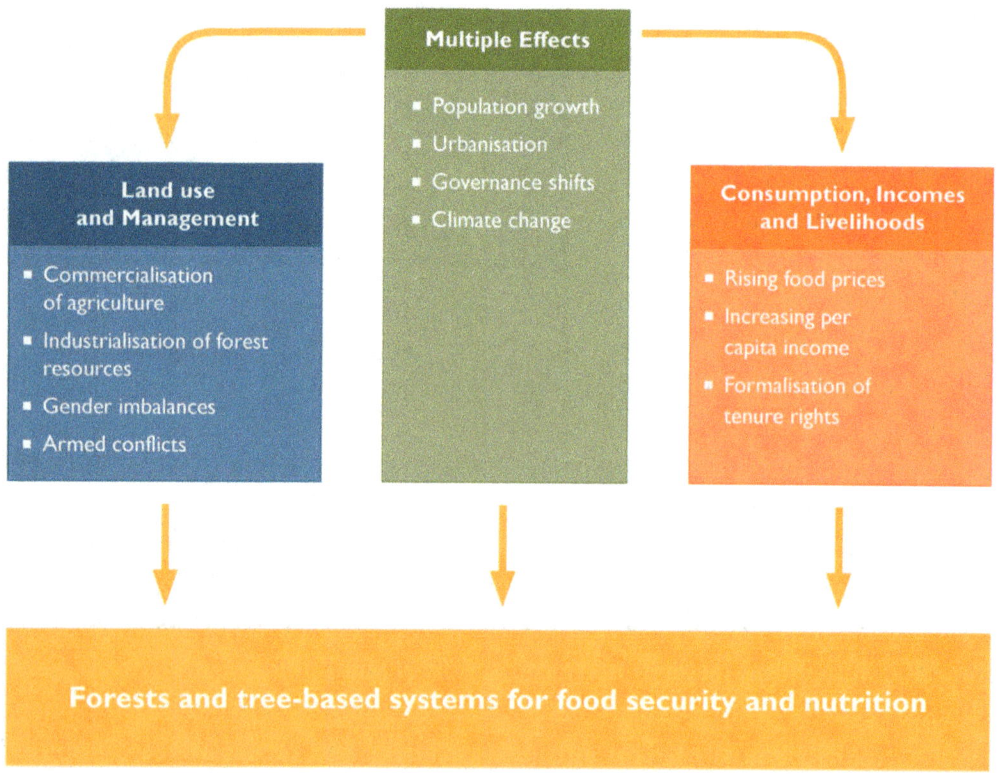

Fig. 4.3 Major drivers affecting forests and tree-based systems for food security and nutrition

The effects of the following drivers on food security and nutrition travel through land use and management as well as through consumption, income and livelihood:

- Population growth places pressure on forests and tree-based systems for food security and nutrition by changing consumptions patterns and by reducing the relative availability of food. Furthermore, population growth leads to changes in land use management forms, resulting in, for example, commercialisation of agriculture and industrialisation of forest resources;
- Urbanisation leads to changes in forest food consumption patterns, with more emphasis on processed products and food safety issues. These changes in demand also lead to changes in land use management, e.g. commercialisation of agriculture. When combined with male migration, urbanisation can lead as well to a change of gender balance in rural areas.
- Governance shifts from state-focused government to multi-sectoral and cross-scale governance present better prospects for integration of different interests and goals related to forest and food systems. The resulting (global) emphasis on ecosystem services can also bring opportunities for improved synergies between forest and food systems, changing management forms and changes of income and livelihood structures. However, when governed by market logics, such valuation poses risks to local control and access over resources.
- Climate change can directly affect the availability and quality of food and nutrition by the appearance of new species. It furthermore impacts forests and tree-based systems for food security and nutrition through forcing changes in land-use and adoption of management forms, and through changes in income from forest products. Climate change consequences are considered not to be gender-balanced and affect vulnerable groups the most.

The following indirect drivers lead to increased *food insecurity* and poor nutrition by forcing changes in land use and management:

- The increasing commercialisation of agriculture to feed a growing (urbanised) population is accelerating forest loss and thereby reducing the availability of forests and tree-based products for food security and nutrition.
- The industrialisation of forest resources (e.g. in plantations) leads to the displacement of local people and undermines the availability of and access to food and nutrition. This change of production format is often based on, and enabled by, weak forest tenure rights.

- Gender imbalances, with male domination, lead to the prioritisation of land uses involving commercial/timber products at the expense of food.
- Conflicts, and in particular armed conflicts, in forest landscapes can lead to exploitation of forest resources and undermine conservation issues. These conflicts often detach households from forests and tree-based food and nutrition. However, armed conflicts weaken institutionalised rules of the game and can also open new (illegal) access to food.

Another set of drivers impacts on forests and tree-based food security and nutrition by changing incomes and livelihoods:

- Formalisation of tenure rights fosters benefit sharing amongst those living in and with the forest. On the other hand, increased formalisation of tenure rights can contribute to increase vulnerability and reduce food security, in particular for the poorest.
- Rising food prices may be less pronounced for many forest foods than for agricultural foods as the former are primarily traded and consumed outside major cities. Data on forest food markets is, however, scant.
- Increasing per capita income changes households' food consumption patterns. This needs to be better understood in relation to ongoing changes in structure and operation of national and regional forest food markets. However, the gathering of forest food will remain important in rural communities with low economic growth.

The range and diversity of drivers demonstrate the interconnectedness between drivers and effects; for example, networked governance leading to gender imbalance can lead to the prioritisation of timber over food. Responding to these messy, interrelated sets of drivers with effective options is a major challenge of our time. This challenge is further exacerbated as the drivers of forests and tree-based systems for food security and nutrition do not allow for a generalisation of causal effects. Social structure influences whether the consequences lead to improvements for food security and nutrition or lead instead to increasing vulnerability. Determining factors are, for example, localities, with urban and rural situations gaining and suffering differently from changes; drivers of change might strive for and achieve positive effects for food security for some groups but result in contradictory effects for the poorest. Hence, responses to drivers need to ensure that they do not only address a relatively small number of elite, but also to find ways to incorporate the aggregated impacts of local, informal responses to drivers. Local stakeholders are in fact not only the most vulnerable, but it can be assumed that they are also the most sensitive to new and innovative response options. The challenge is to maintain the balance to ensure food security and nutrition, and at the same time ensure the sustainability of forests and tree-based systems.

References

Adams, W.M., Aveling, R., Brockington, D., Dickson, B., Elliott, J., Hutton, J., Roe, D., Vira, B. and Wolmer, W., 2004. Biodiversity conservation and the eradication of poverty. *Science* 306: 1146-1149. http://dx.doi.org/10.1126/science.1097920

Adhikari, B., Di-Falco, S. and Lovett, J.C., 2004. Household characteristics and forest dependency: Evidence from common property forest management in Nepal. *Ecological Economics* 48: 245-257. http://dx.doi.org/10.1016/j.ecolecon.2003.08.008

Afreen, S., Sharma, N., Chaturvedi, R.K., Gopalakrishnan, R., and Ravindranath, N.H., 2010. Forest policies and programs affecting vulnerability and adaptation to climate change. *Mitigation and Adaptation Strategies for Global Change* 16(2): 177-197. http://dx.doi.org/10.1007/s11027-010-9259-5

Agarwal, B., 1997. Environmental Action, Gender Equity and Women's Participation. *Development and Change* 28(1): 1-44. http://dx.doi.org/10.1111/1467-7660.00033

Agarwal, B., 2001. Gender and forest conservation: The impact of women's participation in community forest governance. *Ecological Economics* 68(11): 2785-2799. http://dx.doi.org/10.1016/j.ecolecon.2009.04.025

Agarwal, B., 2010. *Gender and Green Governance: The Political Economy of Women's Presence Within and Beyond Community Forestry*. Oxford: Oxford University Press.

Agarwal, B., 2012. *Food Security, Productivity and Gender Inequality*. IEG Working Paper No. 320. Delhi: Institute of Economic Growth, University of Delhi. http://www.binaagarwal.com/downloads/apapers/Food security, productivity, gender inequality.pdf

Agrawal, A., Chatre, A. and Hardin, R., 2008. Changing governance of the world's forests. *Science* 320: 1460-1462. http://dx.doi.org/10.1126/science.1155369

Akter, S., and Basher, S.A., 2014. The impacts of food price and income shocks on household food security and economic well-being: Evidence from rural Bangladesh. *Global Environmental Change* 25: 150-162. http://dx.doi.org/10.1016/j.gloenvcha.2014.02.003

Andersson, K. P., 2004. Who talks with whom? The role of repeated interactions in decentralized forest governance. *World Development* 32: 233-249. http://dx.doi.org/10.1016/j.worlddev.2003.07.007

Angelsen, A. and Kaimowitz, D. (eds.), 2001. *Agriculture Technologies and Tropical Deforestation*. Bogor: Centre for International Forestry Research. http://dx.doi.org/10.1079/9780851994512.0000

Angelsen, A., Jagger, P., Babigumira, R., Belcher, B., Hogarth, N.,Bauch, S., Börner, B., Smith-Hall, C. and Wunder, S., 2014. Environmental income and rural livelihoods: A global-comparative analysis. *World Development* 64: S12-S28. http://dx.doi.org/10.1016/j.worlddev.2014.03.006

Anthias, P., and Radcliffe, S.A., 2013. The ethno-environmental fix and its limits: Indigenous land titling and the production of not-quite-neoliberal natures in Bolivia. *Geoforum (2013)*. http://dx.doi.org/10.1016/j.geoforum.2013.06.007

Arnold, J.E.M., 2008. *Managing Ecosystems to Enhance the Food Security of the Rural Poor*. Gland: IUCN. http://intranet.iucn.org/webfiles/ftp/public/ForumEvents/E1533/Final Document/Situation Analysis Forests and Food Security by M Arnold FINAL DRAFT 30.05.08.pdf

Arnold, M.J.E., Kohlin, G., and Persson, R., 2006. Woodfuels, livelihoods and policy interventions: Changing perspectives. *World Development* 34(3): 596-611. http://dx.doi.org/10.1016/j.worlddev.2005.08.008

Arriagada, R.A., Sills, E.O., Pattanayak, S.K. and Ferraro, P.J., 2009. Combining qualitative and quantitative methods to evaluate participation in Costa Rica's program of payments for environmental services. *Journal of Sustainable Forestry* 28: 343-367. http://dx.doi.org/10.1080/10549810802701192

Arts, B. and Buizer, M., 2009. Forests, discourses, institutions: A discursive-institutional analysis of global forest governance. *Forest Policy and Economics* 11: 340-347. http://dx.doi.org/10.1016/j.forpol.2008.10.004

Auer, M.R., 2012. Group forest certification for smallholders in Vietnam: An early test and future prospects. *Human Ecology* 40: 5-14. http://dx.doi.org/10.1007/s10745-011-9451-6

Baird, I.G. and Shoemaker, B., 2005. *Aiding or Abetting?: Internal Resettlement and International Aid Agencies in the Lao PDR*. Toronto: Probe International.

Baral, N. and Heinmen, J.T., 2006. The Maoist People's War and Conservation in Nepal. *Politics and the Life Sciences* 24(1-2): 2-11. http://dx.doi.org/10.2990/1471-5457(2005)24[2:tmpwac]2.0.co;2

Barney, K., 2008. China and the production of forestlands in Lao PDR—a political ecology of transnational enclosure. In: *Taking Southeast Asia to Market*, edited by J. Nevins and N.L. Peluso. Ithaca: Cornell University Press.

Basnett, B.S., 2013. *Taking Migration Seriously What Are the Implications for Gender and Community Forestry?* InfoBrief No. 65. Bogor: CIFOR. http://dx.doi.org/10.17528/cifor/004183

Benfica, R., 2014. Welfare and distributional impacts of price shocks in Malawi: A non-parametric approach. *Food Security* 6: 131-145. http://dx.doi.org/10.1007/s12571-013-0324-2

Bennett, E.L., 2011. Another inconvenient truth: The failure of enforcement systems to save charismatic species. *Oryx* 45(4): 476-479. http://dx.doi.org/10.1017/s003060531000178x

Bennett, M.T., 2008. China's sloping land conversion program: Institutional innovation or business as usual? *Ecological Economics* 65: 699-711. http://dx.doi.org/10.1016/j.ecolecon.2007.09.017

Berazneva, J. and Lee, D.R., 2013. Explaining the African food riots of 2007-2008: An empirical analysis. *Food Policy* 39: 28-39. http://dx.doi.org/10.1016/j.foodpol.2012.12.007

Bertrand, A., Agbahungba, G.A. and Fandohan, S., 2013. Urbanization and forest foods in Benin. *Unasylva* 241(64): 30-36.

Beswick, D., 2011. Democracy, identity and the politics of exclusion in post-genocide Rwanda: The case of the Batwa. *Democratization* 18: 490-511. http://dx.doi.org/10.1080/13510347.2011.553367

Bharucha, Z. and Pretty, J., 2010. The roles and values of wild foods in agricultural systems. *Philosophical Transactions of the Royal Society B—Biological Sciences* 365(1554): 2913-2926. http://dx.doi.org/10.1098/rstb.2010.0123

Blackman, A. and Rivera, J., 2011. Producer level benefits of sustainability certification. *Conservation Biology* 25: 1176-1185. http://dx.doi.org/10.1111/j.1523-1739.2011.01774.x

Blackman, A., Raimondi, A., and Cubbage, F., 2014. *Does Forest Certification in Developing Countries Have Environmental Benefits? Insights from Mexican Corrective Action Requests*. Washington DC: Resources for the Future. http://dx.doi.org/10.2139/ssrn.2432179

Boucher, D.H., Elias, P., Lininger, K., May-Tobin, C., Roquemore, S. and Saxon, E., 2011. *The Root of the Problem: What's Driving Tropical Deforestation Today?* Cambridge, MA: Union of Concerned Scientists. http://www.ucsusa.org/sites/default/files/legacy/assets/documents/global_warming/UCS_RootoftheProblem_DriversofDeforestation_FullReport.pdf

Boyd, E., 2002. The Noel Kempff project in Bolivia: Gender, power, and decision-making in climate mitigation. *Gender & Development* 10: 70-77. http://dx.doi.org/10.1080/13552070215905

Brody, A., Demetriades, J. and Esplen, E., 2008. *Gender and Climate Change: Mapping the Linkages. A Scoping Study on Knowledge and Gaps.* Brighton: Institute of Development Studies. http://siteresources.worldbank.org/EXTSOCIALDEVELOPMENT/Resources/DFID_Gender_Climate_Change.pdf

Brondizio, E.S., Siqueira, A.D. and Voght, N., 2014. Forest resources, city services: Globalisation, household networks and ubranisation in the Amazon estuary. In: *The Social Lives of Forests: Past, Present and Future of Woodland Resurgence*, edited by S. Hecht, K. Morrisson and C. Padoch. Chicago: University of Chicago Press.

Brown, K. and Corbera, E., 2003. Exploring equity and sustainable development in the new carbon economy. *Climate Policy* 3:S41-S56. http://dx.doi.org/10.1016/j.clipol.2003.10.004

Brown, M.E., Hintermann, B. and Higgins, N., 2009. Markets, climate change and food security in West Africa. *Environmental Science and Technology* 43: 8016-8020. http://dx.doi.org/10.1021/es901162d

Brown, M.E., Tondel, F., Essam, T., Thorne, J.A., Mann, B.F., Leonard, K., Stabler, B. and Eilerts, G., 2012. Country and regional staple food price indices for improved identification of food insecurity. *Global Environmental Change* 22: 784-794. http://dx.doi.org/10.1016/j.gloenvcha.2012.03.005

Büscher, B., Sullivan, S., Neves, K., Igoe, J. and Brockington, D., 2012. Towards a synthesized critique of neoliberal biodiversity conservation. *Capitalism Nature Socialism* 23(2): 4-30. http://dx.doi.org/10.1080/10455752.2012.674149

Cardinale, B.J., Duffy, J.E., Gonzalez, A., Hooper, D.U., Perrings, C., Venail, P., Narwani, A. Mace, G.M., Tilman, D. and Wardle, D.A., 2012. Biodiversity loss and its impact on humanity. *Nature* 486: 59-67. http://dx.doi.org/10.1038/nature11148

Carletto, C., Zezza, A. and Banerjee, R., 2013. Towards better measurement of household food security: Harmonizing indicators and the role of household surveys. *Global Food Security* 2: 30-40. http://dx.doi.org/10.1016/j.gfs.2012.11.006

Carmenta, R., Parry, L., Blackburn, A., Vermeylen, S. and Barlow, J., 2011. Understanding human-fire interactions in tropical forest regions: A case for interdisciplinary research across the natural and social sciences. *Ecology and Society* 16(1): 53. http://www.ecologyandsociety.org/vol16/iss1/art53/

Cassman, K.G., 2012. What do we need to know about global food security? *Global Food Security* 1: 81-82. http://dx.doi.org/10.1016/j.gfs.2012.12.001

Chomitz, K.M., 2007. *At loggerheads? Agricultural Expansion, Poverty Reduction and Environment in the Tropical Forests.* Policy Research Report. Washington DC: World Bank. http://dx.doi.org/10.1596/978-0-8213-6735-3

Chown, S.L., le Roux, P.C., Ramaswiela, T., Kalwij, J.M., Shaw, J.D. and McGeoch, M.A., 2013. Climate change and elevational diversity capacity: Do weedy species take up the slack? *Biological Letters* 9: 20120806. http://dx.doi.org/10.1098/rsbl.2012.0806

Cirera X. and Masset, E., 2010. Income distribution trends and future food demand. *Philosophical Transactions of the Royal Society B—Biological Sciences* 365(1554): 2821-2834. http://dx.doi.org/10.1098/rstb.2010.0164

Coates, J., 2013. Build it back better: Deconstructing food security for improved measurement and action. *Global Food Security* 2: 188-194. http://dx.doi.org/10.1016/j.gfs.2013.05.002

Colchester, M., and Chao, S., 2013. *Conflict or Consent? The Oil Palm Sector at a Crossroads.* Moreton-in-Marsh: Forest Peoples Programme, Sawit Watch and Transformasi untuk Keadilan Indonesia. http://www.forestpeoples.org/sites/fpp/files/publication/2013/11/conflict-or-consentenglishlowres.pdf

Coleman, E.A. and Mwangi, E., 2013. Women's participation in forest management: A cross-country analysis. *Global Environmental Change* 23(1): 193-205. http://dx.doi.org/10.1016/j.gloenvcha.2012.10.005

Collier, P. and Hoeffler, A., 2001. *Greed and Grievance in Civil War*. Policy Research Working Paper 2355. Washington DC: World Bank. http://dx.doi.org/10.1596/1813-9450-2355

Coulibaly-Lingani, P., Tigabu, M., Savadogo, P., Oden, P.-C. and Ouadba, J.-M., 2009. Determinants of access to forest products in southern Burkina Faso. *Forest Policy and Economics* 11(7): 516-524. http://dx.doi.org/10.1016/j.forpol.2009.06.002

Cronkleton, P., Bray, D.B. and Medina, G., 2011. Community forest management and the emergence of multi-scale governance institutions: Lessons for REDD+ development from Mexico, Brazil and Bolivia. *Forests* 2: 451-473. http://dx.doi.org/10.3390/f2020451

Crutzen, P.J., 2006. The "Anthropocene." In: *Earth System Science in the Anthropocene*, edited by E. Ehlers and T. Krafft. Berlin Heidelberg: Springer. http://dx.doi.org/10.1007/3-540-26590-2_3

Dare, M.L., Schirmer, J., and Vanclay, F., 2011. Does forest certification enhance community engagement in Australian plantation management? *Forest Policy and Economics* 13: 328-337. http://dx.doi.org/10.1016/j.forpol.2011.03.011

DeFries, R.S., Rudel, T., Uriarte, M. and Hansen, M., 2010. Deforestation driven by urban population growth and agricultural trade in the twenty-first century. *Nature Geoscience* 3(3): 178-181. http://dx.doi.org/10.1038/ngeo756

De Groot, R.S., Alkemade, R., Braat, L., Hein, L. and Willemen, L., 2010. Challenges in integrating the concept of ecosystem services and values in landscape planning, management and decision making. *Ecological Complexity* 7: 260-272. http://dx.doi.org/10.1016/j.ecocom.2009.10.006

De Jong, W., Donovan, D. and Abe, K.I. (eds.), 2007. *Extreme Conflict and Tropical Forests*. Dordrecht: Springer. http://dx.doi.org/10.1007/978-1-4020-5462-4

De Koning, R., Capistrano, D., Yasmi, Y. and Cerrutti, P.O., 2008. *Forest-related Conflicts: Impacts, Links and Measures to Mitigate*. Washington DC: Rights and Resources Initiative. http://www.rightsandresources.org/documents/files/doc_822.pdf

de Merode, E., Hillman Smith, K., Homewood, K., Pettifor, R., Rowcliffe, M. and Cowlishaw, G., 2007. The impact of armed conflict on protected-area efficacy in central africa. *Biology Letters* 3(3): 299-301. http://dx.doi.org/10.1098/rsbl.2007.0010

de Souza, C.D. and Felfili, J.M., 2006. Uso de plantas medicinais na região de Alto Paraíso de Goiás, GO, Brasil. (Use of medicinal plants in the region of Alto Paraíso de Goiás, GO, Brazil). *Acta Botanica Brasilica* 20: 135-142. http://dx.doi.org/10.1590/s0102-33062006000100013

Debuse, V.J., and Lewis, T., 2014. Long-term repeated burning reduces lantana camara regeneration in a dry eucalypt forest. *Biological Invasions* 16(12): 2697-2711. http://dx.doi.org/10.1007/s10530-014-0697-y

Deere, C.D., 2005. *The Feminization of Agriculture? Economic Restructuring in Rural Latin America*. Policy Report on Gender and Development: 10 Years after Beijing. United Nations Research Institute for Social Development. http://www.unrisd.org/unrisd/website/document.nsf/(httpPublications)/20024EBC6AB9DA45C1256FE10045B101?OpenDocument

Delang, C.O., 2006. The role of wild food plants in poverty alleviation and biodiversity conservation in tropical countries. *Progress in Development Studies* 6(4): 275-286. http://dx.doi.org/10.1191/1464993406ps143oa

DeSchutter, O., 2013. *Gender Equality and Food Security—Women's Empowerment as a Tool against Hunger*. Manila: Asian Development Bank. http://www.fao.org/wairdocs/ar259e/ar259e.pdf

Dewees, P., 2013. Forests, trees and resilient households. *Unasylva* 241(64): 46-53.

Dounias, E. and Froment, A., 2011. From foraging to farming among present-day forest hunter-gatherers: Consequences on diet and health. *International Forestry Review* 13: 294-304. http://dx.doi.org/10.1505/146554811798293818

Dove, M.R., 1986. The practical reason of weeds in Indonesia: Peasant vs. state views of Imperata and Chromolaena. *Human Ecology* 14(2): 163-190. http://dx.doi.org/10.1007/bf00889237

Drimie, S. and McLachlan, M., 2013. Food security in South Africa—first steps toward a transdisciplinary approach. *Food Security* 5: 217-226. http://dx.doi.org/10.1007/s12571-013-0241-4

Dudley, J., Ginsberg, J., Plumptre, A., Hart, J. and Campos, L.C., 2002. Effects of war and civil strife on wildlife and wildlife habitats. *Conservation Biology* 16: 319-329. http://dx.doi.org/10.1046/j.1523-1739.2002.00306.x

Elgert, L., 2012. Certified discourse? The politics of developing soy certification standards. *Geoforum* 43: 295-304. http://dx.doi.org/10.1016/j.geoforum.2011.08.008

Elias, M. and Carney, J., 2007. African shea butter: A feminized subsidy from nature. *Journal of the International African Institute* 77(1): 37-62. http://dx.doi.org/10.3366/afr.2007.77.1.37

Ellis, F. and Manda, E., 2012. Seasonal food crises and policy responses: A narrative account of three food security crises in Malawi. *World Development* 40(7): 1407-1417. http://dx.doi.org/10.1016/j.worlddev.2012.03.005

Eloy, L., Brondizio, E.S. and Do Pateo, R., 2014. New perspectives on mobility, urbanization and resource management in riverine Amazonia. *Bulletin of Latin American Research* 34(1): 3-18. http://dx.doi.org/10.1111/blar.12267

Ewel, J.J., 1999. Natural systems as models for the design of sustainable systems of land use. *Agroforestry Systems* 45: 1-21. http://dx.doi.org/10.1023/a:1006219721151

Fairhead, J. and Leach, M., 1996. *Misreading the African Landscape: Society and Ecology in a Forest-savanna Mosaic.* Cambridge: Cambridge University Press. http://dx.doi.org/10.1017/cbo9781139164023

FAO, 1989. *Forestry and Food Security*. Forestry Paper No. 90. Rome: FAO. http://www.fao.org/docrep/t7750e/t7750e02.htm

FAO, 2008. *The State of Food Insecurity in the World 2008*. Rome: FAO. ftp://ftp.fao.org/docrep/fao/011/i0291e/i0291e00.pdf

FAO, 1997. *Technology Scenarios in the Asia-Pacific Forestry Sector.* Working Paper No: APFSOS/WP/25. Rome and Bangkok: Forestry Policy and Planning Division and Regional Office for Asia and the Pacific.

Ferraro, P.J., 2001. Global habitat protection: Limitations of development interventions and a role for conservation performance payments. *Conservation Biology* 15: 990-1000. http://dx.doi.org/10.1046/j.1523-1739.2001.015004990.x

Ferraro, P.J. and Kiss, A., 2002. Direct payments to conserve biodiversity. *Science* 298: 1718-1719. http://dx.doi.org/10.3126/hjs.v1i2.200

Fox, J., Fujita, Y., Ngidang, D., Peluso, N., Potter, L., Sakuntaladewi, N., Sturgeon, J. and Thomas, D., 2009. Policies, Political-Economy, and Swidden in Southeast Asia. *Human Ecology* 37(3): 305-322. http://dx.doi.org/10.1007/s10745-009-9240-7

Garcia, R.A., Cabeza, M., Rahbek, C., Araújo, M.B., 2014. Multiple dimensions of climate change and their implications for biodiversity. *Science* 344. http://dx.doi.org/10.1126/science.1247579

Gibson, J., 2013. The crisis in food price data. *Global Food Security* 2: 97-103. http://dx.doi.org/10.1016/j.gfs.2013.04.004

Giri, K. and Darnhofer, I., 2010. Outmigrating men: A window of opportunity for women's participation in community forestry? *Scandinavian Journal of Forest Research* 25(9): 55-61. http://dx.doi.org/10.1080/02827581.2010.506769

Glück, P., Angelsen, A., Appelstrand, M., Assembe-Mvondo,S., Auld, G., Hogl, K., Humphreys, D. and Wildburger C., 2010. Core components of the international forest regime complex. In: *Embracing Complexity: Meeting the Challenges of International Forest Governance*, edited by J. Rayner, A. Buck and P. Katila. Vienna: IUFRO World Series No. 28. http://www.iufro.org/science/gfep/forest-regime-panel/report/

Godfray, H.C.J., Beddington, J.R., Crute, I.R., Haddad, L., Lawrence, D., Muir, J.F., Pretty, J., Robinson, S., Thomas, S.M. and Toulmin, C., 2010. Food security: The challenge of feeding 9 billion people. *Science* 327(5967): 812-818. http://dx.doi.org/10.1126/science.1185383

Gómez-Zamalloa, M.G., Caparrós, A. and Ayanz, A.S.M., 2011. 15 years of forest certification in the European Union. Are we doing things right? *Forest Systems* 20: 81-94. http://dx.doi.org/10.5424/fs/2011201-9369

Gorter, H. de, Drabik, D., and Just, D.R., 2013. How biofuels policies affect the level of grains and oilseed prices: Theory, models and evidence. *Global Food Security* 2: 82-88. http://dx.doi.org/10.1016/j.gfs.2013.04.005

Grote, U., 2014. Can we improve global food security? A socio-economic and political perspective. *Food Security* 6: 187-200. http://dx.doi.org/10.1007/s12571-013-0321-5

Gullison, R.E., 2003. Does forest certification conserve biodiversity? *Oryx* 37: 153-165. http://dx.doi.org/10.1017/s0030605303000346

Harvey, C. A., Chacón, M., Donatti, C. I., Garen, E., Hannah, L., Andrade, A., Bede, L., Brown, D., Calle, A. and Chará, J., 2014. Climate smart landscapes: Opportunities and challenges for integrating adaptation and mitigation in tropical agriculture. *Conservation Letters* 7: 77-90. http://dx.doi.org/10.1111/conl.12066

Haug, R. and Hella, J., 2013. The art of balancing food security: Securing availability and affordability of food in Tanzania. *Food Security* 5: 415-426. http://dx.doi.org/10.1007/s12571-013-0266-8

Hecht, S.B. and Saatchi, S.S., 2007. Globalization and Forest Resurgence: Changes in Forest Cover in El Salvador. *BioScience* 57(8): 663-672. http://dx.doi.org/10.1641/b570806

Hecht, S.B., Forthcoming. Trends in migration, urbanization, and remittances and their effects on tropical forests and forest-dependent communities. *CIFOR Working Paper*. Bogor: CIFOR.

Hecht, S.B., Kandel, S., Gomes, I., Cuellar, N. and Rosa, H., 2006. Globalization, forest resurgence, and environmental politics in El Salvador. *World Development* 34(2): 308-323. http://dx.doi.org/10.1016/j.worlddev.2005.09.005

Hertog, W.H. den, and Wiersum, K.F., 2000. Timur (Zanthoxylum armatum) production in Nepal—dynamics in nontimber forest resource management. *Mountain Research and Development* 20(2): 136-145. http://dx.doi.org/10.1659/0276-4741(2000)020[0136:tzapin]2.0.co;2

Hoang, M.H., Do, T.H., Pham, M.T., van Noordwijk, M. and Minag, P.A., 2013. Benefit distribution across scales to reduce emissions from deforestation and forest degradation (REDD+) in Vietnam. *Land Use Policy* 31: 48-60. http://dx.doi.org/10.1016/j.landusepol.2011.09.013

Homma, A.K.O., 1992. The dynamics of extraction in Amazonia: A historical perspective. *Advances in Economic Botany* 9: 23-31.

Hoang, M.H., Do, T.H., Pham, M.T., van Noordwijk, M. and Minang, P.A., 2013. Benefit distribution across scales to reduce emissions from deforestation and forest degradation (REDD+) in Vietnam. *Land Use Policy* 31: 48-60. http://dx.doi.org/10.1016/j.landusepol.2011.09.013

Honey-Roses, J., Lopez-Garcia, J., Rendon-Salinas, E., Peralta-Higuera, A. and Galindo-Leal, C., 2009. To pay or not to pay? Monitoring performance and enforcing conditionality when paying for forest conservation in Mexico. *Environmental Conservation* 36: 120-128. http://dx.doi.org/10.1017/s0376892909990063

Hosonuma N., Herold, M., De Sy V., DeFries R.S., Brockhaus M., Verchot L., Angelsen A. and Romijn, E., 2012. An assessment of deforestation and forest degradation drivers in developing countries. *Environmental Research Letters* 7(4): 044009. http://dx.doi.org/10.1088/1748-9326/7/4/044009

Huggins, C.D., 2014. 'Control Grabbing' and small-scale agricultural intensification: Emerging patterns of state-facilitated 'agricultural investment' in Rwanda. *Journal of Peasant Studies* 41(3): 365-384. http://dx.doi.org/10.1080/03066150.2014.910765

Humphrey, J., 2007. The supermarket revolution in developing countries: Tidal wave or tough competitive struggle? *Journal of Economic Geography* 7: 433-450. http://dx.doi.org/10.1093/jeg/lbm008

Hunsberger, C., Bolwig, S., Corbera, E. and Creutzig, F., 2014. Livelihood impacts of biofuel crop production: Implications for governance. *Geoforum* 54: 248-260. http://dx.doi.org/10.1016/j.geoforum.2013.09.022

Hysing, E., 2009. From government to governance? A comparison of environmental governing in Swedish forestry and transport. *Governance* 22: 647-672. http://dx.doi.org/10.1111/j.1468-0491.2009.01457.x

Ibarra, J.T., Barreau, A., Campo, C.D., Camacho, C.I., Martin, G. and McCandless, S., 2011. When formal and market-based conservation mechanisms disrupt food sovereignty: Impacts of community conservation and payments for environmental services on an indigenous community of Oaxaca, Mexico. *International Forestry Review* 13: 318-337. http://dx.doi.org/10.1505/146554811798293935

Ickowitz, A., Powell, B., Salim, M.A. and Sunderland, T.C.H., 2014. Dietary quality and tree cover in Africa. *Global Environmental Change* 24: 287-294. http://dx.doi.org/10.1016/j.gloenvcha.2013.12.001

IPCC, 2014. Summary for policymakers. In: *Climate Change 2014: Impacts, Adaptation, and Vulnerability. Part A: Global and Sectoral Aspects. Contribution of Working Group II to the Fifth Assessment Report of the Intergovernmental Panel on Climate Change*, edited by C.B. Field, V.R. Barros, D.J. Dokken, K.J. Mach, M.D. Mastrandrea, T.E. Bilir, M. Chatterjee, K.L. Ebi, Y.O. Estrada, R.C. Genova, B. Girma, E. S. Kissel, A.N. Levy, S. MacCracken, P.R. Mastrandrea and L.L. White. Cambridge: Cambridge University Press. http://dx.doi.org/10.1017/cbo9781107415379

Iversen, V, Chhetry, B., Francis, P., Gurung, M., Kafle, G., Pain, A. and Seeley, J., 2006. High value forests, hidden economies and elite capture: Evidence from forest user groups in Nepal's Terai. *Ecological economics* 58(1): 93-107. http://dx.doi.org/10.1016/j.ecolecon.2005.05.021

Jamnadass, R.H., Dawson, I.K., Franzel, S., Leakey, R.R.B., Mithöfer, D., Akinnifesi, F.K. and Tchoundjeu, Z., 2011. Improving livelihoods and nutrition in sub-Saharan Africa through the promotion of indigenous and exotic fruit production in smallholders' agroforestry systems: A review. *International Forestry Review* 13(3): 338-354. http://dx.doi.org/10.1505/146554811798293836

Johns, T., and Eyzaguirre, P.B., 2006. Linking biodiversity, diet and health in policy and practice. *Proceedings of the Nutrition Society* 65: 182-189. http://dx.doi.org/10.1079/pns2006494

Jumbe, C.B.L. and Angelsen, A., 2006. Do the poor benefit from devolution policies? Evidence from Malawi's forest co-management program. *Land Economics* 82(4): 562-581. http://dx.doi.org/10.3368/le.82.4.562

Kaimowitz, D. and Smith, J. 2001. Soybean technology and the loss of natural vegetation in Brazil and Bolivia. In: *Agricultural Technologies and Tropical Deforestation*, edited by A. Angelsen and D. Kaimowitz. Wellington: CABI/CIFOR 195-212. http://dx.doi.org/10.1079/9780851994512.0195

King, C.A., 2008. Community resilience and contemporary agri-ecological systems: Reconnecting people and food, and people with people. *Systems Research and Behavioral Science* 25(1): 111-124. http://dx.doi.org/10.1002/sres.854

Klein, R.J., Schipper, E.L.F. and Dessai, S., 2005. Integrating mitigation and adaptation into climate and development policy: Three research questions. *Environmental Science & Policy* 8: 579-588. http://dx.doi.org/10.1016/j.envsci.2005.06.010

Kröger, M., 2012. The expansion of industrial tree plantations and dispossession in Brazil. *Development and Change* 43: 947-973. http://dx.doi.org/10.1111/j.1467-7660.2012.01787.x

Kull, C.A. and Rangan, H., 2008. Acacia Exchanges: Wattles, Thorn Trees, and the Study of Plant Movements. *Geoforum* 39(3): 1258-1272. http://dx.doi.org/10.1016/j.geoforum.2007.09.009

Lawrence, G. and McMichael, P., 2012. The question of food security. *International Journal of Sociology of Agriculture & Food* 19(2): 135-142.

Lee, J., Martin, A., and Wollenberg, E., Forthcoming. Implications on equity in agricultural carbon market projects: A gendered analysis of access, decision-making, and outcomes. *Environment and Planning A*, under review. http://dx.doi.org/10.1177/0308518x15595897

Lestrelin, G., Castella, J.-C. and Bourgoin, J., 2012. Territorialising sustainable development: The politics of land-use planning in Laos. *Journal of Contemporary Asia* 42: 581-602. http://dx.doi.org/10.1080/00472336.2012.706745

Li, T.M., 2010. Indigeneity, capitalism, and the management of dispossession. *Current Anthropology* 51: 385-414. http://dx.doi.org/10.1086/651942

Li, T.M., 2014. What is land? Assembling a resource for global investment. *Transactions of the Institute of British Geographers* 39: 589-602. http://dx.doi.org/10.1111/tran.12065

Li, T.M., Forthcoming. *Social Impact of Oil Palm: A Gendered Perspective from West Kalimantan*. CIFOR Occasional Paper. Bogor: Center for International Forestry Research. http://dx.doi.org/10.17528/cifor/005579

Locatelli, B., Evans, V., Wardell, A., Andrade, A. and Vignola, R., 2011. Forests and climate change in Latin America: Linking adaptation and mitigation. *Forests* 2: 431-450. http://dx.doi.org/10.17528/cifor/003273

Lujala, P., 2003. *Classification of Natural Resources for Armed Civil Conflict Research*, Paper presented for the European Consortium for Political Research Joint Session of Workshop, Edinburgh, U.K.

Mai, Y.H., Mwangi, E. and Wan, M., 2011. Gender analysis in forestry research: Looking back and thinking ahead. *International Forestry Review* 13(2): 245-258. http://dx.doi.org/10.17528/cifor/003789

Mairena, E., Lorio, G., Hernández, X., Wilson, C., Müller, P. and Larson, A.M., 2012. *Gender and forests in Nicaragua's Autonomous Regions: Community Participation.* CIFOR Infobrief No. 57. Bogor: CIFOR. http://dx.doi.org/10.17528/cifor/004049

Malmer, A., Ardö, J., Scott D. Vignola R. and Xu, J., 2010. Forest water and global water governance. In: *Forests and Society—Responding to Global Drivers of Change*, edited by G. Mery P. Katila G. Galloway, I. Alfaro M. Kanninen, M. Lobovikov and J. Varjo. Vienna: IUFRO World Series No. 25.

Maraseni, T., Neupane, P., Lopez-Casero, F. and Cadman, T., 2014. An assessment of the impacts of the REDD+ pilot project on community forests user groups (CFUGs) and their community forests in Nepal. *Journal of environmental management* 136: 37-46. http://dx.doi.org/10.1016/j.jenvman.2014.01.011

Martin, A., Gross-Camp, N., Kebede, B., McGuire, S. and Munyarukaza, J., 2013a. Whose environmental justice? Exploring local and global perspectives in a payments for ecosystem services scheme in Rwanda. *Geoforum* 28(2014): 216-226. http://dx.doi.org/10.1016/j.geoforum.2013.02.006

Martin, A., McGuire, S. and Sullivan, S., 2013b. Global environmental justice and biodiversity conservation. *The Geographical Journal* 179: 122-131. http://dx.doi.org/10.1111/geoj.12018

Martinez-Alier, J., 2014. The environmentalism of the poor. *Geoforum* 54: 239-241. http://dx.doi.org/10.1016/j.geoforum.2013.04.019

Marx, A. and Cuypers, D., 2010. Forest certification as a global environmental governance tool: What is the macro effectiveness of the Forest Stewardship Council? *Regulation & Governance*, 4: 408-434. http://dx.doi.org/10.1111/j.1748-5991.2010.01088.x

Matulis, B.S., 2014. The economic valuation of nature: A question of justice? *Ecological Economics* 104: 155-157. http://dx.doi.org/10.1016/j.ecolecon.2014.04.010

McAfee, K. and Shapiro, E.N., 2010. Payments for ecosystem services in Mexico: Nature, neoliberalism, social movements, and the state. *Annals of the Association of American Geographers* 100: 579-599. http://dx.doi.org/10.1080/00045601003794833

McAfee, K., 2012. The Contradictory Logic of Global Ecosystem Services Markets. *Development and Change* 43: 105-131. http://dx.doi.org/10.1111/j.1467-7660.2011.01745.x

McDermott, C.L., van Asselt, H., Streck, C., Assembe Mvondo, S., Duchelle, A.E., Haug, C., Humphreys, D., Mulyani, M., Shekhar Silori, C. and Suzuki, R., 2012. Governance for REDD+, forest management and biodiversity: Existing approaches and future options. In: *Understanding Relationships Between Biodiversity, Carbon, Forests and People: The Key to Achieving REDD+ Objectives*, IUFRO World Series No. 31, edited by J.A. Parrotta, C. Wildburger and S. Mansourian. Vienna: International Union of Forest Research Organizations.

McDermott, C.L., Irland, L.C. and Pacheco, P., 2015. Forest certification and legality initiatives in the Brazilian Amazon: Lessons for effective and equitable forest governance. *Forest Policy and Economics* 50: 134-142. http://dx.doi.org/10.1016/j.forpol.2014.05.011

MEA, 2005. *Ecosystems and Human Well-being: Synthesis Report*. Washington DC: World Resources Institute. http://www.millenniumassessment.org/documents/document.356.aspx.pdf

Mehta, A.K. and Shah, A., 2003. Chronic poverty in India: Incidence, causes and policies. In: Special issue on "Chronic Poverty and Development Policy" edited by D. Hulme and A. Shepherd. *World Development* 31(3): 491-511. http://dx.doi.org/10.1016/s0305-750x(02)00212-7

Meidinger, E., 2011. Forest certification and democracy. *European Journal of Forest Research* 130: 407-419. http://dx.doi.org/10.1007/s10342-010-0426-8

Meyfroidt, P. and Lambin, E.F., 2011. Global forest transition: Prospects for an end to deforestation. *Annual Review of Environment and Resources* 36: 343-371. http://dx.doi.org/10.1146/annurev-environ-090710-143732

Milne, S., and Adams, B., 2012. Market Masquerades: Uncovering the politics of community level payments for environmental services in Cambodia. *Development and Change* 43: 133-158. http://dx.doi.org/10.1111/j.1467-7660.2011.01748.x

Minot, N., 2014. Food price volatility in sub-Saharan Africa: Has it really increased? *Food Policy* 45: 45-56. http://dx.doi.org/10.1016/j.foodpol.2013.12.008

Miteva, D.A., Pattanayak, S.K. and Ferraro, P.J., 2012. Evaluation of biodiversity policy instruments: What works and what doesn't? *Oxford Review of Economic Policy* 28: 69-92. http://dx.doi.org/10.1093/oxrep/grs009

Mohai, Paul, David P. and Roberts, J.T., 2009. Environmental Justice. *Annual Review of Environment and Resources* 35: 405-430. http://dx.doi.org/10.1146/annurev-environ-082508-094348

Mukasa, C., Tibazalika, A., Mango, A. and Muloki, H.N., 2012. *Gender and Forestry in Uganda: Policy, Legal and Institutional Frameworks.* CIFOR Working Paper 89. http://dx.doi.org/10.17528/cifor/003795

Muradian, R., Arsel, M., Pellegrini, L., Adaman, F., Aguilar, B., Agarwal, B. Corbera, E., Ezzine de Blas, D., Farley, J., Froger, G., Garcia-Frapolli, E., Gomez-Baggethun E. Gowdy, J., Kosoy, N., Le Coq, J.F., Leroy, P. May, P., Meral, P., Mibielli, P., Norgaard, Ozkaynak, B., Pascual, U., Pengue, W., Perez, M., Pesche, D., Pirard, R., Ramos-Martin, J., Riva, L, Saenz, F., Van Hecken, G., Vatn, A., Vira, B. and Urama, K., 2013. Payments for ecosystem services and the fatal attraction of win-win solutions. *Conservation Letters* 6(4): 274-279. http://dx.doi.org/10.1111/j.1755-263x.2012.00309.x

Murray Li, T., 2007. Practices of assemblage and community forest management. *Economy and Society* 36: 263-293. http://dx.doi.org/10.1080/03085140701254308

Neumann, R.P. and Hirsch, E., 2000. *Commercialisation of Non-timber Forest Products: Review and Analysis of Research.* Bogor and Rome: CIFOR and FAO. http://dx.doi.org/10.17528/cifor/000723

Nevins, J. and Peluso, N.L., 2008. Introduction: Commoditization in Southeast Asia. In: *Taking Southeast Asia to Market,* edited by J. Nevins and N.L. Peluso. London: Cornell University Press, 1-24.

Newton, P., Alves Pinto, H.N. and Pinto, L.F.G., 2014. Certification, forest conservation, and cattle: Theories and evidence of change in Brazil. *Conservation Letters.* http://dx.doi.org/10.1111/conl.12116

Ngaga, Y.M., Munyanziza, E., and Masalu, H.E., 2006. The role of wild mushrooms in the livelihoods of rural people in Kiwele village, Iringa, Tanzania: Implications for policy. *Discovery and Innovation* 18: 246-251. http://dx.doi.org/10.4314/dai.v18i3.15751

Nightingale, A. J., 2002. Participating or just sitting in? The dynamics of gender and caste in community forestry. *Journal of Forestry and Livelihoods* 2(1): 17-24. http://www.forestaction.org/app/webroot/js/tinymce/editor/plugins/filemanager/files/images/stories/pdfs/journal_of_forest_and_livelihood/vol2_1/Participating or just sitting _16_.pdf

Obidzinski, K., Andriani, R., Komarudin, H. and Andrianto, A., 2012. Environmental and social impacts of oil palm plantations and their implivations for biofuel production in indonesia. *Ecology and Society* 17(1):1-19. http://dx.doi.org/10.5751/es-04775-170125

Okojie, C. and Shimeles, A., 2006. *Inequality in Sub-Saharan Africa.* The Inter-Regional Inequality Facility. London: Overseas Development Institute (ODI). http://www.odi.org/sites/odi.org.uk/files/odi-assets/publications-opinion-files/4058.pdf

Olsen, C.S. and Helles, F., 2009. Market efficiency and benefit distribution in medicinal plant markets: Empirical evidence from South Asia. *International Journal of Biodiversity Science & Management* 5(2): 53-62. http://dx.doi.org/10.1080/17451590903063129

Padoch, C. and Pinedo Vasquez, M., 2010. Saving slash and burn to save biodiversity. *Biotropica* 42: 550-552. http://dx.doi.org/10.1111/j.1744-7429.2010.00681.x

Padoch, C., Brondizio, E., Costa, S., Pinedo-Vasquez, M., Sears, R.R. and Siqueira, A., 2008. Urban forest and rural cities: Multi-sited households, consumption patterns and forest resources in amazonia. *Ecology and Society* 13(2): 2. http://www.ecologyandsociety.org/vol13/iss2/art2/

Pagiola, S. and Platais, G., 2007. *Payments for Environmental Services: From Theory to Practice.* Washington DC: World Bank.

Parry, L., Peres, C., Day, B. and Amaral, S., 2010. Rural–urban migration brings conservation threats and opportunities to Amazonian watersheds. *Conservation Letters* 3(4): 251-259. http://dx.doi.org/10.1111/j.1755-263x.2010.00106.x

Pascual, U., Phelps, J., Garmendia, E., Brown, K., Corbera, E., Martin, A. Gomez-Baggethun, E. and Muradian, R., 2014. Social equity matters in payments for ecosystem services. *BioScience* 64: 1027-1036. http://dx.doi.org/10.1093/biosci/biu146

Pattanayak, S.K., Wunder, S. and Ferraro, P.J., 2010. Show me the money: Do payments supply environmental services in developing countries? *Review of Environmental Economics and Policy* 4: 254-274. http://dx.doi.org/10.1093/reep/req006

Pautasso, M., Dehnen-Schmutz, K., Holdenrieder, O., Pietravalle, S., Salama, N., Jeger, M.J., Lange, E. and Hehl-Lange, S., 2010. Plant health and global change—some implications for landscape management. *Biological Reviews of the Cambridge Philosophical Society* 85: 729-755. http://dx.doi.org/10.1111/j.1469-185x.2010.00123.x

Phelps, J., Webb, E.L. and Agrawal, A., 2010. Does REDD+ Threaten to Recentralize Forest Governance? *Science* 328: 312-313. http://dx.doi.org/10.1126/science.1187774

Pimentel, D., McNair, M., Buck, L., Pimentel, M. and Kamil, J., 1997. The value of forests to world food security. *Human Ecology* 25(1): 91-120. http://dx.doi.org/10.1023/a:1021987920278

Pinto, L.F.G. and McDermott, C., 2013. Equity and forest certification—a case study in Brazil. *Forest Policy and Economics* 30: 23-29. http://dx.doi.org/10.1016/j.forpol.2013.03.002

Poppy, G., Chiotha, S., Eigenbrod, F., Harvey, C., Honzák, M., Hudson, M., Jarvis, A., Madise, N., Schreckenberg, K. and Shackleton, C., 2014a. Food security in a perfect storm: Using the ecosystem services framework to increase understanding. *Philosophical Transactions of the Royal Society B: Biological Sciences* 369: 20120288. http://dx.doi.org/10.1098/rstb.2012.0288

Poppy, G., Jepson, P., Pickett, J. and Birkett, M., 2014b. Achieving food and environmental security: New approaches to close the gap. *Philosophical Transactions of the Royal Society B: Biological Sciences* 369: 20120272. http://dx.doi.org/10.1098/rstb.2012.0272

Porras, I.T., Grieg-Gran, M. and Neves, N., 2008. *All that Glitters: A Review of Payments for Watershed Services in Developing Countries.* London: IIED. http://pubs.iied.org/pdfs/13542IIED.pdf

Precillia Ijang, N.T., Sven Walter, S. and Ngueguim, J.R., 2011. An overview of policy and institutional frameworks impacting the use of non timber forest products in central Africa. *International Journal of Social Forestry* 4(1): 63-85. http://www.ijsf.org/dat/art/vol04/ijsf_vol4_no1_04_tata_ijang_ntfp_central_africa.pdf

Pysek, P., and Richardson, D. M., 2010. Invasive species, environmental change and management, and health. *Annual Review of Environment and Resources* 35: 25-55. http://dx.doi.org/10.1146/annurev-environ-033009-095548

Raitio, K., 2008. *You Can't Please Everyone: Conflict Management Cases, Frames and Institutions in Finnish State Forests.* PhD Dissertation, Joensuu, Finland. http://epublications.uef.fi/pub/urn_isbn_978-952-219-117-5/urn_isbn_978-952-219-117-5.pdf

Rasul, G., Choudhary, D., Pandit, B.H. and Kollmair, M., 2012. Poverty and livelihood impacts of a medicinal and aromatic plants project in India and Nepal: An assessment. *Mountain Research and Development* 32(2): 137-148. http://dx.doi.org/10.1659/mrd-journal-d-11-00112.1

Reardon, T. and Hopkins, R., 2007. The supermarket revolution in developing countries: Policies to address emerging tensions among supermarkets, suppliers and traditional retailers. *The European Journal of Development Research* 18(4): 522-545. http://dx.doi.org/10.1080/09578810601070613

Richardson, D.M. and van Wilgen, B.W., 2004. Invasive alien plants in South Africa: How well do we understand the ecological impacts? *South African Journal of Science* 100: 45-52. http://www.capefurniture.za.org/comp_2011/redgum/Working for water invasive alien plant species.pdf

Rigg, J. and Vandergeest, P., 2012. *Revisiting Rural Places: Pathways to Poverty and Prosperity in Southeast Asia.* Honolulu: University of Hawaii Press.

Robbins, P., 2004. Comparing Invasive Networks: Cultural and Political Biographies of Invasive Species. *Geographical Review* 94(2): 139-156. http://dx.doi.org/10.1111/j.1931-0846.2004.tb00164.x

Robledo, C., Clot, N., Hammill, A. and Riché, B., 2012. The role of forest ecosystems in community-based coping strategies to climate hazards: Three examples from rural areas in Africa. *Forest Policy and Economics* 24: 20-28. http://dx.doi.org/10.1016/j.forpol.2011.04.006

Rocheleau, D. and Edmunds, D., 1997. Women, men and trees: Gender, power and property in forest and agrarian landscapes. *World Development* 25(8): 1351-71. http://dx.doi.org/10.1016/s0305-750x(97)00036-3

Rockström, J., Steffen, W., Noone, K., Persson, Å., Chapin, F.S.III, Lambin, E.F., Lenton, T.M., Scheffer, M., Folke, C., Schellnhuber, H.J., Nykvist, B., de Wit, C.A., Hughes, T., van der Leeuw, S., Rodhe, H., Sörlin, S., Snyder, P.K., Costanza, R., Svedin, U., Falkenmark, M., Karlberg, L., Corell, R.W., Fabry, V.J., Hansen, J., Walker, B., Liverman, D., Richardson, K., Crutzen, P. and Foley, J.A., 2009. A safe operating space for humanity. *Nature* 461: 472-475. http://dx.doi.org/10.1038/461472a

Rod, J.K. and Rustad, S.C.A., 2006. Forest resources and conflict and property in Southeast Asia from India. *International Conference on Polarization and Conflict*, Nicosia, Cyprus, 26-29 April 2006.

Romero, C., 2013. Evaluation of forest conservation interventions; the case of forest management certification. *New Frontiers in Tropical Biology: The Next 50 Years (A Joint Meeting of ATBC and OTS)*, 20-30 June 2013.

Romero, C., Putz, F.E., Guariguata, M.R., Sills, E.O., Cerutti, P.O. and Lescuyer, G., 2013. *An Overview of Current Knowledge about the Impacts of Forest Management Certification: A Proposed Framework for its Evaluation.* CIFOR Occasional Paper. Bogor: CIFOR. http://dx.doi.org/10.17528/cifor/004188

Rudel, T.K., Bates, D. and Machinguiashi, R., 2009. A tropical forest transition? Agricultural change, out-migration and secondary forest in the Ecuadorian Amazon. *Annals of the Association of American Geographers* 92(1): 87-102. http://dx.doi.org/10.1111/1467-8306.00281

Salafsky, N. and Wollenberg, E., 2000. Linking livelihoods and conservation: A conceptual framework and scale for assessing the integration of human needs and biodiversity. *World Development* 28: 1421-1438. http://dx.doi.org/10.1016/s0305-750x(00)00031-0

Samii, C., Lisiecki, M., Kulkarni, P., Paler, L. and Chavis, L., 2014. *Effects of Payment for Environmental Services (PES) on Deforestation and Poverty in Low and Middle Income Countries: A Systematic Review.* Campbell Systematic Reviews No. 11.

Sarin, M., 2001. Disempowerment in the name of 'participatory' forestry? Village forests joint management in uttarakhand. *Forests, Trees and People Newsletter* No. 44. http://www.cifor.org/publications/pdf_files/polex/Psarin0301.pdf

Saw, L. G., LaFrankie, J. V., Kochummen, K. M. and Yap, S. K., 1991. Fruit trees in a Malaysian rain forest. *Economic Botany* 45: 120-136. http://dx.doi.org/10.1007/bf02860057

Schlosberg, D., 2013. Theorising environmental justice: The expanding sphere of a discourse. *Environmental Politics* 22: 37-55. http://dx.doi.org/10.1080/09644016.2013.755387

Schmook, B. and Radel, C., 2008. International labour migration from a tropical development frontier: Globalizing households and an incipient forest transition. *Human Ecology* 36: 891-908. http://dx.doi.org/10.1007/s10745-008-9207-0

Schroeder, R.A., 1999. *Shady Practices: Agroforestry and Gender Politics in the Gambia.* Oakland, CA: University of California Press. http://ark.cdlib.org/ark:/13030/ft5n39p01v/

SCR, 2012. *Toward Sustainability: The Roles and Limitations of Certification.* Washington, DC: RESOLVE, Inc. http://www.resolv.org/site-assessment/files/2012/06/Report-Only.pdf

Shackleton, C.M., McGarry, D., Fourie, S., Gambiza, J., Shackleton, S.E. and Fabricius, C., 2006. Assessing the effects of invasive alien species on rural livelihoods: Case examples and a framework from South Africa. *Human Ecology* 35: 113-127. http://dx.doi.org/10.1007/s10745-006-9095-0

Shapiro-Garza, E., 2013. Contesting the market-based nature of Mexico's national payments for ecosystem services programs: Four sites of articulation and hybridization. *Geoforum* 46: 5-15. http://dx.doi.org/10.1016/j.geoforum.2012.11.018

Sheil, D., Casson, A., Meijaard, E., Noordwijk van, M., Gaskell, J., Sunderland-Groves, J., Wertz, K. and Kanninen, M., 2009. *The Impacts and Opportunities of Oil Palm in Southeast Asia: What Do We Know and What Do We Need to Know?* Occasional Paper No. 51. Bogor, Indonesia: CIFOR. http://dx.doi.org/10.17528/cifor/002792

Shrestha, U.B. and Bawa, K.S., 2014. Economic contribution of Chinese caterpillar fungus to the livelihoods of mountain communities in Nepal. *Biological Conservation* 177: 194-202. http://dx.doi.org/10.1016/j.biocon.2014.06.019

Shumsky, S., Hickey, G.M., Johns, T., Pelletier, B. and Galaty, J., 2014. Institutional factors affecting wild edible plant harvest and consumption in semi-arid Kenya. *Land Use Policy* 38: 48-69. http://dx.doi.org/10.1016/j.landusepol.2013.10.014

Siedenburg, J., Martin, A. and McGuire, S., 2012. The power of "farmer friendly" financial incentives to deliver climate smart agriculture: A critical data gap. *Journal of Integrative Environmental Sciences* 9(4): 201-217. http://dx.doi.org/10.1080/1943815x.2012.748304

Sikor, T. and Newell, P., 2014. Globalizing environmental justice? *Geoforum* 54: 151-157. http://dx.doi.org/10.1016/j.geoforum.2014.04.009

Sikor, T. (ed.), 2013. *The Justices and Injustices of Ecosystem Services.* London and New York: Routledge. http://dx.doi.org/10.4324/9780203395288

Sikor, T. and Ngoc Thanh, T., 2007. Exclusive versus inclusive devolution in forest management: Insights from forest land allocation in Vietnam's Central Highlands. *Land Use Policy* 24: 644-653. http://dx.doi.org/10.1016/j.landusepol.2006.04.006

Sikor, T., Auld, G., Bebbington, A.J., Benjaminsen, T.A., Gentry, B.S., Husberger, C., Izac, A.-M., Margulis, M.E., Plieninger, T. and Schroeder, H., 2013. Global land governance: From territory to flow? *Current Opinion in Environmental Sustainability* 5: 522-527. http://dx.doi.org/10.1016/j.cosust.2013.06.006

Silva-Castañeda, L., 2012. A forest of evidence: Third-party certification and multiple forms of proof—a case study of oil palm plantations in Indonesia. *Agriculture and Human Values* 29: 361-370. http://dx.doi.org/10.1007/s10460-012-9358-x

Smith, P., Haberl, H., Popp, A., Erb, K. h., Lauk, C., Harper, R., Tubiello, F. N., Siqueira Pinto, A., Jafari, M. and Sohi, S., 2013. How much land based greenhouse gas mitigation can be achieved without compromising food security and environmental goals? *Global Change Biology* 19: 2285-2302. http://dx.doi.org/10.1111/gcb.12160

Smith-Hall, C., Larsen, H.O. and Pouliot, M., 2012. People, plants and health: A conceptual framework for assessing changes in medicinal plant consumption. *Journal of Ethnobiology and Ethnomedicine* 8: 43. http://dx.doi.org/10.1186/1746-4269-8-43

Strassburg, B., Vira, B., Mahanty, S., Mansourian, S., Martin, A., Dawson, N. Gross-Camp, N., Latawiec, A. and Swainson, L., 2012. Social and economic considerations relevant to REDD+. Understanding relationships between biodiversity, carbon, forests and people: The key to achieving REDD+ objectives. In: *Understanding Relationships between Biodiversity, Carbon, Forests and People: The Key to Achieving REDD+ Objectives*, IUFRO World Series No. 31, edited by J.A. Parrotta, C. Wildburger and S. Mansourian. Vienna: International Union of Forest Research Organizations. http://www.iufro.org/download/file/9212/5303/ws31_pdf/

Strassburg, B.B.N., Latawiec, A.E., Barioni, L.G., Nobre, C.A., Da Silva, V.P., Valentim, J.F., Viana, M. and Assad, E.D., 2014. When enough should be enough: Improving the use of current agricultural lands could meet production demands and spare natural habitats in Brazil. *Global Environmental Change* 28: 84-97. http://dx.doi.org/10.1016/j.gloenvcha.2014.06.001

Sullivan, S., 2013. Banking nature? The spectacular financialisation of environmental conservation. *Antipode* 45: 198-217. http://dx.doi.org/10.1111/j.1467-8330.2012.00989.x

Sunderlin, W.D. and Huynh, T.B., 2005. *Poverty Alleviation and Forests in Vietnam*. Bogor: Center for international forestry research. http://dx.doi.org/10.17528/cifor/001660

Sunderlin, W.D., Angelsen, A., Belcher, B., Burgers, P., Nasi, R., Santoso, L. and Wunder, S., 2005. Livelihoods, Forests and Conservation in Developing Countries: An Overview. *World Development* 33(9): 1382-1402. http://dx.doi.org/10.1016/j.worlddev.2004.10.004

Sunderlin, W.D., Dewi, S., Puntodewo, A., Müller, D., Angelsen, A. and Epprecht, M., 2008. Why forests are important for global poverty alleviation: A spatial explanation. *Ecology and Society* 13(2): 24. http://www.ecologyandsociety.org/vol13/iss2/art24/

Tacconi, L., 2007. Decentralization, forests and livelihoods: Theory and narrative. *Global Environmental Change* 17: 338-348. http://dx.doi.org/10.1016/j.gloenvcha.2007.01.002

Tacconi, L., Mahanty, S. and Suich, H., 2013. The livelihood impacts of payments for environmental services and implications for REDD+. *Society & Natural Resources* 26: 733-744. http://dx.doi.org/10.1080/08941920.2012.724151

Tadesse, G., Algieri, B., Kalkuhl, M. and Braun, J. von, 2014. Drivers and triggers of international food price spikes and volatility. *Food Policy* 47: 117-128. http://dx.doi.org/10.1016/j.foodpol.2013.08.014

Ticktin, T., Ganesan, R., Paramesha, M. and Setty, S., 2012. Disentangling the effects of multiple anthropogenic drivers on the decline of two tropical dry forest trees. *Journal of Applied Ecology* 49(4): 774-784. http://dx.doi.org/10.1111/j.1365-2664.2012.02156.x

Tiwari, P.C. and Joshi, B., 2012. Natural and socio-economic factors affecting food security in the Himalayas. *Food Security* 4: 195-207. http://dx.doi.org/10.1007/s12571-012-0178-z

Tokgoz, S., Zhang, W., Msangi, S. and Bhandary, P., 2012. Biofuels and the future of food: Competition and complementarities. *Agriculture* 2: 414-435. http://dx.doi.org/10.3390/agriculture2040414

Toledo, I. and Burlingame, B., 2006. Biodiversity and nutrition: A common path toward global food security and sustainable development. *Journal of Food Composition and Analysis* 19: 477-483. http://dx.doi.org/10.1016/j.jfca.2006.05.001

Tontisirin, K., Nantel, G. and Bhattacharjee, L., 2002. Food-based strategies to meet the challenges of micronutrient malnutrition in the developing world. *Proceedings of the Nutrition Society* 61: 243-250. http://dx.doi.org/10.1079/pns2002155

Tritsch, I., Marmoex, C., Davy, D., Thibaut, B. and Gond, V., 2014. Towards a Revival of Indigenous Mobility in French Guiana? Contemporary Transformations of the Waypai and Teko Territories. *Bulletin of Latin American Research* 34(1): 19-34. http://dx.doi.org/10.1111/blar.12204

Turnhout, E., Waterton, C., Neves, K. and Buizer, M., 2013. Rethinking biodiversity: From goods and services to "living with". *Conservation Letters* 6: 154-161. http://dx.doi.org/10.1111/j.1755-263x.2012.00307.x

Uberhuaga, P., Larsen, H.O. and Treue, T., 2011. Indigenous forest management in Bolivia: Potentials for livelihood improvement. *International Forestry Review* 13: 80-95. http://dx.doi.org/10.1505/146554811798201134

UN, 2011. *Population, Distribution, Urbanization, Internal Migration and Development: An international Perspective.* New York: United Nations, Department of Economic and Social Affairs Population Division. http://www.un.org/esa/population/publications/PopDistribUrbanization/PopulationDistributionUrbanization.pdf

UN, 2013. *World Population Prospects: The 2012 Revision*. New York: United Nations, Department of Economic and Social Affairs, Population Division.

UN, 2014. *World Urbanization Prospects: The 2014 Revision—Highlights*. New York: United Nations, Department of Economic and Social Affairs, Population Division. http://esa.un.org/unpd/wup/Highlights/WUP2014-Highlights.pdf

UN ECOSOC, 2013. *New Trends in Migrations: Demographic Aspects*. Report of the Secretary General, United Nations Economic and Social Council.

UN Women, 2014. *World Survey on the Role of Women in Development 2014: Gender Equality and Sustainable Development.* United Nations report of the Secretary. New York: UN. http://www.unwomen.org/~/media/headquarters/attachments/sections/library/publications/2014/unwomen_surveyreport_advance_16oct.pdf

UNDP, 2012. *Gender, Climate Change and Food Security*, policy brief. New York: UNDP. http://www.undp.org/content/dam/undp/library/gender/Gender and Environment/PB4_Africa_Gender-ClimateChange-Food-Security.pdf

Van Noordwijk, M., Bizard, V., Wangpakapattanawong, P., Tata, H.L., Villamor, G.B. and Leimona, B., 2014. Tree cover transitions and food security in Southeast Asia. *Global Food Security* 3(3-4): 200-208. http://dx.doi.org/10.1016/j.gfs.2014.10.005

van Wilgen, B.W., Richardson, D.M., Le Maitre, D.C., Marais, C. and Magadlela, D., 2001. The economic consequences of alien plant invasions: Examples of impacts and approaches for sustainable management in South Africa. *Environment, Development and Sustainability* 3: 145-68. http://dx.doi.org/10.1023/a:1011668417953

Vandergeest, P. and Peluso, N. L., 1995. Territorialization and state power in Thailand. *Theory and Society* 24: 385-426. http://dx.doi.org/10.1007/bf00993352

Vanhanen, H., Rayner J. Yasmi, Y. Enters, T. Fabra-Crespo, M. Kanowski, P., Karpinnen, H., Mainusch, J, and Valkeapää, A., 2010. Forestry in Changing Social Landscapes. In: *Forests and Society—Responding to Global Drivers of Change*, edited by G. Mery P. Katila G. Galloway, I. Alfaro M. Kanninen, M. Lobovikov and J. Varjo. Vienna: IUFRO World Series No. 25.

Vermeulen, S.J., Campbell, B.M. and Ingram, J.S.I., 2012. Climate change and food systems. *Annual Review of Environment and Resources* 37: 195-222. http://dx.doi.org/10.1146/annurev-environ-020411-130608

Vinceti, B., Termote, C., Ickowitz, A., Powell, B., Kehlenbeck, K. and Hunter, D., 2013. The contribution of forests and trees to sustainable diets. *Sustainability* 5: 4797-4824. http://dx.doi.org/10.3390/su5114797

Virah-sawmy, M., 2009. Ecosystem management in Madagascar during global change. *Conservation Letters* 2: 163-170. http://dx.doi.org/10.1111/j.1755-263x.2009.00066.x

Welch, J. R., Brondızio, E.S., Hetrick, S.S. and Coimbra Jr, C. E. A., 2013. Indigenous burning as conservation practice: Neotropical savanna recovery amid agribusiness deforestation in central brazil. *PLOS One* 8(12): e81226. http://dx.doi.org/10.1371/journal.pone.0081226

West, P., Igoe, J. and Brockington, D., 2006. Parks and peoples: The social impact of protected areas. *Annual Review of Anthropology* 35: 251-277. http://dx.doi.org/10.1146/annurev.anthro.35.081705.123308

White., A. and Martin, A., 2002. *Who Owns the World's Forests? Forest Tenure and Public Forests in Transition.* Washington, DC: Forest Trends and Center for International Environmental Law.

Williams, M., 2003. *Deforesting the Earth: From Prehistory to Global Crisis*. Chicago: University of Chicago Press.

Williams, M., 2008. A new look at global forest histories of Land Clearing. *Annual Review of Environment and Resources* 33: 345-367. http://dx.doi.org/10.1146/annurev.environ.33.040307.093859

Willis, K.J., Gilson, L. and Brncic, T.M., 2004. How 'Virgin' Is Virgin Rainforest? *Science* 304(5669): 402-403. http://dx.doi.org/10.1126/science.1093991

World Bank, 2003. *World Development Report 2003: Sustainable Development in a Dynamic World: Transforming Institutions, Growth and Quality of Life*. Washington, DC: The World Bank. https://openknowledge.worldbank.org/bitstream/handle/10986/5985/WDR 2003 - English.pdf?sequence=1

Wunder, S., 2001. Poverty alleviation and tropical forests-what scope for synergies? *World Development* 29: 1817-1833. http://dx.doi.org/10.1016/s0305-750x(01)00070-5

Wunder, S., 2005. *Payments for Environmental Services: Some Nuts and Bolts*. Bogor: CIFOR. http://dx.doi.org/10.17528/cifor/001760

Wunder, S., 2013. When payments for environmental services will work for conservation. *Conservation Letters* 6(4): 230-237. http://dx.doi.org/10.1111/conl.12034

Wunder, S., Börner, J., Shively, G. and Wyman, M., 2014. Safety nets, gap filling and forests: A global-comparative perspective. *World Development* 64: 29-42. http://dx.doi.org/10.1016/j.worlddev.2014.03.005

Zbinden, S. and Lee, D.R., 2005. Paying for environmental services: An analysis of participation in Costa Rica's PSA program. *World Development* 33: 255-272. http://dx.doi.org/10.1016/j.worlddev.2004.07.012

Zhou, L. and Veeck, G., 1999. Forest Resource Use and Rural Poverty in China. *Forestry Economics* 4 (1): 80-92.

Zulu, C.L., 2010. The forbidden fuel: Charcoal, urban woodfuel demand and supply dynamics, community forest management and woodfuel policy in Malawi. *Energy Policy* 38(7): 3717-3730. http://dx.doi.org/10.1016/j.enpol.2010.02.050

5. Response Options Across the Landscape

Coordinating lead author: *Terry Sunderland*

Lead authors: *Frédéric Baudron, Amy Ickowitz, Christine Padoch, Mirjam Ros-Tonen, Chris Sandbrook and Bhaskar Vira*

Contributing authors: *Josephine Chambers, Elizabeth Deakin, Samson Foli, Katy Jeary, John A. Parrotta, Bronwen Powell, James Reed, Sarah Ayeri Ogalleh, Henry Neufeldt and Anca Serban*

This chapter presents potential landscape-scale responses that attempt to reconcile the oft-competing demands for agriculture, forestry and other land uses. While there is no single configuration of land-uses in any landscape that can optimise the different outcomes that may be prevalent within a particular landscape, there are options for understanding and negotiation for the inherent trade-offs that characterise such outcomes. With increasing pressure on biodiversity and ecosystem services across many landscapes from the growing impact of human activities, hard choices have to be made about how landscapes could and should be managed to optimise outcomes. In a context where views on landscape-scale management options are often deeply entrenched and conflicts of interest are difficult to reconcile, consensus on what constitutes "success" may be difficult to achieve. Political economy and wider governance issues have often meant that a theoretically optimal landscape is unrealistic or unachievable on the ground. However, in this chapter we attempt to provide an over-arching framework for landscape approaches and how such approaches can contribute to both conservation and the achievement of food security and nutrition goals.

 http://dx.doi.org/10.11647/OBP.0085.05

5.1 Introduction

Habitat loss, largely driven by agricultural expansion, has been identified as the single largest threat to *biodiversity*[1] (Newbold et al., 2014) worldwide. Agricultural activities are intensifying, and particularly in the tropics (Laurance, et al., 2014; Shackelford et al., 2015) due to increasing global demands for food, fibre and biofuels (OECD/FAO, 2011). As such, "global food security is increasingly trading off food for nature" Lambin (2012). This habitat loss is further compounded by *land degradation* and competition from other land uses such as urbanisation (Ellis et al., 2010). Between 2000 and 2010, in the developing world alone, it is estimated that land degradation and urbanisation consumed between 2.6 and 6.2 million hectares of arable land (Lambin and Meyfroidt, 2011).

The tropics host the majority of biodiversity-rich areas on the planet (Myers et al., 2000), and the realisation that we may be witnessing a sixth mass extinction (Barnosky et al., 2011) has been answered by a call to expand the extent of protected areas, particularly in tropical regions. Consequently tropical land is increasingly subject to competing claims (Giller et al., 2008) and reconciling these claims presents what are sometimes referred to as "wicked problems" (Rittel and Webber, 1973). A range of concepts and frameworks for implementation are now being discussed which aim to consider land-use change in forested *landscapes* in such a way that competing demands for food, commodities and forest services may be, hopefully, reconciled (e.g. Pirard and Treyer, 2010). There is abundant theory to underpin the desirability of establishing landscape "mosaics" (Naveh, 2001; Sunderland et al., 2008), where such competing demands are addressed in a more holistic, integrated manner.

"*Landscape approaches*" to achieving food production, natural resource conservation and *livelihood* security goals seek to better understand and recognise interconnections between different land uses and the stakeholders that derive benefits from them (Milder et al., 2012). Such approaches also aim to reconcile competing land uses and to achieve conservation, production and socio-economic outcomes (Sayer et al., 2013) and as such are now ubiquitous paradigms in the natural resource management discourse (DeFries and Rosenzweig, 2010). Furthermore, the environmental services that support the sustainability of agriculture are also sought through landscape approaches (Scherr and McNeely, 2008; Brussaard et al., 2010; Foli et al., 2014). However, the very complexity of landscape approaches defies definition (Reed et al., 2015), despite the clarion calls for such clarification.

In parallel, both in the North and in the South, industrial agriculture, the ultimate legacy of the Green Revolution, is being questioned as a model to achieve global *food security* sustainably (McLaughlin, 2011). This model may have been appropriate to the context of the 1960s and 1970s, when reducing hunger was the main goal, when water and nutrients were abundant, energy was cheap, and when *ecosystems* were able to detoxify agricultural pollutants. The global context today is very different

1 All terms that are defined in the glossary (Appendix 1), appear for the first time in italics in a chapter.

with the growing scarcity of cheap energy (Day et al., 2009), water (Wallace, 2000) and nutrients (e.g. phosphorus, Cordell et al., 2009). The adoption of large-scale industrial agriculture has resulted in negative impacts on the environment (Conway, 1997; Cassman et al., 2003), public health (Fewtrell, 2004; Bandara et al., 2010) and even *nutrition* (Ellis et al., 2015), suggesting the paradigm itself needs to be challenged (Tilman and Clark, 2014).

In addition, industrial agriculture, with its narrow focus on a few crops (Sunderland, 2013; Khoury et al., 2014), has proven to be highly susceptible to shocks such as drought, flooding, pests and disease outbreaks, and market vagaries (Holling and Meffe, 1996; Swinnen and Squicciarini, 2012). In response to these challenges, various approaches have emerged using ecological concepts and principles to design sustainable agricultural systems (Gliessman, 1997). These approaches are based on the assumption that chemical and mechanical inputs can be replaced (at least partially) by biological functions (Doré et al., 2011; Cumming et al., 2014). Such functions are performed by the planned biodiversity (e.g. managed diversity of crop and livestock species), but also by the unplanned biodiversity (e.g. pollination or biological pest control), which is often crucial in these agroecological systems (Klein et al., 2007). The maintenance of unplanned biodiversity in agricultural landscapes is often due to dispersion from nearby (undisturbed) natural patches (Blitzer et al., 2012; Tscharntke et al., 2012). Natural areas may also provide nutrient subsidies to agricultural lands. For example, birds can be important vectors of nutrient subsidies from natural areas to agricultural lands (Young et al., 2010). This suggests the importance of landscape approaches not only for biodiversity conservation, but also for the design of sustainable agricultural systems.

Finally, non-intensive agricultural land may host significant biodiversity within a given landscape (Benton et al., 2003; Clough et al., 2011). Multifunctional landscapes are often described as patches of natural habitat embedded in an agricultural matrix (Fischer et al., 2006). Implicitly, this division assumes that patches are biodiversity-rich whilst the matrix is depleted in biodiversity (Tscharntke et al., 2005). However, the matrix may be part of the habitat of several species (Wright et al., 2012). This is particularly the case if the matrix is structurally similar to the native vegetation, for example, tropical agroforests (Clough et al., 2011). In addition, in human-dominated landscapes, agriculture is often the dominant force maintaining open patches on which many species depend (Arnold et al., 2014). This is the case for example of open-habitat bird species, which have become totally dependent on agricultural land in many areas (Wright et al., 2012). In tropical *forests*, traditional *shifting cultivation* agricultural practices create patches of open grassy fallow in an otherwise homogeneous forest cover. The resulting landscape mosaic may be beneficial for several species. For example, shifting cultivation systems in Sri Lanka were found to provide a key food source to populations of endangered Asian elephant (Wikramanayake et al., 2004), but also led to serious issues of crop raiding (Mackenzie and Ahabyona, 2012).

Despite the utility of landscape approaches for both sustainable agriculture and biodiversity conservation, it should however not be seen as a prescriptive approach to spatial planning. Published principles for landscape approaches (Fischer et al., 2006; Lindenmayer et al., 2008; Sayer et al., 2013) should not be seen as a set of boxes to be ticked in the search for an agreed spatial plan but rather as a framework of approaches from which practitioners may draw in order to solve real problems on the ground. There are fundamental difficulties in identifying and agreeing on metrics to measure progress in solving wicked problems particularly if opinions differ on the optimal solution to a problem when no single metric can measure, or even define, "success", particularly when trade-offs are the norm (Sunderland et al., 2008). National level reviews of landscape and ecosystem approaches to *forest management* have revealed that this is still very much work in progress (Sayer et al., 2014). The application of landscape principles might eventually lead to a spatial plan accepted by stakeholders but landscapes are constantly changing under the influence of multiple drivers and end points in the form of long-term plans appear to be the exception rather than the rule (Carrasco et al., 2014).

Much of the theory and practice of landscape approaches is underpinned by the assumption that facilitation and negotiation will eventually allow for a consensus on a desired outcome. However, in reality there are often entrenched views, conflicts of interest and power plays as a result of which, true consensus is rarely achievable (Colfer and Pfund, 2010). Conflict between agriculture, at both industrial and small scales, conservation and other competing land uses (e.g. industry, urbanisation, tourism, recreation, dams, reservoirs) is often the subject of strongly contested activism with highly polarised positions (Sunderland et al., 2008). Landscape approaches sometimes appear to be advocated on the assumption that they can resolve these fundamental differences in a way that will avoid conflict, particularly with regard to achieving both food and nutritional security. In reality, any intervention will bring "winners" and "losers" as any rural community – including "traditional societies" living in or on the edge of forest habitats – is heterogeneous and characterised by various internal conflicts. Ignoring this heterogeneity and these internal conflicts may weaken local communities against the influence of new powerful stakeholders, for example logging and mining concessions (Giller et al., 2008).

Boys with *Parkia biglobosa* pods, Labe, Guinea. Photo © Terry Sunderland

This chapter seeks to highlight the options related to the integration of agriculture, forestry and other land uses (Sayer et al., 2013; Sunderland et al., 2013). The intention is to identify landscape-scale policies, interventions and actions that may achieve this integration through land use change, recognising subsequent implications of forest loss and degradation on food security and nutrition. We also look at landscape configuration (including management systems, land sharing/sparing, intensification, productive landscapes, eco-agriculture etc.) and necessary synergies and trade-offs between different land uses (crops, livestock etc. but also other sectors), and *forests and tree-based systems*. Finally, we look at integrated and cross-sectoral options (that include forests and tree-based systems) for food security.

5.2 The Role of Landscape Configurations

5.2.1 Temporal Dynamics within Landscapes

Landscapes change over time and the spatial configuration of land uses is rarely static. Such changes are not only a result of anthropogenic pressures (such as *deforestation*), but can also be caused by natural ecological dynamics (e.g. Vera, 2000). Failure to understand these dynamics and their origins can lead to misguided management interventions, as in the case of Sahelian forest dynamics where it was assumed incorrectly that people were responsible for forest loss (Fairhead and Leach, 1996). Given this dynamism, in many forest landscapes it may be inappropriate to permanently delineate land uses in fixed spatial patches – often referred to as "zoning". However, finding workable alternative *governance* arrangements in such systems can be very difficult (Scott, 1999).

Mosaic of agriculture, agroforestry systems and forest in Chittagong, Bangladesh. Photo © Terry Sunderland

In some cases, particular configurations of the landscape level social-ecological system, containing multiple different patches of land uses, may be more or less sustainable in the long term. For example, the best configuration to maximise production of a particular commodity (such as a *tree crop* like oil palm) in the short term may be a large monoculture, but this might degrade the productivity of the land and other *ecosystem services* in the long term. Similarly, the best configuration to maximise the abundance of a given species of interest today may be very different from the best configuration to maximise the abundance of the same species in a couple of decades, as *climate change* is driving shifts in species ranges (Parmesan and Yohe, 2003). The optimum configuration to produce the same desired outcome in the longer term might look very different. The fact that multifunctional landscapes are "moving targets' with "multiple futures" calls for adaptive management approaches (Holling and Meffe, 1996).

Related to adaptive management is the concept of *resilience*: "the capacity of a system to continually change and adapt yet remain within critical thresholds" (Stockholm Resilience Centre, 2014). Some landscape configurations may be better able to cope with emerging pressures in the future, such as anthropogenic climate change. A considerable literature argues that landscapes containing diverse social and ecological systems (multifunctional landscapes) are likely to be more resilient to change than more simple systems (e.g. Elmqvist et al., 2003; Tscharntke et al., 2005). Production landscapes that are configured to maximise resilience by mimicking the structure of natural ecosystems are sometimes referred to as "eco-agricultural" landscapes (Scherr and McNeely, 2008). In addition, the numerous ecological interactions between cultivated and natural patches of vegetation in landscape mosaics (see above) result in complex ecological networks and stabilise the functions of these landscapes. In comparison, ecological interactions in more homogeneous landscapes are limited, and the functions of such landscapes (including agricultural productivity) are more vulnerable to shocks (e.g. extreme climatic events) (Loeuille et al., 2013). Forests and tree-based landscapes also sustain the resilience of social systems: forest products are consumed more frequently in times of food scarcity and can provide crucial livelihood safety nets (Johns and Eyzaguirre, 2006; Powell et al., 2013)

5.2.2 Trade-offs and Choices at the Landscape Scale

Landscapes are complex systems that generate a range of social and ecological outcomes over time. These outcomes are not limited to food; they include biodiversity conservation, sources of income, provision of cultural, regulatory and social services and a host of other benefits. Different landscapes produce different combinations of these elements, dependent on biophysical (such as soils and rainfall) and social conditions (such as who has the right to manage and harvest what).

There is no single configuration of land-uses in any landscape that can provide all the different outcomes that people might find desirable. For example, the "best" landscape configuration for biodiversity conservation might include large areas of forest strictly protected from human use, but this might support the livelihood needs of only a very small human population or even displace previously resident people (Brockington and Igoe, 2006). This has often been the case in the establishment of protected areas in many parts of the world (West et al., 2006). For example, in Madagascar the expansion of protected forest areas has alienated people from previously common lands, a phenomenon that can restrict community access to forest resources, including food (Corson, 2011). In contrast, the "best" landscape for cereal production might contain very little or no forest at all. Other desirable outcomes, such as malaria mitigation (Mendenhall et al., 2013) or food security (Thrupp, 2000; Chappell et al., 2013; Sunderland et al., 2013) may be best provided by more diverse landscapes.

Box 5.1 Novel technologies

New applications of technologies such as remote sensing and mobile phones, also contribute to improving the integration of agriculture and forest conservation within landscapes. A few examples have been collected:

- The recently launched Soil Moisture Active Passive Observatory (SMAP) will be used in designing global early-warning systems and improving the precision of crop suitability maps (NASA website). This technology can improve climate and weather forecasts, allowing scientists to monitor floods and droughts and therefore better predict crop yields.
- In Kenya, through the Kilimo Salama initiative of Syngenta Foundation, farmers are able to purchase insurance via their mobile phone messaging service, which lowers the cost of insurance provision. With their crops insured, farmers can more readily experiment with higher-risk, higher-yield crops and stay assured that regardless of the weather, they will be able to feed their families (Rojas-Ruiz and Diofasi, 2014).
- In India, studies revealed that the introduction of mobile technology enhanced farmers' awareness of markets and prices and improved decision-making with regard to technology adoption. Challenges to further increase the adoption and utility of mobile technology include availability of content in local languages, compatibility of these languages with the handsets, overall literacy, retrieval costs of voice messages and the lack of transmission masts in remote areas (Mittal et al., 2010; Mittal, 2012).
- In East Africa, researchers linked scientists with a private sector communications firm that produces Shamba Shape-Up (SSU), a farm reality TV show broadcast in Kenya, Tanzania and Uganda. The show seeks and presents climate-smart agriculture (CSA) information, reaching an average monthly viewership of 9 million people across East Africa. Research shows a trend of increasing uptake of CSA practices, with an average of 42 percent of SSU viewers changing their practices, as well as benefitting Kenya's GDP through net soil fertility and net dairy production increase. In a further development, the company is expanding CSA platforms by linking SSU to a mobile/SMS/internet service allowing farmers to ask questions and receive technical advice from experts. (http://ccafs.cgiar.org/blog/communicating-behavior-change-how-kenyan-tv-show-changing-rural-agriculture).

With increasing anthropogenic and biophysical pressures on biodiversity and ecosystem services across many landscapes, choices have to be made about what is desirable and how landscapes should be managed (MEA, 2005; Laurance et al., 2014). Management regimes can serve to optimise trade-offs and synergies among different outcomes (Naidoo et al., 2006; DeFries and Rosenzweig, 2010), but there are always likely to be some trade-offs and opportunity costs (McShane et al., 2011; Leader-Williams et al., 2011). To address this problem, increasing attention has been given by researchers to the question of how to resolve trade-offs at the landscape scale to produce desirable outcomes for both biodiversity and production goals (e.g. Polasky et al., 2008).

5.3 Land Sparing and Land Sharing

The *land sharing/land sparing* framework is potentially useful for considering trade-offs between agriculture and biodiversity conservation (Balmford et al., 2005; Green et al., 2005; Garnett et al., 2013). One rationale for accepting the negative ecological consequences of land-use intensification on existing farmland is that natural habitats can be "spared" from further expansion of agriculture and as such will be sufficient for the maintenance of biological communities and ecosystem services. Meanwhile, integrating agricultural production and conservation on the same land ("land sharing" or "wildlife-friendly farming"), coupled with the likelihood of further expansion acts as an alternative solution for balancing trade-offs between production and conservation. However, the central question in the land sparing/land sharing debate is whether it is more favourable for biodiversity if desired increases in agricultural production are met by increasing the area of low yield farmland (land sharing) or by increasing the intensity of farming on existing farmland (land sparing).

To answer this question it is necessary to understand the relationship between biodiversity and agricultural production in landscapes. Empirical fieldwork in Ghana and India (Phalan et al., 2011), Uganda (Hulme et al., 2013) and Malaysia (Edwards et al., 2014) has consistently found that land sparing is the "better" strategy for reconciling biodiversity and food production targets, because many species cannot survive in farming systems of even the lowest management intensity (Ewers et al., 2009; Phalan et al., 2011; Balmford et al., 2012). More recently, it has been shown that with relatively modest and sustainable increases in productivity on existing farmland, Brazil could reduce deforestation caused by agriculture to zero (Strassburg et al., 2014). Pretty and Barucha (2014) also conclude that *sustainable intensification* can result in desirable outcomes both for enhanced food yields and improved environmental goods and services, yet Phelps et al. (2013) suggest that with intensification, productivity increases could incentivise further clearance of forest for agriculture. The majority of farmers in developing countries also lack the necessary capital to either intensify their farming systems or spare land for nature (Bennett and Franzel, 2013). Box 5.1

highlights some examples of novel technologies applied to better integrate agriculture, forest and food security in a landscape.

The land sparing/sharing framework and associated research have consequently generated some debate (Perfecto and Vandermeer, 2010). Critics of the land sparing approach argue that the intensification of agriculture has a negative impact on biodiversity and ecosystem services, and that for "sparing" to work, intensification of agriculture in one place must be explicitly coupled with protection of natural habitat elsewhere, which rarely happens in practice (Chappell and LaValle, 2009; Perfecto and Vandermeer, 2010; Angelsen, 2010; Tilman et al., 2011). Links between the intensification of agricultural systems (through increased fertiliser application, pesticide use, animal stocking rates and irrigation) and *in situ* declines of biodiversity on farmland have been well documented (Green et al., 2005; Kleijn et al., 2009), even though biodiversity loss need not necessarily accompany increased agricultural yields across all systems (Clough et al., 2011). Meanwhile, the potential ecological impacts of "spillover" effects (Matson and Vitousek, 2006; Didham et al., 2015), from the agricultural matrix into adjacent natural systems (e.g. inputs of nutrient subsidies through fertiliser drift and down-slope leaching (Duncan et al., 2008), livestock access (Didham et al., 2009) and the spillover of predator or consumer organisms (Blitzer et al., 2012)) could likely compromise the effectiveness of land sparing strategies.

Proponents of land sharing advocate the creation of multi-functional agricultural landscapes that generate and utilise natural ecological processes within a social and cultural context (Bolwig et al., 2006; Perfecto and Vandermeer, 2008; Knoke et al., 2009; Barthel et al., 2013). In turn, this approach has been criticised for promoting lower yields and therefore leading to further forest clearance for agriculture. It is also claimed that land sharing is only suitable for conserving only those species able to survive in human-dominated landscapes, namely generalist or common species (Kleijn et al., 2006; Jackson et al., 2007; Phalan et al., 2011). Meanwhile, others have criticised the entire framing of land sparing/sharing on the basis that it fails to consider broader social and ecological complexities such as other ecosystem services, food security and poverty (Fischer et al., 2014).

In reality the choice and distinction between land sparing and land sharing, while context dependent, is unclear. For example, what appears to be sharing at the landscape scale may look more like sparing at the local scale (Grau et al., 2013; Baudron and Giller, 2014). The framework offers a useful tool for thinking about choices in landscapes, but policymakers should recognise that there are important limitations to its use in real world situations (Perfecto and Vandermeer, 2010; Fischer et al., 2014). Furthermore, such "landscape design" thinking might be intuitively appealing, but it faces a number of limitations in practice:

- Trade-off analyses tend to be incomplete, meaning that they neglect important issues (Fischer et al., 2014). For example, the "best" landscape for balancing forest conservation and food production may be very different from the "best" landscape to balance conservation, food and space for urban expansion.

- Results may be affected by the spatial scale of analysis. The "best" landscape configuration at one scale may be different at a larger scale. Additionally, landscape analyses often fail to incorporate flows of people and materials between landscapes (Phalan et al., 2011; Seto et al., 2012; Grau et al., 2013).
- The concept of idealised landscape design ignores the social and political realities on the ground (Fischer et al., 2014). Who owns what within the landscape, and who gets to decide what happens? Who benefits or loses from particular choices? What is the history and current status of the landscape? These political economy issues may mean that a theoretically optimal landscape configuration is unrealistic or unachievable on the ground.

The research reviewed in this section demonstrates the importance of thinking beyond the site scale by taking into account broader interactions between land-uses within landscapes. However, it also highlights the inherent complexity in any such analysis, and the trade-offs that are likely to exist between the desired outcomes of different stakeholders. Research at this scale is in its infancy, and faces daunting data and analytical deficiencies. Addressing these challenges will be a priority for the coming years.

A broader question is how far research can go in providing useful information about relationships between forest food systems and other land-uses at the landscape scale.

5.4 Landscapes and Localised Food Systems

Landscape approaches offer promise for solving some food-related problems that have proved to be more intractable than the basic task of producing more calories, such as improving access to food and nutrition through the provision of a diversity of products, and thus improving diets (Scherr and McNeely, 2008; Ickowitz et al., 2014).

Landscape approaches, especially those that are developed locally, are often more suitable for lands where previous agricultural intensification has been unsuccessful, for example on sloping lands and other areas that are marginal for conventional approaches. The diverse production activities that such systems comprise are often well adapted to the panoply of environmental, demographic, social, political and economic changes that is sweeping across much of the less-developed world. Diverse, locally-adapted production and resource management systems tend to increase the resilience of rural households in the face of such changes (Padoch and Sunderland, 2014).

It is estimated that 40 percent of all food in the less-developed world, and up to 80 percent if solely focusing on Africa and Asia (FAO, 2012), originates from smallholder systems, and many of these systems depend essentially on diverse landscape systems (Godfray et al., 2010). Smallholder farmers worldwide and throughout history have managed landscapes for food and other livelihood needs. Forests, woodlots, parklands, swidden-fallows and other tree-dominated areas are integral parts of many smallholder landscapes and household economies (Agrawal et al., 2013).

The greatest obstacle to including shifting cultivation in the new landscape paradigm, in the eyes of both development professionals and conservationists, is not necessarily the illegibility of its patchy landscapes or the complexity of its management, but its inherent dynamism. Change is what defines a system as shifting cultivation: annual crops are moved from plot to plot every year or two; as forests regenerate in one area, they are felled in another. Can so much dynamic change be tolerated in a "sustainable" landscape? (Scott, 1999). Can shifting cultivation be considered sustainable if it includes slashing and burning woody vegetation? These questions are inherent in complex socio-ecological systems and landscape dynamics and can only be addressed at a landscape level through an adaptive approach that is based on continual learning – two essential features of a landscape approach (Sayer et al., 2014; Holling and Meffe, 1996).

Many shifting cultivation systems worldwide have adapted successfully to larger human populations, new economic demands and the directives of anti-slash-and-burn policies and conservation prohibitions. Such adaptation has taken a large number of pathways, of which the more active management of fallows has perhaps been the most important. Examples include the management of rich mixtures of marketable fruits and fast-growing timbers in Amazonia and the production of rubber and rattans in Southeast Asia (Sears and Pinedo-Vasquez, 2004; Cairns, 2007). These adaptations suggest that the sustainability of shifting cultivation systems emerges when it is seen at broader spatial and longer temporal scales: shifting cultivation, in common with many smallholder-influenced landscapes, is constantly mutable.

As exemplified in the case study in Box 5.2, productive, complex and dynamic landscapes in the Lao People's Democratic Republic and elsewhere, lend flexibility to household economies and contribute to appropriate responses to climatic and economic perturbations. Programmes of directed change, such as the one promoted by the Lao government, attempt to create distinct zones for agricultural intensification and forest conservation, but until now have failed to enhance sustainable resource management or local livelihoods.

Box 5.2 The long-term benefits of shifting agriculture: a case study from Lao PDR

An important study (Castella et al., 2013) analysed changes in the patterns of field-forest landscapes that occurred as environmental and socio-economic change transformed the territories of seven villages in the northern uplands of the Lao People's Democratic Republic over a period of 40 years. In this region, where a tradition of shifting cultivation had created intricately-patterned landscapes of forest, fallows and farms, such landscapes are now being radically altered by policies aimed at increasing forest cover and promoting intensive commercial farming. Shifting cultivation, with its complex landscapes, is deliberately being replaced with a land sparing model of agriculture. This is because the segregation of land uses is perceived as most efficient for achieving multiple objectives in the context of a growing population, and shifting cultivation is widely viewed as "primitive" by government and other institutions.

Based on extensive field research, however, Castella et al. (2013) found that by imposing strict boundaries between agricultural and forest areas, interventions in the name of land-use planning have had significant negative impacts on the well-being of rural communities and especially on their ability to adapt to change. Farm and forest products that previously were "intricately linked at both landscape and livelihood levels, are now found in specialized places, managed by specialized households" (i.e. the domestication of non-wood forest products) and collected by specialised traders. The authors argued that "this trend may have negative consequences for the resilience of the overall landscape as it reduces its biological and socio-economic diversity and therefore increases vulnerability to external shocks" (Castella et al., 2013).

5.5 "Nutrition-sensitive" Landscapes

Nutrition-sensitive approaches to agriculture and food security are gaining increasing acceptance as an important dimension of global food security policy (Ruel and Alderman, 2013; Pinstrup-Andersen, 2013), recognising that the ultimate solution to *malnutrition* lies in the consumption of sufficient quantities of nutritious foods (Burchi et al., 2011). While protein and calorie deficiencies are still widespread, the prevalence of micronutrient deficiencies outweighs that of hunger, and should be a public health, food security and agricultural priority (Allen, 2002). Most of the discourse surrounding nutrition-sensitive approaches focuses on the role of monoculture agriculture, overlooking the role of agroecological systems, wild foods and forests in contributing to nutrition and dietary adequacy (Powell et al., in press). Some recent work, however, suggests that the contribution of forests and tree-based agriculture to nutrition in particular may be substantial (Golden et al., 2011; Ickowitz et al., 2014; also see discussion in Chapter 2).

Malnutrition, including under-nutrition and over-nutrition together with the concomitant increases of non-communicable diseases in poor and middle-income countries are key developmental and political challenges for donors, governments and smallholders (Frison et al., 2011). Direct pathways to malnutrition include poor diet and infection often combined with lifestyle factors, which are determined by personal factors (e.g. physiology, psychology and knowledge), household factors (such as quantity, quality, seasonality and use of own food production, income and education), as well as broader structural social, cultural, political and environmental factors (such as inequality and access to productive resources, information etc.). Indirect pathways to malnutrition are important, operating through income, education, equity and other factors that can have sustained and longer-term impacts.

The best way to address the challenge of under-nutrition and malnutrition is to coordinate activities across different sectors and different levels of scale: a more holistic "systems approach" (Frison et al., 2011; Powell et al., in press). There is a bidirectional link: while landscapes have an influence on the nutrition and health of the communities that depend on them (Golden et al., 2011; Ickowitz et al., 2014), the behaviour of people can also have an influence on the very well-being and long-term sustainability of integrated landscape systems themselves.

Market sellers at the roadside, Nyimba, Zambia. Photo © Terry Sunderland

A number of landscape level factors lead to insufficient production, sale and use of nutritious food. These include internal factors such as poor productivity of the agricultural, aquatic and forestry systems; loss of *agricultural biodiversity* of the systems; access to markets and lack of knowledge and awareness on healthy diets (e.g. Powell et al., 2014); but also external drivers of land use and landscape change including environmental, institutional, social and political factors. A better understanding of these factors would help to reduce their impact on food security and nutrition.

While there is evidence that increased income and improved food security are correlated at the national scale, evidence is beginning to emerge showing that incomes from diverse landscapes may be used in a nutritionally-sensitive manner (Ickowitz et al., 2014). The interactions between urban and rural populations have profound implications on livelihoods, markets and wellbeing. The layers of these relationships need to be understood and supported when positive, and mitigated when shown to reduce resilience.

5.6 Landscape Governance

There are diverse uses and understandings across disciplines of the term "governance" (Kozar et al., 2014). At its core, the term denotes the inclusion of multiple non-state actors in deliberating and deciding society's most pressing issues and their solutions, and refers to new spaces where increasingly complex problems can be solved by multiple types of actors (Kozar et al., 2014). Landscape governance is thereby concerned with the institutional arrangements, decision-making processes, policy instruments and underlying values in the system by which multiple actors pursue their interests in sustainable food production, biodiversity and ecosystem service provision and livelihood security in multifunctional landscapes.

As people living in and around a particular landscape seek from it a wide range of qualities and benefits, the divergent values and interests of multiple types of actors at different levels create new challenges for landscape governance. Throughout the world, innovative efforts are being pursued to couple the sustainable governance of ecological resources and human activity within a common framework. These efforts seek to realise multiple ecosystem services and livelihood benefits for diverse stakeholders within the same geographic location. At the same time, advances in the study of socio-ecological systems (Liu et al., 2007) and the corresponding practice of integrated landscape governance (FAO, 2005; Scherr et al., 2013) is rooted in the growing recognition that nature conservation need not necessarily pose a trade-off with development.

Rather, investments in conservation, restoration and sustainable ecosystem use are increasingly viewed as potentially synergistic in generating ecological, social and economic benefits and therefore providing solutions to the "wicked" problems identified earlier in this chapter (de Groot et al., 2010; see also discussion in Chapter 6).

As inhabitants of landscapes and other practitioners continue to experiment and innovate with the scaling-up of landscape approaches from their diverse entry points, emerging institutional issues of multi-level and multi-actor governance and their incongruity within administrative and jurisdictional boundaries pose an imminent challenge to successfully realising multiple outcomes from multi-functional landscapes.

Consensus across multiple fields, spanning ecological, political and geographical disciplines, concludes that a core challenge for addressing complex problems bridging social and ecological systems is effective governance at multiple levels. Yet the inhabitants of landscapes and other practitioners struggling to implement landscape approaches often focus on one level, whether international, national, regional or local (Nagendra and Ostrom, 2012). Multilevel decision-making for the governance of landscapes helps to link actors and address the complex issues that arise in governing social-ecological systems (Görg, 2007). However, the way in which the issues of scale and multi-actor governance are conceptualised and the manner in which solutions for viable governance systems are designed are both emergent and variant.

Effective governance structures in multifunctional landscapes remain elusive, giving rise to questions such as: what functions will be located where, what rules determine who has rights to what resources at what time, and how to enforce those rules. Who decides such questions based on what values, and who is included and excluded from activities and benefits linked to different functions are also key challenges within the management of complex landscapes.

Decision-making processes that can accommodate diverse values, interests and knowledge while balancing the influence and power among different types of actors can help to formulate a common vision and maintain it in the face of dynamic socio-ecological change in the landscape. Robust institutions capable of traversing scales and levels can contribute to providing the mechanisms and incentives by which public, private and civic sector actors can cooperate to realise their desired outcomes.

Colfer and Pfund (2010) identified recurring issues that are likely to impinge on any efforts to work collaboratively with tropical forest communities and landscapes. These include, governmental policies with complex, diverse and often unpredictable effects, varying interfaces between customary and formal legal systems, differences in the use and governance of agricultural production and *non-timber forest products (NTFPs)*, and the potential even within collaborative governance for harm (win-win solutions are unlikely always to be an option and many argue that trade-offs are the norm) (Giller et al., 2008).

Based on a comparative study of pantropical landscapes, Colfer and Pfund (2010) conclude that there are six key issues that represent governance constraints at the landscape scale: 1. the powerful duo of government and industry (for example, oil palm expansion); 2. risks linked to national policies (for example, the focus on men and timber in forest management, without complimentary income-generating and gender-balanced activities); 3. complexities of pluralistic governance (such as differing relations between hinterland groups and governments); 4. differences in cultural significance and governance of NTFPs and other forest products, including differentiation in roles between sexes and among social groups; 5. discontinuity between national laws and swidden *agroforestry* systems; and 6. new potential dangers for hinterland people from international sources (such as risks of exclusion linked to international encouragement of proliferation of protected areas).

Cattle grazing in *Borassus aethiopium* savannah, Senegal. Photo © Terry Sunderland

Most of these issues demonstrate the global variety and variation over time in contexts, peoples, and regimes governing natural resources. Such diversity and dynamism reinforces the desirability of: a) strengthening and supporting their involvement in their own governance and b) tailoring any interventions to the specificities of any locale. Indeed, implementation of the latter probably requires the implementation of the former. Thus formal governmental shortcomings strengthen the argument for stronger citizen involvement, to serve as monitor and ultimately provide some constraint on such power.

5.7 Conclusions

The ability to create change in policy and practice in the context of landscape approaches to land management is currently impaired by a dearth of scientific evidence. While there is a growing body of evidence, our understanding of how forests and landscapes with tree cover contribute to food security and nutrition and the provisioning of healthy and nutritious foods to local and global food systems remains limited. Greater attention to the production of and access to nutrient-dense foods is needed in the debate on the respective benefits of land sharing versus land sparing which has focused to date on the impacts of staple crop yields (one important aspect of food security) on biodiversity and forest conservation.

Future work on forests, and food security and nutrition should also focus on linking the health of forests and landscapes to *food sovereignty* (which encompasses food security, the right to food and healthy diets, as well as the right to control over one's own *food system* (Pimbert, 2009) to help mitigate nutrition transitions while contributing to sustainable management of wildlands. The concept of food sovereignty has been widely accepted by many indigenous groups (e.g. http://www.indigenousfoodsystems.org/food-sovereignty), and it is seen as a potential mechanism and argument to enhance greater autonomy of indigenous communities over their local food and agricultural systems as well as their wider landscapes and bio-cultural environments.

The need for local food systems is clearly demonstrated by the fact that current global food production is more than adequate to feed the entire global population, at least in terms of calories (Stringer, 2000; Chappell and LaValle, 2011), while more than 800 million people are undernourished (FAO, 2009). Clearly, producing large amounts of food in the North is not enough to guarantee food security in the South. A main reason for this is that the agricultural production from the North is subject to multiple demands, not only from the food sector, but also from the livestock (Goodland, 1997) and energy sectors (OECD-FAO, 2011).

Enhanced food sovereignty will help ensure local people have control over their own diets and are engaged in efforts to improve the nutritional quality of their diets. Such community level engagement will be particularly important for those people facing a nutrition transition and the burden of malnutrition. Community level engagement with local food and agricultural systems additionally creates a setting ideal for engaging communities for more sustainable management of these food and agricultural systems and the wider landscapes in which they reside.

Although food security is dependent on issues of sustainability, availability, access and utilisation, and not production alone, it is evident that a "new agriculture" (Steiner, 2011) needs to be found to feed the world's population both efficiently and equitably. It needs to produce food where it is needed i.e. in areas where agriculture is dominated by small farms (e.g. two thirds of African farms are smaller than two hectares (Altieri, 2009)) and where negligible quantities of external inputs are used (agriculture "organic by default", Bennett and Franzel, 2013). Thus, *agroecology* (i.e. the application of ecological concepts and principles in the design of sustainable agricultural systems, Gliessman, 1997) appears well suited to these geographies. As such, the United Nations' (2011) vision of an "agro-ecological" approach that combines biodiversity concerns, along with food production demands, provides a more compelling vision of future food production.

The integration of biodiversity conservation and agricultural production goals must be a first step, whether through land sharing or land sparing, or a more nuanced, yet complex, multi-functional integrated landscape approach. However, conservation and restoration in human dominated ecosystems must strengthen connections between agriculture and biodiversity (Novacek and Cleland, 2001). In such landscapes, characterised by impoverished biodiversity and in particular "defaunated", depopulated of their medium and large size vertebrates (Galetti and Dirzo, 2013), agriculture may represent an opportunity, and not necessarily a threat, for conservation and ecosystem restoration. When native large vertebrates are lost, several ecological functions such as the maintenance of habitat heterogeneity, nutrient cycling and seed dispersion are impaired (Owen-Smith, 1988; Hansen and Galetti, 2009). Domestic livestock may mimic ecosystem functions once provided by wild herbivores (Wright et al., 2012), and restore the ecological integrity of landscape mosaics. In extreme cases, domestic livestock has been used to restore biodiversity and ecosystem functions of landscapes that previously lost large native vertebrates, most famously in the Oostvaardersplassen in the Netherlands (Vera, 2009).

Managing landscapes on a multi-functional basis that combines food production, biodiversity conservation and the maintenance of ecosystem services should be at the forefront of efforts to achieve food security (Godfray, 2011). In order for this to happen, knowledge from biodiversity science and agricultural research and development need to be integrated through a systems approach at a landscape scale. This provides a unique opportunity for forestry and agricultural research organisations to coordinate efforts at the conceptual and implementation levels to achieve more sustainable agricultural systems. As such, a clear programme of work on managing landscapes and ecosystems for biodiversity conservation, agriculture, food security and nutrition should be central to development aid.

References

Agrawal, A., Cashore, B., Hardin, R., Shepherd, G., Benson, C. and Miller, D., 2013. *Economic Contributions of Forests*. Background Paper 1, United Nations Forum on Forests (UNFF), 10th Session, Istanbul, Turkey. http://www.un.org/esa/forests/pdf/session_documents/unff10/EcoContrForests.pdf

Allen, L.H., 2002. Iron supplements: Scientific issues concerning efficacy and implications for research and programs. *The Journal of Nutrition* 132: 813S-819S.

Altieri, M.A., 2009. Agroecology, small farms, and food sovereignty. *Monthly Review* 61: 102-113. http://dx.doi.org/10.14452/mr-061-03-2009-07_8

Angelsen, A., 2010. Policies for reduced deforestation and their impact on agricultural production. *Proceedings of the National Academy of Sciences of the USA* 107: 19639-19644. http://dx.doi.org/10.1073/pnas.0912014107

Balmford, A., Green, R. and Phalan, B., 2012. What conservationists need to know about farming. *Proceedings of the Royal Society B: Biological Sciences* 279: 2714-2724. http://dx.doi.org/10.1098/rspb.2012.0515

Balmford, A., Green, R.E. and Scharlemann, J.P.W., 2005. Sparing land for nature: Exploring the potential impact of changes in agricultural yield on the area needed for crop production. *Global Change Biology* 11: 1594-1605. http://dx.doi.org/10.1111/j.1365-2486.2005.001035.x

Bandara, J.M.R.S., Wijewardena, H.V.P., Liyanege, J., Upul, M.A. and Bandara, J.M.U.A., 2010. Chronic renal failure in Sri Lanka caused by elevated dietary cadmium: Trojan horse of the green revolution. *Toxicology Letters* 198: 33-39. http://dx.doi.org/10.1016/j.toxlet.2010.04.016

Barthel, S., Crumley, C. and Svedin, U., 2013. Bio-cultural refugia—safeguarding diversity of practices for food security and biodiversity. *Global Environmental Change* 23: 1142-1152. http://dx.doi.org/10.1016/j.gloenvcha.2013.05.001

Baudron, F. and Giller, K.E., 2014. Agriculture and nature: Trouble and strife? *Biological Conservation* 170: 232-245. http://dx.doi.org/10.1016/j.biocon.2013.12.009

Barnosky A.D., Matzke, N., Tomiya, S., Wogan, G.O.U., Swartz, B., Quental, T.B., Marshall, C., McGuire, J.L., Lindsey, E.L., Maguire, K.C., Mersey, B. and Ferrer, E.A. 2011. Has the Earth's sixth mass extinction already arrived? *Nature* 471: 51-57. http://dx.doi.org/10.1038/nature09678

Bennett, M. and Franzel, S., 2013. Can organic and resource-conserving agriculture improve livelihoods? A synthesis. *International Journal of Agricultural Sustainability* 11(3): 193-215. http://dx.doi.org/10.1080/14735903.2012.724925

Benton, T.G., Vickery, J.A. and Wilson, J.D., 2003. Farmland biodiversity: Is habitat heterogeneity the key? *Trends in Ecology & Evolution* 18: 182-188. http://dx.doi.org/10.1016/s0169-5347(03)00011-9

Blitzer, E.J., Dormann, C.F., Holzschuh, A., Klein, A.M., Rand, T.A. and Tscharntke, T., 2012. Spillover of functionally important organisms between managed and natural habitats. *Agriculture Ecosystem and Environment* 146: 34-43. http://dx.doi.org/10.1016/j.agee.2011.09.005

Bolwig, S., Pomeroy, D., Tushabe, H. and Mushabe, D., 2006. Crops, trees, and birds: Biodiversity change under agricultural intensification in Uganda's farmed landscapes. *Geografisk Tidsskrift, Danish Journal of Geography* 106(2): 115-130. http://dx.doi.org/10.1080/00167223.2006.10649561

Brockington, D. and Igoe, J., 2006. Eviction for conservation: A global overview. *Conservation and Society* 4: 424.

Brussaard, L., Caron, P., Campbell, B., Lipper, L., Mainka, S., Rabbinge, R., Babin, D. and Pulleman, M., 2010. Reconciling biodiversity conservation and food security: Scientific challenges for a new agriculture. *Current Opinion in Environmental Sustainability* 2(1): 34-42. http://dx.doi.org/10.1016/j.cosust.2010.03.007

Burchi, F., Fanzo, J. and Frison, E., 2011. The role of food and nutrition system approaches in tackling hidden hunger. *International Journal of Environmental Research in Public Health* 8(2): 358-373. http://dx.doi.org/10.3390/ijerph8020358

Cairns, M. (ed.), 2007. *Voices from the Forest: Integrating Indigenous Knowledge into Sustainable Upland Farming*. Washington, DC: Resources for the Future.

Carrasco, R., Nghiem, T., Chisholm, R.A., Sunderland, T. and Koh, L.P., 2014. Mapping ecosystem service values in tropical forests: A spatially explicit regression meta-analysis. *Biological Conservation* 178: 163-170.

Cassman, K.G., Dobermann, A., Walters D.T. and Yang, H., 2003. Meeting cereal demand while protecting natural resources and improving environmental quality. *Annual Review of Environment and Resources*. 28: 315-358. http://digitalcommons.unl.edu/cgi/viewcontent.cgi?article=1317&context=agronomyfacpub

Castella, J.-C., Lestrelin, G., Hett, C., Bourgoin, J., Fitriana, Y.R., Heinimann, A. and Pfund, J.-L., 2013. Effects of landscape segregation on livelihood vulnerability: Moving from extensive shifting cultivation to rotational agriculture and natural forests in Northern Laos. *Human Ecology* 41(1): 63-76. http://dx.doi.org/10.1007/s10745-012-9538-8

Chappell, M.J., Wittman, H., Bacon, C.M., Ferguson, B.G., Barrios, L.G., Barrios, R.G., Jaffee, D., Lima, J., Méndez, V.E., Morales, H., Soto-Pinto, L., Vandermeer, J. and Perfecto, I., 2013. Food sovereignty: An alternative paradigm for poverty reduction and biodiversity conservation in Latin America. *F1000Research* 2: 235. http://dx.doi.org/10.12688/f1000research.2-235.v1

Chappell, M.J. and LaValle, L.A., 2009. Food security and biodiversity: Can we have both? An agroecological analysis. *Agriculture and Human Values* 28: 3-26. http://dx.doi.org/10.1007/s10460-009-9251-4

Clough, Y., Barkmann, Y., Juhrbandt, J., Kessler, M., Cherico Wanger, T., Anshary, A., Buchori, D., Cicuzza, D., Darras, K., Dwi Putrak, D., Erasmi, S., Pitopang, R., Schmidt, C., Schulze, C.H., Seidel, D., Steffan-Dewenter, I., Stenchly, K., Vidal, S., Weist, M., Christian Wielgoss, A. and Tscharntke, T., 2011. Combining high biodiversity with high yields in tropical agroforests. *Proceedings of the National Academy of Sciences of the USA* 108: 8311-8316. http://dx.doi.org/10.1073/pnas.1016799108

Colfer, C. and Pfund, J.-L. (eds.), 2010. *Collaborative Governance of Tropical Landscapes*. London: Earthscan. http://dx.doi.org/10.4324/9781849775601

Conway, G., 1997. *The Doubly Green Revolution: Food for All in the Twenty-First Century*. Ithaca, New York: Comstock Publishing Associates.

Cordell, D., Drangert, J.-O. and White, S., 2009. The story of phosphorus: Global food security and food for thought. *Global Environmental Change* 19: 292-305. http://dx.doi.org/10.1016/j.gloenvcha.2008.10.009

Corson, C., 2011. Territorialization, enclosure and neoliberalism: Non-state influence in struggles over Madagascar's forests. *Journal of Peasant Studies* 38: 703-726. http://dx.doi.org/10.1080/03066150.2011.607696

Cumming, G., Buerkert, A., Hoffman, E., Schlect, E., von Cramon-Taubadel, S. and Tsharntke, T., 2014. Implications of agricultural transitions and urbanization for ecosystem services. *Nature* 515: 50-57. http://dx.doi.org/10.1038/nature13945

Day, J.W. Jr., Hall, C.A., Yáñez-Arancibia, A. and Pimentel, D., 2009. Ecology in times of scarcity. *BioScience* 59: 321-331. http://dx.doi.org/10.1525/bio.2009.59.4.10

DeFries, R. and Rosenzweig, C., 2010. Toward a whole-landscape approach for sustainable land use in the tropics. *Proceedings of the National Academy of Sciences of the USA*. 107(46): 19627-19632. http://dx.doi.org/10.1073/pnas.1011163107

De Groot, R.S., Alkemade, R., Braat, L., Hein, L. and Willemen, L., 2010. Challenges in integrating the concept of ecosystem services and values in landscape planning, management and decision making. *Ecological Complexity* 7(3): 260-272. http://dx.doi.org/10.1016/j.ecocom.2009.10.006

Didham, R.K., Barker, G.M., Bartlam, S., Deakin, E.L., Denmead, L.H., Fisk, L.M., Peters, J.M., Tylianakis, J.M., Wright, H.R. and Schipper, L.A., 2015. Agricultural intensification exacerbates spillover effects on soil biogeochemistry in adjacent forest remnants. *PLoS ONE* 10(1): e0116474. http://dx.doi.org/10.1371/journal.pone.0116474

Didham, R.K., Barker, G.M., Costall, J.A., Denmead, L.H., Floyd, C.G. and Watts, C.H., 2009. The interactive effects of livestock exclusion and mammalian pest control on the restoration of invertebrate communities in small forest remnants. *New Zealand Journal of Zoology* 36: 135-163. http://dx.doi.org/10.1080/03014220909510148

Doré, T., Makowski, D., Malézieux, E., Munier-Jolain, N., Tchamitchian, M. and Tittonell, P., 2011. Facing up to the paradigm of ecological intensification in agronomy: Revisiting methods, concepts and knowledge. *European Journal of Agronomy* 34: 197-210. http://dx.doi.org/10.1016/j.eja.2011.02.006

Duncan, D.H., Dorrough, J., White, M. and Moxham, C., 2008. Blowing in the wind? Nutrient enrichment of remnant woodlands in an agricultural landscape. *Landscape Ecology* 23: 107-119. http://dx.doi.org/10.1007/s10980-007-9160-0

Edwards, D.P., Gilroy, J.J., Woodcock, P., Edwards, F.A., Larsen, T.H., Andrews, D.J.R., Derhé, M.A., Docherty, T.D.S., Hsu, W.W., Mitchell, S.L., Ota, T., Williams, L.J., Laurance, W.F., Hamer, K.C. and Wilcove, D.S., 2014. Land-sharing versus land-sparing logging: Reconciling timber extraction with biodiversity conservation. *Global Change Biology* 20: 183-191. http://dx.doi.org/10.1111/gcb.12353

Ellis, A., Myers, S. and Ricketts. T., 2015. Do pollinators contribute to nutritional health? *PLoS One* 10(1): e114805. http://dx.doi.org/10.1371/journal.pone.0114805

Ellis, E., Goldewijk, K., Siebert, S., Lightman, D. and Ramankutty, N., 2010. Anthropogenic transformation of the biomes, 1700-2000. *Global Ecology and Biogeography* 19: 589-606. http://dx.doi.org/10.1111/j.1466-8238.2010.00540.x

Elmqvist, T., Folke, C., Nyström, M., Peterson, G., Bengtsson, J., Walker, B. and Norberg, J., 2003. Response diversity, ecosystem change, and resilience. *Frontiers in Ecology and the Environment* 1: 488-494. http://dx.doi.org/10.2307/3868116

Ewers, R., Scharlemann, J., Balmford, A. and Green, R., 2009. Do increases in agricultural yield spare land for nature? *Global Change Biology* 15: 1715-1726. http://dx.doi.org/10.1111/j.1365-2486.2009.01849.x

Fairhead, J. and Leach, M., 1996. *Misreading the African Landscape: Society and Ecology in a Forest Savanna Mosaic*. Cambridge: Cambridge University Press.

FAO, 2012. *Roles of Forests in Climate Change*. Rome: FAO. http://www.fao.org/forestry/climatechange/53459/en/

FAO, 2009. *The State of Food Insecurity in the World. Economic Crises: Impacts and Lessons Learned.* Rome: FAO. ftp://ftp.fao.org/docrep/fao/012/i0876e/i0876e.pdf

FAO, 2005. *An Approach to Rural Development: Participatory and Negotiated Territorial Development (PNTD)*. Rome: FAO. http://www.fao.org/3/a-ak228e.pdf

Fewtrell, L., 2004. Drinking-water nitrate, methemoglobinemia, and global burden of disease: A discussion. *Environmental Perspective* 112: 1371-1374. http://dx.doi.org/10.1289/ehp.7216

Fischer, J., Abson, D.J., Butsic, V., Chappell, M.J., Ekroos, J., Hanspach, J., Kuemmerle, T., Smith, H.G. and von Wehrden, H., 2014. Land sparing versus land sharing: Moving forward. *Conservation Letters* 7: 149-157. http://dx.doi.org/10.1111/conl.12084

Fischer, J., Lindenmayer, D.B. and Manning, A.D., 2006. Biodiversity, ecosystem function, and resilience: Ten guiding principles for commodity production landscapes. *Frontiers in Ecology and the Environment* 4(2): 80-86. http://dx.doi.org/10.1890/1540-9295(2006)004[0080:befart]2.0.co;2

Foli, S., Reed, J., Clendenning, J., Petrokofsky, G., Padoch, C. and Sunderland, T., 2014. Exploring the dynamics between forests, ecosystem services and food production: A systematic review protocol. *Environmental Evidence*. http://www.environmentalevidencejournal.org/content/3/1/15

Frison, E., Cherfas, J. and Hodgkin, T., 2011. Agricultural biodiversity is essential for a sustainable improvement in food and nutritional security. *Sustainability* 3: 238-253. http://dx.doi.org/10.3390/su3010238

Galetti, M. and Dirzo, R., 2013. Ecological and evolutionary consequences of living in a defaunated world. *Biological Conservation* 163: 1-6. http://dx.doi.org/10.1016/j.biocon.2013.04.020

Garnett, T., Appleby, M., Balmford, A., Bateman, I., Benton, T., Bloomer, P. Burlingame, B., Dawkins, M., Dolan, L., Fraser, D., Herroro, M., Hoffman, I. Smith, P., Thornton, P., Toulmin, C., Vermeulen, S. and Godfray, C., 2013. Sustainable intensification in agriculture: Premises and policies. *Science* 341: 33-34. http://dx.doi.org/10.1126/science.1234485

Giller, K.E., Leeuwis, C., Andersson, J.A., Andriesse, W., Brouwer, A., Frost, P., Hebinck, P., Heitkönig, I., Van Ittersum, M. K. and Koning, N., 2008. Competing claims on natural resources: What role for science. *Ecology and Society* 13(2): 34. http://www.ecologyandsociety.org/vol13/iss2/art34/

Gliessman, S.R., 1997. *Agroecology: Ecological Processes in Sustainable Agriculture*. New York: CRC Press, Taylor & Francis.

Godfray, C.J. 2011. Food and biodiversity. *Science* 333: 1231-2132. http://dx.doi.org/10.1126/science.1211815

Godfray, H.C.J., Beddington, J.R., Crute, I.R., Haddad, L., Lawrence, D., Muir, J.F., Pretty, J., Robinson, S., Thomas, S.M. and Toulmin, C., 2010. Food security: The challenge of feeding 9 billion people. *Science* 327: 812-818. http://dx.doi.org/10.1126/science.1185383

Golden, C., Fernald, L., Brashares, J., Rasolofoniaina, B. and Kremen, C., 2011. Benefits of wildlife consumption to child nutrition in a biodiversity hotspot. *Proceedings of the National Academy of Sciences of the USA*. 108(49): 19653-19656. http://dx.doi.org/10.1073/pnas.1112586108

Goodland, R., 1997. Environmental sustainability in agriculture: Diet matters. *Ecological Economics* 23: 189-200. http://dx.doi.org/10.1016/s0921-8009(97)00579-x

Görg, C., 2007. Landscape Governance: The "politics of scale" and the "natural" conditions of places. *Geoforum* 38(5): 954-66. http://dx.doi.org/10.1016/j.geoforum.2007.01.004

Grau, R., Kuemmerle, T. and Macchi, L., 2013. Beyond "land sparing versus land sharing": Environmental heterogeneity, globalization and the balance between agricultural production and nature conservation. *Current Opinion in Environmental Sustainability* 5: 477-483. http://dx.doi.org/10.1016/j.cosust.2013.06.001

Green, R.E., Cornell, S.J., Scharlemann, J.P.W. and Balmford, A. 2005. Farming and the fate of wild nature. *Science* 307: 550-555. http://dx.doi.org/10.1126/science.1106049

Hansen, D.M. and Galetti, M., 2009. The forgotten megafauna. *Science* 324: 42-43. http://dx.doi.org/10.1126/science.1172393

Holling, C.S. and Meffe, G.K., 1996. Command and control and the pathology of natural resource management. *Conservation Biology* 10: 328-337. http://dx.doi.org/10.1046/j.1523-1739.1996.10020328.x

Hulme, M.F., Vickery, J.A., Green, R.E., Phalan B., Chamberlain, D.E., Pomeroy, D.E., Nalwanga, D., Mushabe, D., Katebaka, R., Bolwig, S. and Atkinson, P.W., 2013. Conserving the Birds of Uganda's Banana-Coffee Arc: Land Sparing and Land Sharing Compared. *PLoS ONE* 8(2): e54597. http://dx.doi.org/10.1371/journal.pone.0054597

Ickowitz, A., Powell, B., Salim, A., and Sunderland, T., 2014. Dietary quality and tree cover in Africa. *Global Environmental Change* 24: 287-294. http://dx.doi.org/10.1016/j.gloenvcha.2013.12.001

Jackson, L.E., Pascual, U. and Hodgkin, T. 2007. Utilizing and conserving agrobiodiversity in agricultural landscapes. *Agriculture, Ecosystems and Environment* 121: 196-210. http://dx.doi.org/10.1016/j.agee.2006.12.017

Johns, T. and Eyzaguirre, P., 2006. Linking biodiversity, diet and health in policy and practice. *Proceedings of the Nutrition Society* 65: 182-189. http://dx.doi.org/10.1079/pns2006494

Khoury, C., Bkorkman, A., Dempewolf, H., Rimirez-Villegas, J., Guarino, L., Jarvis, A., Rieseberg, L. and Struik, P., 2014. Increasing homogeneity in global food supplies and the implications for food security. *Proceedings of the National Academy of Science of the USA* 111(11): 4001-4006. http://dx.doi.org/10.1073/pnas.1313490111

Kleijn, D., Kohler, F., Baldi, A., Batary, P., Concepcion, E.D., Clough, Y., Diaz, M., Gabriel, D., Holzschuh, A., Knop, E., Kovacs, A., Marshall, E. J. P., Tscharntke, T. and Verhulst, J., 2009. On the relationship between farmland biodiversity and land-use intensity in Europe. *Proceedings of the Royal Society B: Biological Sciences* 276: 903-909. http://dx.doi.org/10.1098/rspb.2008.1509

Kleijn, D., Baquero, R.A., Clough, Y., Díaz, M., De Esteban, J., Fernández, F., Gabriel, D., Herzog, F., Holzschuh, A., Jöhl, R., Knop, E., Kruess, A., Marshall, E.J., Steffan-Dewenter, I., Tscharntke, T., Verhulst, J., West, T.M. and Yela, J.L., 2006. Mixed biodiversity benefits of agri-environment schemes in five European countries. *Ecology Letters* 9: 243-254. http://dx.doi.org/10.1111/j.1461-0248.2005.00869.x

Klein, A.-M., Vaissière, B.E., Cane, J.H., Steffan-Dewenter, I., Cunningham, S.A., Kremen, C. and Tscharntke, T., 2007. Importance of pollinators in changing landscapes for world crops. *Proceedings of the Royal Society B: Biological Sciences* 274: 303-313. http://dx.doi.org/10.1098/rspb.2006.3721

Knoke, T., Calvas, B., Aguirre, N., Román-Cuesta, R.M., Günter, S., Stimm, B., Weber, M. and Mosandl, R., 2009. Can tropical farmers reconcile subsistence needs with forest conservation? *Frontiers in Ecology and the Environment* 7: 548-554. http://dx.doi.org/10.1890/080131

Kozar, R., Buck, L.E., Barrow, E.G., Sunderland, T.C.H., Catacutan, D.E., Planicka, C., Hart, A.K., and Willemen, L., 2014. *Toward Viable Landscape Governance Systems: What Works?* Washington, DC: EcoAgriculture Partners, on behalf of the Landscapes for People, Food, and Nature Initiative. http://peoplefoodandnature.org/publication/toward-viable-landscape-governance-systems-what-works/

Lambin, E., 2012. Global land availability: Malthus versus Ricardo. *Global Food Security* 1: 83-87. http://dx.doi.org/10.1016/j.gfs.2012.11.002

Lambin, E.F. and Meyfroidt, P., 2011. Global land use change, economic globalization, and the looming land scarcity. *Proceedings of the National Academy of Science of the USA* 108: 3465-3472. http://dx.doi.org/10.1073/pnas.1100480108

Laurance, W.F., Sayer, J. and Cassman, K.G., 2014. Agricultural expansion and its impacts on tropical nature. *Trends in Ecology and Evolution* 29: 107-116. http://dx.doi.org/10.1016/j.tree.2013.12.001

Leader-Williams, N., Adams, W.M. and Smith, R.J., 2011. *Trade-offs in Conservation: Deciding What to Save.* Oxford: John Wiley and Sons.

Lindenmayer, D., Hobbs, J.R., Montague Drake, R., Alexandra, J., Bennett, A., Burgman, M., Cale, P., Calhoun, A., Cramer, V. and Cullen, P., 2008. A checklist for ecological management of landscapes for conservation. *Ecology Letters* 11(1): 78-91. http://dx.doi.org/10.1111/j.1461-0248.2007.01114.x

Liu, J., Dietz, T., Carpenter, S. R., Alberti, M., Folke, C., Moran, E., Pell, A.N., Deadman, P., Kratz, T., Lubchenco, J., Ostrom, E., Ouyang, Z., Provencher, W., Redman, C.L., Schneider, S.H. and Taylor, W.W., 2007. Complexity of coupled human and natural systems. *Science* 317 (5844): 1513-1516. http://dx.doi.org/10.1126/science.1144004

Loeuille, N., Barot, S., Goergelon, E., Kylafis, G. and Lavigne, C., 2013. Eco-evolutionary dynamics of agricultral networks for sustainable management. *Advances in Ecological Research* 49: 339-435. http://dx.doi.org/10.1016/b978-0-12-420002-9.00006-8

Mackenzie, C.A. and Ahabyona, P., 2012. Elephants in the garden: Financial and social costs of crop raiding. *Ecological Economics* 75: 72-82. http://dx.doi.org/10.1016/j.ecolecon.2011.12.018

Matson, P.A. and Vitousek, P.M., 2006. Agricultural intensification: Will land spared from farming be land spared for nature? *Conservation Biology* 20: 709-710. http://dx.doi.org/10.1111/j.1523-1739.2006.00442.x

MEA, 2005. *Ecosystems and Human Well-Being: Biodiversity Synthesis.* World Resources Institute: Washington, DC. http://www.millenniumassessment.org/documents/document.354.aspx.pdf

McLaughlin, D., 2011. Land, food and biodiversity. *Conservation Biology* 6: 1117-1120. http://dx.doi.org/10.1111/j.1523-1739.2011.01768.x

McShane, T.O., Hirsch, P.D., Trung, T.C., Songorwa, A.N., Kinzig, A., Monteferri, B., Mutekanga, D., Thang, H.V., Dammert, J.L., Pulgar-Vidal, M., Welch-Devine, M., Brosius, J.P., Coppolillo, P. and O'Connor, S., 2011. Hard choices: Making trade-offs between biodiversity conservation and human well-being. *Biological Conservation* 144: 966-972. http://dx.doi.org/10.1016/j.biocon.2010.04.038

Mendenhall, C.D., Archer, H.M., Brenes, F.O., Sekercioglu, C.H. and Sehgal, R.N.M., 2013. Balancing biodiversity with agriculture: Land sharing mitigates avian malaria prevalence. *Conservation Letters* 6: 125-131. http://dx.doi.org/10.1111/j.1755-263x.2012.00302.x

Milder, J.C., Buck, L.E., DeClerck, F. and Scherr, S.J., 2012. Landscape Approaches to Achieving Food Production, Natural Resource Conservation, and The Millennium Development Goals. In: *Integrating Ecology and Poverty Reduction*, edited by J.C. Ingram, F. De Clerck and C. Rumbaitis del Rio. New York: Springer, 77-108. http://dx.doi.org/10.1007/978-1-4419-0633-5_5

Mittal, S., 2012. *Modern ICT for Agricultural Development and Risk Management in Smallholder Agriculture in India.* CIMMYT, Socioeconomics Working Paper No. 3. Mexico: International Maize and Wheat Improvement Center.

Mittal, S., Gandhi, S. and Tripathi, G., 2010. *Socio-economic Impact of Mobile Phone on Indian Agriculture*. ICRIER Working Paper No. 246. New Dehli: International Council for Research on International Economic Relations. http://www.eaber.org/sites/default/files/documents/Socio-Economic Impact of Mobile Phones on Indian Agriculture.pdf

Myers, N., Mittermeier, R.A., Mittermeier, C.G., da Fonseca, G.A.B. and Kent, J., 2000. Biodiversity hotspots for conservation priorities. *Nature* 403: 853-858. http://dx.doi.org/10.1038/35002501

Nagendra, H. and Ostrom, E., 2012. Polycentric governance of forest resources. *International Journal of the Commons* 6: 104-133. http://www.thecommonsjournal.org/index.php/ijc/article/view/321/270

Naidoo, R., Balmford, A., Ferraro, P.J., Polasky, S., Ricketts, T.H. and Rouget, M., 2006. Integrating economic costs into conservation planning. *Trends in Ecology and Evolution* 21: 681-687. http://dx.doi.org/10.1016/j.tree.2006.10.003

NASA website. Soil Moisture Active Passive. http://smap.jpl.nasa.gov

Naveh, Z., 2001. Ten major premises for a holistic conception of multifunctional landscapes. *Landscape and Urban planning* 57(3): 269-284. http://dx.doi.org/10.1016/s0169-2046(01)00209-2

Newbold, T., Huson, L., Phillips, H., Hill, S., Contu, S., Lysenko, I., Blandon, A., Butchart, S., Booth, H., Day, J., de Palma, A., Harrison. M., Kirkpatrick, L., Pynegar, E., Robinson, A., Simpson, J., Mace, G., Scharlemann, J. and Purvis, A., 2014. A global model of the response of tropical and sub-tropical biodiversity to anthropogenic pressures. *Proceedings of the Royal Society B: Biological Sciences* 281: 20141371. http://dx.doi.org/10.1098/rspb.2014.1371

Novacek, M.J. and Cleland, E.E., 2001. The current biodiversity extinction event: Scenarios for mitigation and recovery. *PNAS USA* 98: 5466-5470. http://dx.doi.org/10.1073/pnas.091093698

OECD/FAO, 2011. *OECD-FAO Agricultural Outlook 2011-2020*. Paris: OECD Publishing. http://dx.doi.org/10.1787/agr_outlook-2011-en

Owen-Smith, N., 1988. *Megaherbivores; The Influence of Very Large Body Size on Ecology*. Cambridge: Cambridge University Press. http://dx.doi.org/10.1017/cbo9780511565441

Padoch, C. and Sunderland. T., 2014. Managing landscapes for food security and enhanced livelihoods: Building upon a wealth of local experience. *Unasylva* 241(64) http://www.fao.org/docrep/019/i3482e/i3482e01.pdf

Parmesan, C. and Yohe, G., 2003. A globally coherent fingerprint of climate change impacts across natural ecosystems. *Nature* 421: 37-42. http://dx.doi.org/10.1038/nature01286

Perfecto, I. and Vandermeer, J., 2010. The agroecological matrix as alternative to the land-sparing/agriculture intensification model. *Proceedings of the National Academy of Sciences of the USA* 107: 5786-5791. http://dx.doi.org/10.1073/pnas.0905455107

Perfecto, I. and Vandermeer, J. 2008. Biodiversity conservation in tropical agroecosystems: A new conservation paradigm. *Annals of the New York Academy of Sciences* 1134: 173-200. http://dx.doi.org/10.1196/annals.1439.011

Phalan, B., Onial, M., Balmford, A. and Green, R.E., 2011. Reconciling food production and biodiversity conservation: Land sharing and land sparing compared. *Science* 333: 1289-1291. http://dx.doi.org/10.1126/science.1208742

Phelps, J., Carrasco, R., Webb, E., Koh, L.P. and Pascual, U., 2013. Agricultural intensification escalates future conservation costs. *Proceedings of the National Academy of Sciences of the USA* 10(19): 7601-7606. http://dx.doi.org/10.1073/pnas.1220070110

Pimbert, M.P., 2009. *Towards Food Sovereignty. Reclaiming Autonomous Food Systems.* London: IIED. http://pubs.iied.org/pdfs/14585IIED.pdf

Pinstrup-Andersen, P., 2013. Can Agriculture Meet Future Nutrition Challenges? Special Debate Section. *European Journal of Development Research* 25(1): 5-12. http://dx.doi.org/10.1057/ejdr.2012.44

Pirard, R. and Treyer, S., 2010. *Agriculture and Deforestation: What Role Should REDD+ and Public Support Policies Play?* Working Papers No. 10/2010. IDDRI. http://www.iddri.org/publications/agriculture-and-deforestation-what-role-should-redd+-and-public-support-policies-play

Polasky, S., Nelson, E., Camm, J., Csuti, B., Fackler, P., Lonsdorf, E., Montgomery, C., White, D., Arthur, J., Garber-Yonts, B., Haight, R., Kagan, J., Starfield, A., and Tobalske, C., 2008. Where to put things? Spatial land management to sustain biodiversity and economic returns. *Biological Conservation* 141: 1505-1524. http://dx.doi.org/10.1016/j.biocon.2008.03.022

Powell, B., Ickowitz, A., Termote, C., Thilsted, S., Sunderland, T. and Herforth, A., [in press] Strategies to improve diets with wild and cultivated biodiversity from across the landscape. *Food Security*. http://dx.doi.org/10.1007/s12571-015-0466-5

Powell, B., Ouarghidi, A., Johns, T., Tattou, M.I. and Eyzaguirre, P., 2014. Wild leafy vegetable use and knowledge across multiple sites in Morocco: A case study for transmission of local knowledge? *Journal of Ethnobiology and Ethnomedicine* 10: 34. http://dx.doi.org/10.1186/1746-4269-10-34

Powell, B., Ickowitz, A., McMullin, S., Jamnadass, R., Padoch, C., Pinedo-Vasquez, M. and Sunderland, T., 2013. *The Role of Forests, Tress and Wild Biodiversity for Nutrition-sensitive Food Systems and Landscapes*. Expert background paper for the International Conference on Nutrition 2. Rome: FAO. http://www.fao.org/fileadmin/user_upload/agn/pdf/2pages_Powelletal.pdf

Pretty, J. and Barucha, Z.P., 2014. Sustainable intensification in agricultural systems. *Annals of Botany*. http://dx.doi.org/10.1093/aob/mcu205

Reed, J., Deakin, L. and Sunderland, T., 2015. Integrated Landscape Approaches: A systematic map of the evidence. *Environmental Evidence* 4: 2. http://www.environmentalevidencejournal.org/content/4/1/2/abstract

Rittel, H. W. and Webber, M.M., 1973. Planning Problems are Wicked. *Polity* 4: 155-169.

Rojas-Ruiz, J. and Diofasi, A., 2014. Upwardly Mobile: How Cell Phones are Improving Food and Nutrition Security. *The Hunger and Undernutrition Blog*. http://www.hunger-undernutrition.org/blog/2014/04/upwardly-mobile-how-cell-phones-are-improving-food-and-nutrition-security.html

Ruel, M., and Alderman, H., 2013. Nutrition-Sensitive Interventions and Programs: How Can They Help Accelerate Progress in Improving Maternal and Child Nutrition? *The Lancet* 382 (9891): 536-551. http://dx.doi.org/10.1016/s0140-6736(13)60843-0

Sayer, J., Margules, C., Boedhihartono, A.K., Dale, A. and Sunderland, T., 2014. Landscape approaches: What are the pre-conditions for success? *Sustainability Science*. http://dx.doi.org/10.1007/s11625-014-0281-5

Sayer, J., Sunderland, T., Ghazoul, J., Pfund, J.-L., Sheil, D., Meijaard, E., Venter, M., Boedhihartono, A.K., Day, M., Garcia, C., van Oosten, C. and Buck, L., 2013. The landscape approach: Ten principles to apply at the nexus of agriculture, conservation and other competing land-uses. *Proceedings of the National Academy of Sciences of the USA*. 110(21): 8345-8348. http://www.pnas.org/content/110/21/8349.full.pdf

Shackelford, G., Steward, P., German, R., Sait, S. and Benton, T., 2015. Conservation planning in agricultural landscapes: Hotspots of conflict between agriculture and nature. *Diversity and Distributions* [early on-line publication] http://dx.doi.org/10.1111/ddi.12291

Scherr, S. J. and McNeely, J.A., 2008. Biodiversity conservation and agricultural sustainability: Towards a new paradigm of "ecoagriculture" landscapes. *Philosophical Transactions of the Royal Society B: Biological Sciences* 363(1491): 477-494. http://dx.doi.org/10.1098/rstb.2007.2165

Scherr, S., Shames, S. and Friedman, R., 2013. *Defining Integrated Landscape Management for Policy Makers*. Washington, DC: Ecoagriculture Partners. http://ecoagriculture.org/wp-content/uploads/2015/08/DefiningILMforPolicyMakers.pdf

Scott, J., 1999. *Seeing Like a State: How Certain Schemes to Improve the Human Condition Have Failed*. New Haven, CT: Yale University Press.

Sears, R.R., and Pinedo-Vasquez, M., 2004. Axing the trees, growing the forest: Smallholder timber production on the Amazon varzea. In: *Working Forests in the American Tropics: Conservation through Sustainable Management?* Edited by D. Zarin, J. Alavalapatti, F.E. Putz and M.C. Schmink. New York: Columbia University Press.

Seto, K.C., Reenberg, A., Boone, C.G., Fragkias, M., Haase, D., Langanke, T., Marcotullio, P., Munroe, D.K., Olah, B. and Simon, D., 2012 Urban land teleconnections and sustainability. *Proceedings of the National Academy of Sciences of the USA* 109(2): 7687-7692. http://dx.doi.org/10.1073/pnas.1117622109

Steiner, A., 2011. Conservation and farming must learn to live together. *New Scientist*. http://dx.doi.org/10.1016/s0262-4079(11)60853-1

Stockholm Resilience Centre, 2014. *What is Resilience?* http://www.stockholmresilience.org/21/research/what-is-resilience.html

Strassburg, B.B.N., Latawiec, A.E., Barioni, L.G., Nobre, C.A., da Silva, V.P., Valentim, J.F., Vianna, M. and Assad, E.D. 2014. When enough should be enough: Improving the use of current agricultural lands could meet production demands and spare natural habitats in Brazil. *Global Environmental Change* 28: 84-97. http://dx.doi.org/10.1016/j.gloenvcha.2014.06.001

Stringer, R. 2000. *Food Security in Developing Countries*. Policy Discussion Paper No. 11. Adelaide, Australia: Centre for International Economic Studies, University of Adelaide. http://dx.doi.org/10.2139/ssrn.231211

Sunderland, T.C.H., Powell, B., Ickowitz, A., Foli, S., Pinedo-Vasquez, M., Nasi, R. and Padoch, C., 2013. *Food Security and Nutrition: The Role of Forests*. Discussion Paper. Bogor, Indonesia: CIFOR. http://www.cifor.org/online-library/browse/view-publication/publication/4103.html

Sunderland, T.C.H., 2011. Food security: Why is biodiversity important? *International Forestry Review* 13(3): 265-274. http://dx.doi.org/10.1505/146554811798293908

Sunderland, T.C.H., Ehringhaus, C. and Campbell, B.M., 2008. Conservation and development in tropical forest landscapes: A time to face the trade-offs? *Environmental Conservation* 34(4): 276-279. http://dx.doi.org/10.1017/s0376892908004438

Swinnen, J. and Squicciarini, P., 2012. Mixed messages on prices and food security. *Science* 335: 405-406. http://dx.doi.org/10.1126/science.1210806

Thrupp, L.A., 2000. Linking agricultural biodiversity and food security: The valuable role of agrobiodiversity for sustainable agriculture. *International Affairs* 76: 283-297. http://dx.doi.org/10.1111/1468-2346.00133

Tilman, D. and Clark, M., 2014. Global diets link environmental sustainability and human health. *Nature* 515: 518-522. http://dx.doi.org/10.1038/nature13959

Tilman, D., Balzer, C., Hill, J. and Befort, B., 2011. Global food demand and the sustainable intensification of agriculture. *Proceedings of the National Academy of Science* 108(50): 20260-20264. http://dx.doi.org/10.1073/pnas.1116437108

Tscharntke, T., Clough, Y., Wanger, T.C., Jackson, L., Motzke, I., Perfecto, I., Vandermeer, J. and Whitbread, A., 2012. Global food security, biodiversity conservation and the future of agricultural intensification. *Biological Conservation* 151: 53-59. http://dx.doi.org/10.1016/j.biocon.2012.01.068

Tscharntke, T., Klein, A.M., Kruess, A., Steffan-Dewenter, I. and Thies, C., 2005. Landscape perspectives on agricultural intensification and biodiversity—ecosystem service management. *Ecology Letters* 8: 857-874. http://dx.doi.org/10.1111/j.1461-0248.2005.00782.x

United Nations, 2011. Report submitted by the Special Rapporteur on the right to food, Olivier De Schutter. http://www.srfood.org/images/stories/pdf/officialreports/20110308_a-hrc-16-49_agroecology_en.pdf

Vera, F., 2000. *Grazing Ecology and Forest History*. West Berkshire, UK: CABI Publishing. http://dx.doi.org/10.1079/9780851994420.0000

Vera, F., 2009. Large-scale nature development—the Oostvaardersplassen. *British Wildlife* 20(5): 28-36. http://www.lhnet.org/assets/pdf/britishwildlifevera.pdf

Wallace, J.S., 2000. Increasing agricultural water use efficiency to meet future food production. *Agriculture, Ecosystems and Environment* 82: 105-119. http://dx.doi.org/10.1016/s0167-8809(00)00220-6

West, P., Igoe, J. and Brockington, D., 2006. Parks and Peoples: The social impact of protected areas. *Annual Review of Anthropology* 35: 251-277. http://dx.doi.org/10.1146/annurev.anthro.35.081705.123308

Wikramanayake, E.D., Hathurusinghe, H.S., Janaka, H. K., Jayasinghe, L.K.A., Fernando, R., Weerakoon, D.K. and Gunawardene, M.D., 2004. The human-elephant conflict in Sri Lanka: Lessons for mitigation, management and conservation from traditional land-use patterns. In: *Endangered Elephants: Past, Present, and Future: Symposium on Human-Elephant Relationships and Conflicts*, Sri Lanka, September 2003. Biodiversity and Elephant Conservation Trust.

Wright, H.L., Lake, I.R. and Dolman, P.M., 2012. Agriculture: A key element for conservation in the developing world. *Conservation Letters* 5: 11-19. http://dx.doi.org/10.1111/j.1755-263x.2011.00208.x

Young, H.S., McCauley, D.J., Dunbar, R.B. and Dirzo, R., 2010. Plant cause ecosystem nutrient depletion via the interruption of bird-derived spatial subsidies. *Proceedings of the National Academy of Sciences of the USA* 107: 2072-2077. http://dx.doi.org/10.1073/pnas.0914169107

6. Public Sector, Private Sector and Socio-cultural Response Options

Coordinating lead author: *Henry Neufeldt*
Lead authors: *Pablo Pacheco, Hemant R. Ojha, Sarah Ayeri Ogalleh, Jason Donovan and Lisa Fuchs*
Contributing authors: *Daniela Kleinschmit, Patti Kristjanson, Godwin Kowero, Vincent O. Oeba and Bronwen Powell*

This chapter focuses on political, economic and social response options at national to supranational scales to drivers of unsustainable management of forests and tree-based landscapes and their effects on food security and nutrition. Three different angles are considered: a) policy responses to enhance linkages between food security and forests with a focus on setting up the right institutional and governance structures and addressing the important issue of forest tenure reform; b) market-based response options that focus on global processes for supporting sustainable supply, and innovative corporate and multi-actor initiatives to support inclusive value chains of forest and tree products; and c) socio-cultural response options to enhance food security where the focus is on: changing urban demand; education to change behaviour and improve dietary choices; reducing inequalities and promoting gender-responsive interventions; and social mobilisation for food security.

For the public sector, a central governance issue is how and to what extent policy and regulatory frameworks help ensure that the most vulnerable groups, in particular the poorest members of society and women, have equitable access and rights to food security and nutrition from forests and tree-based systems. To this end, it is important to include relevant actors, from local communities to government departments, and initiate tenurial reform, devolution of decision-making to sub-national levels and a strengthening of institutional capacity at local levels.

For the private sector, sustainability standards supported by multi-stakeholder processes, complement policy frameworks and offer opportunities for change on the ground, particularly if these can include smallholders. In addition, pledges by corporate actors to zero deforestation and sustainable supply will likely have significant influence in shaping future production practices and business models if they include benefits for smallholder rural populations. Co-regulatory approaches that involve both public and private sector actors to achieve more inclusive food systems through innovations and greater valuation of local practices, management systems and knowledge, may in the future further enhance the governance of food systems.

At the level of social responses, education plays a pivotal role in empowering rural populations and has the potential to generate tangible benefits for households and communities in achieving food security and nutrition, sustainable forest and landscape management, and improved health. Targeting women and other vulnerable groups is particularly important to enable greater inclusiveness in decision-making and benefit sharing in forests and tree-based systems. Behavioural change that is often driven by social movements toward the consumption of food with lower environmental impact, particularly in growing urban areas, can have significant positive impacts on rural populations if the value chains necessary to meet the demand are set up to include smallholders and marginalised groups.

 http://dx.doi.org/10.11647/OBP.0085.06

6.1 Introduction

Food security[1] has become a matter of global concern, in particular since the last food price spikes in 2008 and 2010 (Beddington et al., 2012). FAO projections suggest that food production must rise by 60 percent by 2050 if a growing and increasingly more affluent population of over 9 billion is to be fed (Alexandratos and Bruinsma, 2012). At the same time, our environmental footprint which is leading to large scale soil *degradation*, *deforestation*, loss of *biodiversity*, crop varieties and *ecosystem services*, must be reduced as our current mode of operation is inconsistent with the planet's long-term provisioning capacity (IAASTD, 2009). All current trajectories imply that humanity is moving farther away from safe spaces (Rockström et al., 2009). *Climate change* is further compounding the challenge, for instance by undermining gains in crop productivity through increased floods and droughts, but also through longer-term shifts in temperature and rainfall distribution (IPCC, 2014; Nelson et al., 2010). This highlights the need for more sustainable agricultural methods for food production while knowledge gaps regarding trade-offs arising from competing economic and environmental goals, and key biological, biogeochemical and ecological processes involved in more sustainable food production systems remain (Tilman et al., 2002).

There is now growing recognition for the urgent need to act more decisively against these trends (Beddington et al., 2012). The revived attention to food security and *nutrition* is already leading to more sustained national and international efforts to increase food production and productivity, particularly in developing countries. Several countries, such as Mexico, India and South Africa have enshrined national food security in their constitutions and the United Nations Framework Convention on Climate Change is addressing issues related to the sustainable management of land through a number of frameworks such as REDD+ (Reducing Emissions from Deforestation and forest Degradation, conservation and sustainable management of forests, and enhancement of forest carbon stocks), the Ad-Hoc Durban Platform, and technology transfer (Campbell et al., 2014). At the same time there is a better understanding that food production must rise while enhancing climate *resilience* and lowering agriculture's greenhouse gas emissions' intensity (FAO, 2013a). To provide national and international support to this idea the global Alliance for Climate-Smart Agriculture, a voluntary association of national governments, intergovernmental organisations, development banks, private sector, civil society and research organisations, was launched at the United Nations Climate Summit in September 2014 (GACSA, 2014). It remains to be seen if climate-smart agriculture can deliver on the triple win, and this platform for action can indeed mobilise the financial, political, social and research resources necessary to significantly influence our current trajectories (Neufeldt et al., 2013).

1 All terms that are defined in the glossary (Appendix 1), appear for the first time in italics in a chapter.

The growing demand for food, fibre, energy and other products from the land often leads to market pressures for exploitation that can lead to *forest* destruction. Perverse incentives, for instance subsidies that have been set up to address the demand for cheap food without considering environmental externalities, may aggravate these pressures. These and other drivers affect the contribution of *forests and tree-based systems* to food security and nutrition as many drivers of deforestation and forest degradation lie outside the *landscapes* in which they manifest themselves. For example, agriculture is believed to be the driver of up to 80 percent of current deforestation, which often is resulting from national agricultural development policies intended to boost oil palm, cattle or soybean production (Kissinger et al., 2012). While an increase in agricultural productivity can potentially reduce the pressure on forests and other natural *ecosystems*, focusing on one outcome at the expense of others will often lead to sub-optimal results for overall sustainability (Sayer et al., 2013). Taking a landscape perspective that integrates across agriculture, forests and other land uses rather than considering different land use sectors in isolation is increasingly understood as crucial to long-term sustainability and food security and nutrition (Padoch and Sunderland, 2013; Scherr et al., 2012; and see Chapter 5).

This chapter focuses on political, economic and social response options to drivers at national to supranational scales that lead to *food insecurity* and negative nutrition outcomes due to the degradation of forests and tree-based systems. While they often support the sustainable management of land-based natural resources at landscape scales, many of these responses lie outside the land sectors altogether. The chapter addresses the topic from three different angles: a) policy responses to enhance linkages between food security and forests with a focus on setting up the right institutional and *governance* structures and addressing the important issue of forest *tenure* reform; b) market-based response options that focus on global processes for supporting sustainable supply, and innovative corporate and multi-actor initiatives to support inclusive value chains of forest and tree products; and c) socio-cultural response options to enhance food security where the focus is on: changing urban demand; education to change behaviour and improve dietary choices; reducing inequalities and promoting *gender*-responsive interventions; and social mobilisation for food security. Together, they cover a wide range of response options that are available to governmental, corporate and social agents. While these areas are presented separately here, they are strongly interlinked. For example, market forces require national rules and regulations to govern them in ways that are consistent with sustainable development goals but also social and cultural norms and values, which in turn shape the forms that institutions and governance structures take. Therefore topics from different sections within the chapter frequently touch upon each other. The chapter concludes by summarising the different lessons drawn from each of the three areas.

6.2 Governance Responses to Enhance Linkages between Forests and Tree-based Systems and Food Security and Nutrition

6.2.1 Introduction

Given the diverse roles of forests and tree-based systems for food security and nutrition (see Chapter 2), governance responses need to be understood in the widest sense. In this section, we discuss three governance response options: forest tenure reforms, decentralisation and market regulation. This is followed by a review of lessons on catalysing governance reform drawn broadly from the field of innovation studies and governance reform experiences (see Figure 6.1).

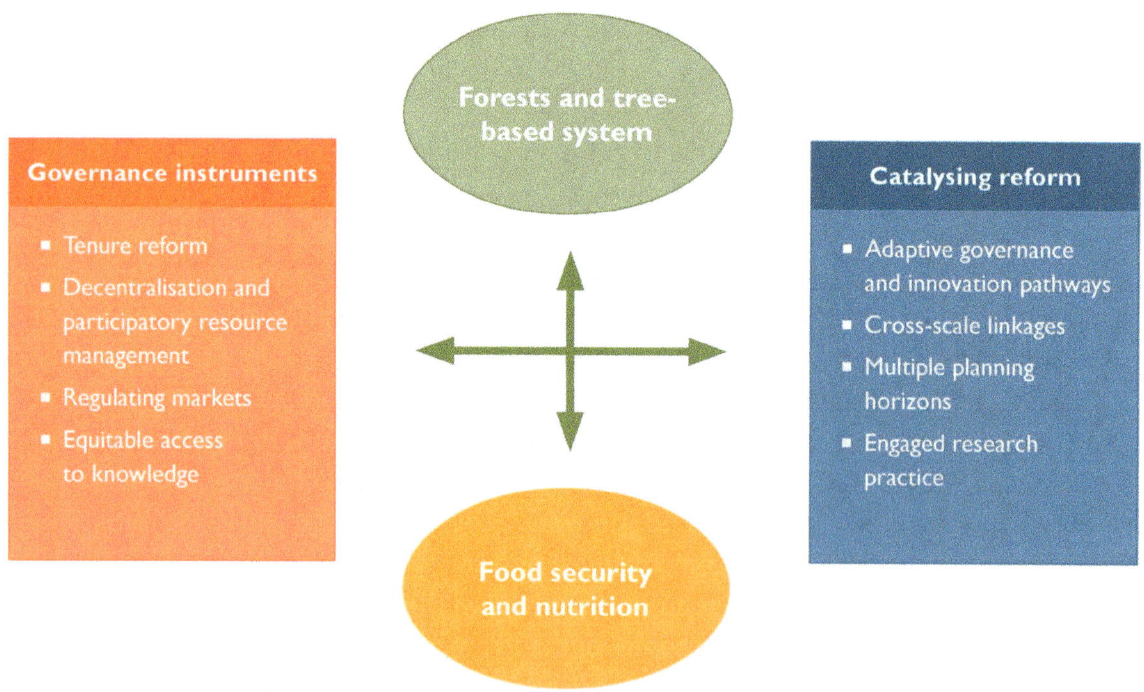

Fig. 6.1 Governance responses linking forests and tree-based systems with food security and nutrition

Forest governance has historically been a highly contested field, often very different from the agriculture sector which is governed in a more decentralised way (Colfer, 2013). While the forest sector is conventionally governed either for biodiversity conservation or timber production (Kennedy et al., 2001), a shifting emphasis on *non-timber forest products* and participatory conservation has given way to more food-friendly *forest management* practices (Belcher et al., 2005). A key manifestation of these shifts is rising concerns for food and nutritional security, highlighting the need for more proactive measures to reorient forest governance to address these *livelihood* priorities (Sunderland et al., 2013).

6.2.2 Reforms Related to Tenure and Resource Rights

Who controls forests significantly determines what they are managed for and who benefits from them, both outcomes having profound implications for food security. For example, globally an estimated 13 percent of all forests are officially protected for conservation values (FAO, 2010), but nearly half of these legally protected areas are heavily used (usually illegally) for agriculture and forest product extraction (Scherr et al., 2004). Forest tenure is also linked to land use policy that shapes how benefits can be optimised at the level of land use. Such land ownership issues have gained prominence in the past two decades, resonating Sen's argument that "entitlement" is more critical than production in reducing hunger at the global scale (Sen, 1999). Over the past three decades, forest tenure reforms have seen major strides globally, as manifested in increased recognition of the rights of local communities and/or local governments. Such reforms range from the titling of land parcels to indigenous communities to sharing timber revenues (Larson et al., 2010). At least five forms of tenurial reform can be identified: a) state-community collaborative or joint management, empowering communities to secure their livelihood interests, including meeting their food and nutrition needs, in forest management plans (Sundar, 2000; Bampton et al., 2007); b) formal community rights supported by concurrent reforms in state institutions (Bray and Merino-Pérez, 2002); c) national laws granting rights to communities for forest management, but still focusing narrowly on subsistence use, as in the case of Nepal (Sunam et al., 2013); d) pro-poor forest tenure reforms (leasehold forestry) allowing poor households to grow annual and perennial crops (Thoms et al., 2006); and e) institutional arrangements for enhancing the access of indigenous people to land resources (e.g. indigenous forest rights in Mexico (Toledo et al., 2003)).

However, tenure reforms are frequently insufficient to secure livelihood benefits, including food security. As Larson et al. (2010) argue, "new statutory rights do not automatically result in rights in practice, however, nor do local rights necessarily lead to improvements in livelihoods or forest condition". This can be seen for example in Nepal despite the country having granted clear legislative rights to communities (Ojha et al., 2014; Sunam et al., 2013). A wave of recentralisation is reported from cases elsewhere in the world (Ribot et al., 2006). Even in areas with significant formal devolution of forest authority, many communities have limited rights in practice (Larson et al., 2010).

Recognising the issue of intra-community equity, pro-poor tenure reforms have been initiated within community-based forest management, with explicit rights to grow food and cash crops in forest areas granted to the poorest members of society (Bhattarai et al., 2007). Nevertheless, even in countries promoting participatory or community-based forest management, many policy responses and forest laws intended to support smallholders and the poorest of society still contain restrictive provisions. As such they fail to authorise food cultivation or other means of enhancing food benefits from forests by smallholders, as is the case for example with India's Forest Right Act. Equally, in Nepal, where community forestry has come of age, with the establishment of successful local institutions, several forest ecosystem services are not yet defined in the tenure policy, thus creating a sense of tenurial insecurity (Sharma and Ojha, 2013).

Overall, forest tenure reform has emerged as an important governance response in relation to linking forest management with food security, despite varied and diverse experiences across the globe. The challenge is often that, even when tenure is redefined, a supportive institutional system – including capacity and political will – to translate the reform into practice remains absent. More attention is thus needed to how local innovations in resource access and control are linked effectively to an enabling policy and institutional environment.

6.2.3 Decentralisation and Community Participation in Forest Management

Another important forest governance response with profound implications on food security is decentralisation of authority (Colfer and Capistrano, 2004). While tenure reforms seek to transfer resource rights, decentralisation has involved much broader processes including institutional reform, power sharing and accountability. Indeed, the past three decades have seen a tidal wave of decentralisation in developing and transitional economies driven by diverse forces: loss of legitimacy of the centralised state (Bardhan, 2002), demands for a greater role of the market and for deregulation (Mohan, 1996), escalating concerns for poverty reduction (Crook, 2003), environmental conservation (Agrawal, 2001) and heightened demands for citizen participation in governance (Ribot, 2003; Ribot, 2007; Fung and Wright, 2001). Decentralisation endeavours entail a varying mix of activities aimed at empowering either communities of citizens, elected local government bodies, or other forms of quasi-political and administrative institutions, and involve political, administrative and fiscal measures depending on the context. Further, decentralisation responses are linked to a variety of ideas that have influenced governance practices such as deliberation (Dryzek, 2010), interactive governance (Kooiman et al., 2008), empowered participation (Fung and Wright, 2001), as well as representation and multi-stakeholder involvement (Hemmati, 2002; Vallejo and Hauselmann, 2004).

Although practices vary, the idea underlying decentralisation is to engage local actors in decision-making through locally-elected authorities to ensure accountable governance. In the forest sector, three forms of decentralisation have been found: transfer of rights to locally-elected government (democratic decentralisation), transfer of power to local offices of the national government ("deconcentration", as seen in Senegal for example (Ribot, 2006)), and transfer of rights to local communities (devolution, as seen in Nepal (Pokharel et al., 2008)).

However, there is no consensus that decentralisation leads to better outcomes in terms of local livelihood impacts and environmental sustainability. Questions of accountability and legitimacy in the exercise of power have become more critical than in the past (Lund, 2006; Mwangi and Wardell, 2012), challenging conventional forest governance authorities. Policies of decentralisation, while intended to "include" communities in multi-level participation, are often distorted in practice (Ribot et al., 2006; Head, 2007). Problems of participatory exclusion persist even in pro-poor environment and development programmes (Agarwal, 2001), and development practice continues to remain separated from politics (Hickey and Mohan, 2005).

Despite Ostrom's seminal work refuting Hardin's "tragedy of the commons" (Ostrom, 1990) and promoting the evolution of common property institutions, community action has tended to be smaller in scale, involving face-to-face channels of communication and coordination in practice. Such small-scale approaches are perceived to provide easier solutions for countries with weak and unaccountable governments (Blaikie, 2006). The limits of small-scale, community-based approaches to decentralisation have manifested themselves in various forms. Field experiences demonstrate that beyond a certain point, community effectiveness cannot improve unless supported by larger systems of local governance (Ojha, 2014). Communities may not necessarily be inclusive and accountable internally (Benjamin, 2008; Blaikie, 2006). In Nepal, the successful development of community forestry systems is meeting challenges of internal exclusion, market manipulation, elite domination and timber smuggling – issues that cannot be left entirely to community-level decision-makers (Mohan and Stokke, 2000). In India, the Forest Rights Act aimed to empower local forest-dependent people, but it was not effectively implemented due to inadequate local capacity (Springate-Baginski et al., 2013). In the Philippines, national policy entrusts local communities with rights to manage forests, but actual implementation has remained ineffective due to bureaucracy (Pulhin et al., 2007; Dahal and Capistrano, 2006).

Decentralisation responses should also be seen in the context of the growing consensus that forest governance has become a multi-level process (Mwangi and Wardell, 2012; Ojha, 2014). A multi-scale approach to governance may help to enhance food security by overcoming policy barriers and ensuring policy coherence from production to consumption, to eliminate poor policies (e.g. distorting trade) and to put in place positive ones (e.g. overcoming food waste) (Brooks, 2014).

Decentralisation and community participation remain important tools of forest governance reform to contribute to food security by: a) fostering local level decision-making and land use planning; b) resolving conflicts among different types of forest users; c) forging an effective interface between local knowledge and science; and d) enhancing the sustainability of innovation processes. These are linked to inclusive, accountable and transparent decision-making and equitable benefit sharing arrangements at the local level. In particular, evidence suggests that women's presence in decision-making has helped to improve forest conservation outcomes (Agarwal, 2009). Wider decentralised responses are important to address intra-community heterogeneity and equity issues, as people who depend on forests for their livelihoods and for food are also the ones who suffer the problems of inequity and injustice most (Mahanty et al., 2006). This is particularly critical in view of the findings that despite significant rights offered to local communities across the globe, inclusion remains elusive (Agrawal and Ribot, 1999). It is thus important to enhance decentralisation in such a way that it has equitable impacts on the community, while making sure that there is a concurrent reform in governance at multiple scales to support decentralisation.

6.2.4 Regulating Markets

Resource tenure reform is not enough to gain benefits from the market, as illustrated by the case of Mexican forestry wherein 80 percent of forest is owned by communities while they possess only five percent of total processing capacity (Scherr et al., 2004). As Scherr et al. (2004) argue, there is a need to "re-think the potential contributions of small-scale forest producers to commercial production and conservation goals, and ensure that a much higher share of the profits needs to go to local people rather than central governments or private interests". Stringent regulatory reforms are needed on the sale of forest products from production systems managed by local communities, local governments or state-community partnerships, such that significant incomes can reach poor rural households (Grieg-Gran et al., 2005). Even when communities are given rights to market forest products, the poor are not likely to benefit without regulatory arrangements to mandate community groups to spend the money for the benefit of the poor (Iversen et al., 2006). In Nepal's community forestry for example, a government directive requires community forestry groups to spend at least 35 percent of community revenue in projects directly related to the livelihoods of the poor (Nepal DoF, 2009).

Box 6.1 Regulatory constraints to community benefits from marketing of forest products

- Tenure rights of local and indigenous people remain weak
- Use rights frequently limit harvested products to those for subsistence use
- Decisions to harvest products commercially are limited by stringent requirements
- Inhibitive regulatory requirements for non-forest sectors (such as transport)
- Policies tend to favour industrial scale logging over community scale operations
- Onerous taxes and fees for forest products at various stages of the value chain
- Requirement for special permits to harvest forest products
- Requirement for special permits to transport goods to market
- Weak governance often leads to lack of transparency along the value chain
- Uncertainty about how to address legal issues including taxation
- Resistance by government officials to relinquish control over forests
- Government officials' demands for unofficial incentives to provide permits

Source: Gilmour, 2011.

As the markets for environmental goods and services increase globally, benefit sharing has become a crucial question for communities, generating a wide-range of policy and practical responses (Antinori and Bray, 2005; Pandit et al., 2009). At stake are the crucial questions of whether and how communities interact and negotiate with market players, and what agency they wield in these relationships of economic exchange and sometimes political contestations (Pacheco and Paudel, 2010). Studies in forest markets show that communities can benefit only when they have capabilities, necessary support services and suitable regulatory arrangements in their favour (Pacheco and Paudel, 2010). A recent review and analysis has identified several regulatory and governance constraints that prevent communities from benefitting from the marketing of forest products (Box 6.1).

Small-scale producers flourish primarily where there are fewer regulations and subsidies to large industry, and where there are secure forest rights (Scherr et al., 2004). For them, appropriate market regulations include: a) low regulatory costs of market entry (e.g. no registration fees, low cost management plans, no bribes required); b) no producer/consumer subsidies (and hence greater competitiveness for small-scale producers); c) a low-cost regulatory environment (e.g. few permits required); and d) secure local rights for forest products and environmental services (Scherr et al., 2004). These factors are critical in enhancing the commercial use of forest resources for local livelihoods even in situations where formal resource tenure exists.

In recent years, non-state market regulatory arrangements have also emerged such as certification mechanisms (Cashore, 2002; Durst et al., 2006) and payments for environmental services (PES) (also see Section 6.3). Certification chiefly consists in harnessing demand for sustainably-harvested products (including timber and food). In recent years, within both state and non-state frameworks, a policy agenda to support PES has emerged (Wunder et al., 2008) but challenges persist in relation to monitoring and verification. Concerns have also surfaced about ensuring the control of smallholders on genetic resources while encouraging the private sector to deliver improved seeds and technologies. The concern that markets do not favour the poor has inspired a series of instruments such as safeguards and free, prior, informed consent of indigenous peoples in commercial projects (Pimbert, 2012).

Interactions with the market are now inevitable for improving rural livelihoods, and the agenda of enhancing food security from forest cannot ignore this. It is also clear that "laissez faire" approaches to market development neither ensure equitable access, nor are likely to create sufficient conditions for the sustainable management of resources. Hence, regulated markets are an important governance response where a number of issues such as capacity, equity, marketability, fund management and planning, decision-making and others are directly regulated through different forms of governance instruments, while also ensuring ample entrepreneurial freedom and incentives.

6.2.5 Catalysing Governance Reform

In effecting the required change in policy and practice, quite often the issue is more about how a particular process of change emerges or is catalysed by some champions of change, triggered by particular sets of drivers, and sustained by an effective interplay between science and policy deliberations. Scholars and practitioners have considered catalysing changes in governance in the fields of forestry, environment and development in a variety of ways, as shown by a review of approaches (see Box 6.2).

Box 6.2 Approaches to catalyse changes in governance

- **Innovation system approach** (Hall, 2002) emphasises linking research, practice, policy together as essential for improving systems and practices. The approach emerged with industrial innovations in the West, followed by agricultural extension in the developing world, such as in India.
- **Social learning approaches** (Schusler et al., 2003) emphasise open communication, engagement and co-learning as necessary for changing systems. Examples can be found across both Western and developing countries.
- **Participatory research** (Pretty, 1995) holds that research can make a difference when conducted in close engagement with the subjects or local communities. This is applied widely in agriculture and natural resource management in the developing world.
- **Critical action research** (Ojha, 2013) emphasises the role of locally-engaged researchers in catalysing change by acting at different levels to generate alternative and critical knowledge. Examples can be found in developing countries – mainly in South America and South Asia.
- **Knowledge brokering** (Meyer, 2010) and using research as capacity building (Hall et al., 2003) are also emerging tools of innovation. Here, the role of new and hybrid actors as knowledge brokers is important in linking policy, practice and research groups. This idea has emerged in both the West and in the South.
- **Transformative innovation** "needs to give far greater recognition and power to grassroots innovation actors and processes, involving them within an inclusive, multi-scale innovation politics" (Leach et al., 2012).
- Participatory technology development (Schot, 2001) emphasises that technology and institutions co-evolve over time.
- **Adaptive collaborative approaches** (Colfer, 2005; Ojha and Hall, 2012) emphasise that management actions are experiments for learning and conflict management, as problem systems are always emergent and dynamic. Evidence is generated from across Asia, Africa and South America.

In more practical terms, we identify the following strategies to catalyse forest governance reforms so as to enhance food security outcomes:

1. **Reframing the facilitative regime**. Learning and innovation can be seen as the property of a system to self-organise and evolve, but this can be catalysed much faster and with much better results, in terms of fairness and equity

through appropriate mechanisms to inform, support, nurture, enable, capacitate and strengthen relevant groups and organisations involved in innovation development. Such facilitative arrangements are particularly crucial in the forestry sector in which governance has historically been organised around a "command and control" model. Even decentralised systems of forest governance face recentralisation threats (Ribot, 2006; Sunam et al., 2013). Options for forest governance to be more food-friendly include for example establishing demonstration landscape sites, creating incentives and offering subsidies for provisioning services.

2. **Conceptualising cross-scale linking.** Cross-scale linkages involve a diversity of transactions or interactions, and require building coalitions that go beyond technological innovations (Biggs and Smith, 2003). Such cross-scale forums can be harnessed for their potential to generate innovation, enable negotiations, manage conflicts etc. This means for example, inviting forest and food actors together along with farmers and public officials to open up informal spaces to explore and negotiate opportunities to enhance forest-food linkages. The Rights and Resources Initiative (RRI), a research-policy group working on forest rights globally, also promotes such policy fora at national, regional and global levels (RRI, 2014).
3. **Adopting multiple planning horizons.** Conceiving, facilitating and supporting multiple and overlapping planning processes, including forests, landscapes, and subnational and national levels, can help to facilitate change simultaneously at different temporal and spatial scales (Biggs and Smith, 2003). For example, a community forest user group can focus on a 3-5 year planning cycle, while district or landscape level plans traditionally require more time. Similarly, monitoring systems can also be tailored to the needs of decision-makers at different scales of governance, without overburdening local households and communities to gather information that is not immediately relevant to them.
4. **Cultivating local champions of change.** Many success stories in forest governance – and more generally in environment and development – around the world are linked to the strong role of a few passionately engaged agents of change. Identifying and nurturing such champions can be part of the broader strategy of reaching transformative change in forest governance for food security (World Bank, 2003).

6.3 Private Sector-driven Initiatives for Enhancing Governance in Food Systems

6.3.1 Introduction

The global *food system* is undergoing important changes which are associated with a reorganisation of value chains that are becoming increasingly global, the adoption of improved policy frameworks aimed at regulating food production and markets, and the emergence of private sector-driven initiatives to promote the adoption of sustainable practices in the supply of agricultural commodities (e.g. grains, palm oil, beef). The food system is characterised by increased vertical integration from the local to the global level and the development of large and complex value chains. The architecture underpinning the global food system is growing in complexity with an increasing role of the private sector, mainly large-scale corporate and transnational groups, in organising value chains, as well as multi-stakeholder processes with an active role of civil society groups influencing the governance of value chains at different levels (Margulis, 2013).

With growing foreign investment not only in processing but in upstream production, global value chains are speeding up concentration and technological change. This is stimulated by global traders and transnational companies that are seeking to enhance their economies of scale in both supply and marketing, which ultimately tends to displace local farmers who are integrated into more traditional food production systems (Page, 2013). Retailers and supermarkets also tend to impose higher quality standards to suppliers in order to meet more demanding consumption patterns, mainly in urban markets (Reardon et al., 2003). Nonetheless, in spite of growing interconnections between rural economies and urban markets, several market failures and asymmetries persist. These failures often lead to undesired environmental and social outcomes. Main environmental impacts relate to deforestation, soil erosion and water pollution, while social ones are the exclusion of traditional farmers from the value chains due to their more limited capacities to compete – in terms of costs and quality – in more demanding markets leading to unequal distribution of economic benefits from food markets (United Nations, 2014).

This section examines the main institutional initiatives aimed at building a more sustainable and inclusive food supply, with a focus on those driven by the private sector that is expanding its influence in the governance of value chains as part of new modes of governance that increasingly adopt the form of "hybrid" institutional arrangements in which state regulations and market-based mechanisms interact (Djama, 2011; Marsden et al., 2009). It includes an overview of the main challenges to achieve sustainability and inclusiveness in the global food systems, as well as the scale, scope and potential of different governance instruments that are in place. Some of the main global processes that are emerging in order to support sustainable supply are examined. An overview is included of corporate sector initiatives and commitments towards sustainable supply while reducing deforestation and protecting local people's rights, including "hybrid" models where both the public and private sectors collaborate to build sustainable value chains.

6.3.2 The Challenges of Sustainability and Inclusiveness in Food Supply

Our analysis focuses on the mechanisms, initiatives and processes, located at different levels and driven by non-state actors that are aimed at promoting sustainable food supply. Particular emphasis is given to large-scale investors and initiatives related to the corporate sector. Figure 6.2 shows the main mechanisms and processes in the institutional architecture that shape sustainable food supply, with a focus on agro-industrial value chains. This diagram is not exhaustive. It shows some of the main initiatives undertaken at the global level, supported by company associations such as codes of conduct, and multilateral organisations such as guidelines for responsible investment and land governance. It also refers to labelling and certification associated with specific production standards and third-party certification to promote the adoption by companies, on a voluntary basis, of standards for sustainable crop production, and other social safeguards. In addition, there are emerging corporate initiatives expressed in the form of commitments to adopt deforestation-free supply chains. Combinations of these different mechanisms with specific state public policy lead to so-called "hybrid mechanisms" which can take different forms in practice.

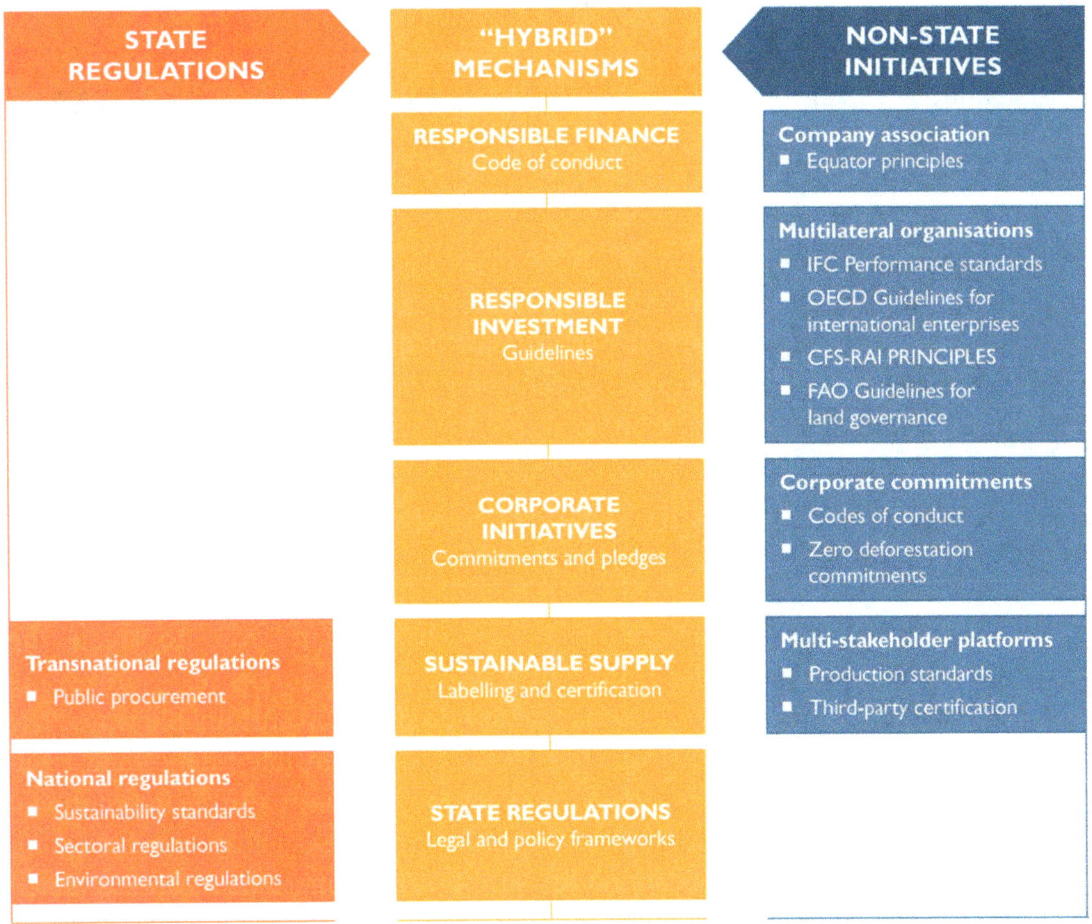

Fig. 6.2 State and non-state instruments shaping food systems

The most relevant instruments adopted to enhance private sector performance in food supply comprise responsible investment instruments, codes of conduct, sustainability standards, and certification and labelling (Candel, 2014). Some of these measures focus on finance and investments, while others concentrate on the production and trade realms (van Gelder and Kouwenhoven, 2011). Incentives for adoption are related to managing social risks by reducing civil society pressure and improving relations with communities, as well as reducing the implications that reputational risk can have on financial risks for company operations (Campbell, 2007). In other cases, adopting codes of conduct, sustainability standards and certification schemes may enable access to more discerning and specialised markets as well as optimising harvesting and production processes (Page, 2013). While these instruments provide little scope for public actors' participation in their design, the implementation phase provides more opportunities for achieving synergies and complementarities among different actors (Pacheco et al., 2011).

Processing shea butter (*Vittelaria paradoxa*), Labé, Guinea. Photo © Terry Sunderland

6.3.3 Global Initiatives to Support Sustainable Finance and Supply

Different mechanisms and instruments have emerged in order to create the conditions and mechanisms for upstream producers and downstream processors to target markets which demand goods that are produced in sustainable ways. The most relevant initiatives are instruments promoting responsible finance and large-scale investment developed by multilateral organisations (e.g. the International Finance Corporation – IFC) and multi-stakeholder processes such as roundtables for certification and traceability of commodity supply. These instruments are explained in more detail below, with a particular focus on what are the distinctive features that make them innovative. While the levels of adoption of these instruments are limited, they tend to expand slowly over time.

Initiatives and processes to promote responsible finance

The most important and well-known collective responsibility investment policy is the Equator Principles (EP), which is a financial industry benchmark for determining, assessing and managing social and environmental risk in project financing (Equator Principles, 2014). By 2014, 80 financial institutions had adopted the EP (Equator Principles, 2014). Signatories of the EP commit to adhere to the environmental and social guidelines (Performance Standards) of the IFC when providing project finance or related advisory services for projects costing USD 10 million or more. The Performance Standards of the IFC address a wide range of social and environmental risks, such as protection of human rights, protection and conservation of biodiversity, use and management of dangerous substances, impacts on affected communities and indigenous peoples, labour rights, pollution prevention and waste minimisation. There is important variation in the way in which these principles are implemented. In practice, the IFC's actual policy prescriptions tend to vary, such as happened in the palm oil sector (van Gelder and Kouwenhoven, 2011). Nonetheless, knowledge of the deficiencies in following IFC standards led the World Bank to revisit its strategy for engagement in the palm oil sector in 2011 (World Bank, 2011).

Initiatives shaping large-scale investments

Large-scale foreign direct investments (FDI) in land acquisition have expanded in developing countries, mainly in sub-Saharan Africa, with negative impacts on local livelihoods (Cotula et al., 2011; Deininger, 2011). While these investments can contribute to economic development in hosting countries, they also take advantage of relatively favourable economic and regulatory conditions, thus mechanisms are needed in order to maximise their benefits while minimising their adverse social and environmental impacts (Haberli and Smith, 2014). Several international initiatives, including statements of principles and voluntary codes of conduct, have emerged in response to the need for transparency, sustainability, involvement of local stakeholders and recognition of their interests, emphasising concerns about deforestation, domestic food security and rural development (Hallam, 2011).

Among these initiatives the three most relevant ones are all led by international organisations. There is a first draft for negotiation of the World Committee on Food Security (CFS) associated with the "Principles for Responsible Agricultural Investment" (CFS-RAI Principles) (CFS, 2014). Two other initiatives include the FAO-led "Voluntary Guidelines on the Responsible Governance of Tenure of Land, Fisheries and Forests in the Context of National Food Security" endorsed by the CFS in 2012 (CFS-FAO, 2012), 2), and the OECD's "Guidelines for Multinational Enterprises" (OECD, 2008). In addition to those mentioned above, some human rights commitments have been included in the voluntary guidelines on the right to food (FAO, 2004). These different principles and guidelines are all so-called soft law instruments and thus not legally binding. The first two address adverse effects associated with agri-FDI, while the others seek to prevent human rights violations by investors.

Certification schemes and voluntary sustainability standards

Voluntary sustainability standards provide assurance that a project, process or service conforms to a set of criteria defining good social and environmental practices. Specific schemes cover production (e.g. organic certification), the relations between chain actors (e.g. Fairtrade), and some cover both production and chain relations (e.g. the Forest Stewardship Council – FSC) (van Dam et al., 2008). In some cases, downstream supply chain actors (e.g. retailers, processors) impose standards on their suppliers as a way to inform consumers of their commitment to environmental and social objectives. In the case of Fairtrade, for example, buyers have sought collaborative business relationships with cooperatives in order to increase access to high-quality coffee (Raynolds, 2009). In other cases, upstream chain actors (e.g. cooperatives and privately-owned businesses) seek out certification on their own for the purposes of obtaining higher prices from the sale of their raw material. Non-governmental organisations (NGOs) and government agencies often support farmers in their efforts to obtain certification under the assumption that certification contributes to environmental and social goals. In some cases, certification can become a prerequisite for producers to access markets (Donovan, 2011). Box 6.3 examines a case of coffee certification in Nicaragua.

Box 6.3 Fairtrade coffee certification in Nicaragua

In Nicaragua, researchers have focused considerable energy on the issue of access to certified coffee markets and related implications for coffee supplies and rural development. In the late 1990s, governments and donors supported certification in Nicaragua in response to the dramatic and sustained reduction in price for coffee, with the expectation that access to markets for certified coffee would offer economic benefits over the short and long term (USAID, 2003; Varangis et al., 2003). Considerable investments were made by NGOs and donors to build local capacities for increasing coffee quality, obtaining certification and enhancing smallholder supply capacity. In many cases, cooperatives played a critical role in upgrading production capacities and in building relations with buyers and credit providers. However, in practice the results have been mixed. Arguments explaining these outcomes have centred on the persistence of low yields and relatively high labour requirements (Valkila, 2009; Barham et al., 2011; Beuchelt and Zeller, 2011), declining prices relative to conventional coffee (Weber, 2011) and the inability of smallholders to intensify coffee systems given their livelihood insecurities and rising production and household consumption costs (Mendez et al., 2010; Wilson, 2010; Donovan and Poole, 2014). There appears to be a growing consensus that smallholders in Nicaragua were probably too poor to be able to respond to the demands of buyers and certification systems.

An interesting case of voluntary sustainability standards is the so-called commodity roundtables, specifically the Roundtable for Sustainable Palm Oil (RSPO) and the Roundtable on Responsible Soy (RTRS), as mechanisms for certification of supply based on agreed sustainable production standards (Schouten and Glasbergen, 2011). These roundtables have been established to include different stakeholders along the value chain such as government, NGOs, industry, importers and exporters. The process has not been exempt from tensions, particularly in the context of building the legitimacy of the certification mechanism in the eyes of affected private sector actors, including industry and traders (WWF, 2010). In the case of RSPO, the adoption of sustainability standards by company members has been relatively slow, but it tends to expand over time. To date, 11.95 million tonnes of palm oil are certified, covering a total of 3.16 million hectares, and accounting for 18 percent of total global production (RSPO, 2014). Potts et al. (2014) provide a comprehensive review of the status and progress achieved by the implementation of a diverse set of voluntary sustainability standards – including FSC, RSPO, Fairtrade, Bonsucro, among several others.

6.3.4 Emerging Corporate Sustainability Initiatives

The corporate sector has a decisive role in shaping social and environmental outcomes associated with food supply in the context of current, globally-integrated food systems (Magdoff et al., 2000). Transnational corporations are central actors in the development of the global food system since they tend to dominate production and trade, and constitute important players in the processing, distribution and retail sectors (Clapp and Fuchs, 2009). Financial institutions and investors are also key actors in the food value chains. The most significant initiatives involving these actors are revised below.

Efforts towards the adoption of responsible financial investments

The adoption of policies and practices for due diligence, mandated and voluntary environmental and social risk management, and preferential green investments, such as those developed by IFC and a few commercial banks all contribute to the adoption of responsible finance. Responsible investment policies need to contain well-defined, verifiable criteria – preferably derived from internationally-recognised standards – that the financial institution can use to evaluate the proposed investment. Many financial institutions have set up their own benchmarks that meet these criteria, but there are also collective responsible investment policies undersigned by a group of financial institutions. Over the past ten years, more and more financial institutions have developed their own responsible investment policies for various sectors and sustainability issues (Perez, 2007). Leading this development was the World Bank Group. Its private-sector subsidiary, the IFC, has over two decades of experience with assessing investment proposals against its Performance Standards on Environmental and Social Sustainability, which define criteria on a broad range of social and environmental issues (IFC, 2012). Some public banks have followed this trend such as the Brazilian Development Bank (BNDES) that has also adopted similar guidelines (BNDES, 2014).

As the issues and sectors for which banks have developed policies or benchmarks vary, the number of banks that have developed benchmarks relevant to the agricultural sector is relatively more limited. A BankTrack study comprising 49 large international banks indicated that 16 institutions had developed a forestry policy and nine had developed an agricultural policy (van Gelder et al., 2010). This suggests that there is scope for adoption of policies by financial institutions that can lead to more responsible investments.

Voluntary commitments by the corporate sector for sustainable supply

Many corporate groups involved in supply, processing and retailing are adopting commitments, some with well-defined targets, for achieving their projected production goals with lower negative social and environmental impacts. On the supply side, these are made by corporate groups developing their operations in landscapes where there is a high risk of environmental impacts (e.g. peatlands in Indonesia, tropical forests in Brazil). On the demand side, these commitments are made by consumer goods companies that are well positioned in the markets responding to social pressure on corporate social and environmental performance (Baron et al., 2009). For example, the Consumer Goods Forum (CGF) has adopted a "Global Social Compliance Programme" aimed at improving social and environmental sustainability in the supply chains by harmonising existing efforts (GSCP, 2014).

Specifically, Wilmar International Ltd., the largest palm-oil trader, committed in late 2013 to ensure that its plantations and suppliers protect certain forests and abstain from using fire to clear land, and also banned development on high-carbon-stock landscapes including peatlands (Wilmar International, 2014). Unilever, the second-largest manufacturer of consumer goods, also committed in late 2013 to purchasing all palm oil from sustainable sources by 2015, and that all palm oil would be certified and come from traceable sources by 2020 (Unilever, 2014). Some end-user companies of palm oil, notably Starbucks agreed in early 2013 to source 100 percent of their palm oil from certified sustainable suppliers by 2015, which was a response to a shareholder resolution filed by an environmental mutual fund (Starbucks, 2014). Additional commitments to source sustainable palm oil have been made by some other consumer goods companies such as McDonalds, Walmart and Nestlé.

On the supply and processing side, in part as a result of the commitments made by consumer goods companies, five of the world's largest palm oil companies (Asian Agri, IOI Corporation Berhad, Kuala Lumpur Kepong Berhad, Musim Mas Group and Sime Darby Plantation) together with Cargill, subscribed to the "Oil Palm Manifesto" in July 2014 (HCSS, 2014). This manifesto aims to achieve three specific objectives: 1) build traceable and transparent supply chains, 2) implement the conservation of high carbon stock (HCS) forests and the protection of peat areas regardless of depth, and 3)

Washing vegetables in the river in Ouagadougou, Burkina Faso. Photo © Terry Sunderland

increase benefit sharing while ensuring a positive social impact on people and communities. Furthermore, in September 2014, three companies (Cargill, Golden Agri-Resources GAR and Wilmar) subscribed to the "Indonesian Palm Oil Pledge" known also as the KADIN pledge. These companies commit to achieving the following: 1) adopt and promote sustainable production practices based on acceptable methods of classifying HCS forests, and sustainable supply chain management and processing, 2) work with the Indonesian Chamber of Commerce (KADIN) to engage the Government of Indonesia to encourage development of policy frameworks that promote the implementation of the pledge, 3) expand the social benefits from palm oil production, and 4) improve the competitiveness of Indonesian palm oil.

During the UN Climate Summit in New York City, held in September 2014, about 150 different governments, businesses and NGOs joined forces to announce "The New York Declaration on Forests" under which these different groups committed to cutting forest loss in half by 2020, and ending it by 2030. This declaration also calls for eliminating forest loss from agricultural commodity supply chains by 2020 and restoring at least 350 million hectares of degraded forestlands by 2030. This declaration was signed by some of the major players of the palm oil industry, including palm oil traders (APP, Cargill, GAR and Wilmar) and consumer companies (Kellogg's, General Mills, Nestlé and Unilever), which complements their no-deforestation commitments for palm oil sourcing.

6.3.5 "Hybrid" Models for Sustainable and Inclusive Supply

Several initiatives have emerged in both consumer and producer countries to promote trade of commodities in national and international markets that originate from more sustainable sources, or that place reduced impacts on local people, which somehow adopt the form of "hybrid" models since they tend to articulate public regulations with private standards in different ways. Two such experiences are described here. While these experiences are still in their infancy, they may have potential to develop into more consolidated initiatives that could lead to improved outcomes in both inclusiveness and sustainability, which continues to be an elusive goal in many cases.

Linking international standards and national regulations

In Indonesia, a mandatory government-led standard, labelled "Indonesian Sustainable Palm Oil" (ISPO) has been issued for the production of palm oil in addition to the RSPO, a relatively consolidated voluntary, market-based certification system (described above). The main rationale for setting up the ISPO is that only large-scale palm oil companies are members of RSPO, whereas medium-scale companies account for a large share of the palm oil sector (Indonesia Ministry of Agriculture, 2012). Furthermore, only a small portion of certified palm oil suppliers have established markets in developed countries, which tend to demand sustainably-produced palm oil, while a significant portion of demand lies in less demanding markets, such as China and India that received 38 percent of total crude palm oil exports in 2013. When including other developing countries this segment of the market makes up 70 percent of the total (COMTRADE, 2014). The Indonesian government has acted carefully with regard to the RSPO (Rayda, 2012), and after initially joining the organisation, the Indonesian Palm Oil Association (GAPKI) withdrew its membership in 2011, stating its intention to fully support the recently announced ISPO. The latter is conceived as a way to improve the adoption of standards in palm oil in Indonesia by complying with existing regulations, which have been hard to enforce. This may lead to reducing existing gaps in the adoption of good practices in palm oil production.

Linking sustainable value chains to jurisdictional approaches

The "Green Municipalities" is an initiative emerging in the state of Pará in the Brazilian Amazon, associated with public-private arrangements to improve sustainable supply and enhance landscape management, particularly in the cattle beef value chains. This initiative has developed as part of a broader, relatively complex, institutional arrangement involving the public sector to halt deforestation and promote forest regeneration. The arrangements are enforced by the federal government with the assistance of environmental agencies (IBAMA) and contributions of municipal governments. In addition, banks have been mandated by state regulations to limit commercial loans to farmers that are not able to comply with environmental regulations, and to voluntarily subscribe to a Rural Environmental Cadastre (Whately and Campanili, 2014). This institutional scheme now plays a crucial role in voluntary actions from supermarkets that are mainly located in the highly populated urban centres of southern Brazil, which have thus banned beef originating from illegally deforested lands in the Amazon (Pacheco and Poccard-Chapuis, 2012). Furthermore, as a response to changes in demand, farmers have engaged in improving practices of herd and pasture management by implementing standards developed by the state research agency (EMBRAPA) as a way to intensify cattle production and reduce pressures on conversion of forests. However, these practices are mainly adopted by medium- and large-scale cattle beef systems, while smallholders with more diversified farming systems face the risk of being left behind with limited options.

Table 6.1 Instruments with potential to contribute to sustainable commodity supply.
Adapted from **Pacheco et al. (2011).**

Instrument	Market Mechanism	Influence on	Incentive	Stakeholders	Role of actors in instrument design and/or implementation	
					Public	Private
Responsible investment instruments	Anticipated benefits associated with a good corporate image encourages financiers to invest only in those corporate actors whose practices are considered sustainable or low-risk	Finance and investments	- Reduced civil society pressure - Reduced risk of investments - Improved public image	- Financiers (design and application of instrument) - Corporate actors (complying with criteria so as to access finance) - Civil society (lobbying for responsible investment practices)	- None	- Establishing criteria - Application of instrument - Verification of compliance - Implementation of actions to meet criteria
Criteria self-developed by corporations/ codes of conduct	Anticipated benefits associated with a good corporate image encourages practices deemed to be environmentally and/or socially beneficial or benign by key stakeholders	Production and processing	- Improved relations with local communities - Reduced civil society pressure - Reduced risk to operations - Marketing tool	- Affected communities - Civil society in developed and developing countries - Government in developing countries - Shareholders	- None	- Establishing and implementing policies - Voluntary reporting to shareholders
Sustainability standards	Anticipated benefits associated with a good corporate image encourages portfolio of land uses to shift to align with standards	Mostly production	- Access to certain markets - Optimisation of production processes	- All actors along the supply chain	- Definition of standards	- Negotiating standards - Sometimes driven by actors further downstream
Certification and labelling	Strict standards for accessing benefits encourages portfolio of land uses to shift to provide uses or services required by the instrument	Production and markets	- Access to niche markets - Price premium	- All actors along the supply chain - Consumers	- Identification of effective criteria - Setting targets	- Independent verification

6.4 Socio-cultural Response Options

6.4.1 Introduction

Social and socio-cultural response options to enhance food security by influencing forest and tree-based systems are manifold. Addressing social drivers requires a more nuanced approach as these drivers are strongly influenced by cultural differences and, due to their frequently informal nature, are not always easy to grasp, to categorise and to quantify. Macro-scale responses addressed in this section encompass opportunities and challenges of changing urban demand, nutrition education and behavioural changes, reducing inequalities and promoting gender-responsive interventions and policies, as well as social mobilisation for sustainable food security.

A homestead lunch, Cat Ba, Vietnam. Everything on the table is sourced from the forest, farm or nearby water courses. Photo © Terry Sunderland

6.4.2 Changing Urban Demand

Cities are centres of creativity, power and wealth. Understanding the dynamics, growth and organisation of cities, using a sustainability lens, is important for food security and environmental sustainability (Bettencourt, 2013; Bettencourt et al., 2007). With more than half of the world's population currently living in cities with a continued rise predicted, securing adequate food supply for city dwellers will be even more crucial than it is today. Growing urbanisation often calls for increased food production in surrounding rural areas, but also raises pressures to convert agricultural land in the wake of urban development. In order to address the complexity of divergent priorities, there is a need for planning alternatives, policies and incentives that aim to reconcile growth, management, food security and sustainable diets, and the enhancement of agriculture (Forster and Escudero, 2014).

Urban consumers are increasingly aware of the fact that modern agriculture can have negative environmental externalities, for example through the use of agricultural biocides and synthetic fertilisers and the concentration on few crop varieties (Badgley and Perfecto, 2007). These have led to eutrophication of aquifers, soil degradation, loss of biodiversity or reduction of genetic diversity, among others, and highlight the need for more sustainable agricultural methods for food production. However, knowledge gaps regarding trade-offs arising from competing economic and environmental goals, and key biological, biogeochemical and ecological processes involved in more sustainable food production systems remain (Tilman et al., 2002).

These negative environmental externalities have prompted environmentalists and others to support more sustainable methods of food production and to advocate for a shift in dietary choices (Halberg et al., 2005). In Western countries, advocacy for organic agriculture and vegetarianism are two of the most prominent responses to such criticism.

Organic agriculture has gained recognition as having an important potential to help feed the world and restore biodiversity and landscape richness at the same time (Badgley and Perfecto, 2007; Fuller et al., 2005). Health, ecology, care and fairness principles form the core of the organic agriculture vision, all working towards supporting the health and integrity of ecological systems and cycles in a sustainable manner (IFOAM, 2005). Similarly, urban (relatively wealthy) consumers actively seek and pay more for food labelled or certified as "environmentally-friendly" or "pesticide free", characteristics that attract them to organic foods (Dimitri and Greene, 2002). Today, organic farming is practised in 162 countries, and organic food and drink sales worldwide reached almost USD 63 billion in 2011 (Soil Association, 2014). This has been achieved through a change in perceptions of natural food from being a prerogative for alternative lifestyles to what is consensually understood as being healthy. Whereas in 1997 organically produced food was primarily sold in natural food stores, almost half of it was purchased in chain supermarkets in 2008. At the same time, the number of farmers' markets, where organic farmers sell their products directly to end-users quadrupled from less than 2,000 in 1994 to more than 8,000 in 2013 (Alkon, 2014).

The second shift in urban demand that we address here concerns meat-free food choices. While full or partial vegetarianism is part of the history and culture of many people in the world, most vegetarians in Western societies are not life-long practitioners but converts (Beardsworth and Keil, 1992). European and North American campaigns promoting a shift from meat consumption to vegetarian diets are nowadays often associated with a desire to reduce the ecological footprint of food production. For example, feed grains given to animals for human consumption in urban areas contribute largely to the overall urban footprint, and the corresponding intensive livestock systems are often blamed for forest loss, reduced water quality and diseases (Forster and Escudero, 2014). Furthermore, through feed, enteric fermentation manure management and post-production processes, livestock production is among the largest emitters of greenhouse gases in agriculture (Smith et al., 2014). Modelling results suggest that dietary shifts to lower meat consumption and a healthy diet could result in agricultural emission reductions of 34 to 64 percent by 2050 (Stehfest et al., 2009).

Both vegetarianism and organic food production are conducive to improving the livelihoods of rural populations. Many of the products that can supplement basic staple foods and substitute meat are grown on trees (Jamnadass et al., 2013) providing significant opportunities for *agroforestry* systems if appropriate value chains are established (see Section 6.3). The lower demand for land to feed livestock can also contribute to reduced deforestation (Smith et al., 2014). However, in spite of an increase in vegetarianism and the perception that sustainable forms of agricultural production are needed, and despite widespread knowledge of the adverse effects of excessive meat consumption (Bender, 1992), meat, dairy and poultry consumption continues to rise with growing affluence (Alexandratos and Bruinsma, 2012), in turn aggravating the environmental impacts associated with livestock production (Westhoek et al., 2014). Asia, in particular, is among the regions with the highest increase in meat consumption and requires heavy investments in education to change behaviours and improve dietary choices.

6.4.3 Behaviour Change and Education to Improve Dietary Choices

Dietary choices depend on access, availability and affordability (Dibsdall et al., 2002) but also socio-cultural and environmental factors (Sobal et al., 2014; Kuhnlein and Receveur, 1996; Fischler, 1988). Even small changes in access or prices can have significant impacts on diets (Glanz et al., 2005; Story et al., 2008). While reformed political and market frameworks can enhance access to food and stabilise its prices, *malnutrition* also needs to be addressed on an interpersonal level. Better knowledge about healthy diets through nutrition education can therefore play an important role (FAO, 1997; Jamnadass et al., 2011). A revalorisation of knowledge on the origins and properties of food items and effects of these food items on human health can potentially lead to increased interest in *traditional ecological knowledge* about forest and tree foods. Such interest can be an important counter-movement to the rapidly progressing loss of traditional and indigenous knowledge, widely attributed to social change and modernisation (Keller et al., 2006; Lykke et al., 2002; Ogoye-Ndegwa and Aagaard-Hansen, 2003; also see Chapter 3).

Nutrition and health education in its broadest sense has three components: providing information through communication strategies (e.g. information campaigns, dietary advice in health service settings), providing skills that enable consumers to act on the information provided (e.g. meat preparation, food preservation) and providing an enabling food environment (e.g. marketing to children, making different foods available) (Hawkes, 2013). Nutrition and health education can translate greater food availability at the household level into healthier diets by targeting women, men and children in the households with tailored messages about improved food choices (McCullough et al., 2004), for example, through optimal feeding and care practices for infants, young children and women of reproductive age. In terms of agroforestry and food security, there is, for example, a need to understand how best to educate consumers on the benefits of eating fruits, many of which are tree products (Jamnadass et al., 2013).

There are numerous examples documenting successful nutrition education campaigns. A particular nutrition education programme in Thailand's Kanchanaburi

Province, for instance, considers children as effective agents of change in societies and therefore, teaching them about agriculture and nutrition is seen to be a wise investment (Jamnadass et al., 2011; Sherman, 2003). In this example, education for sustainable development involved teaching school children how to sustainably harvest forest foods, how to plant, cultivate and harvest their local traditional agricultural crops in village areas, and prepare healthy meals for their families. By targeting children from a young age, foundations for behavioural changes in entire communities are laid (Burlingame and Dernini, 2012). Numerous forest conservation programmes also attempt to introduce agroforestry practices and integrate dietary concerns in their environmental conservation efforts. In the Kenyan Mau Forest, for instance, a number of conservation projects promote the establishment of school gardens, planting of fruit trees and integrated agroforestry practices alongside conventional tree planting drives (NECOFA, 2013). Nutrition education has also gained momentum in education systems of developed countries, mainly through practical application of knowledge in school gardens, cooking classes and sometimes specialised students' clubs (Alkon, 2014). In the USA and other developed countries, apart from rising awareness of malnutrition and hunger, nutrition education is popularly used to fight against obesity – hunger and obesity being two sides of the same coin (Patel and McMichael, 2009; Zerbe, 2010).

Despite important positive contributions, the effectiveness of education-based initiatives for sustainability and food security needs to be further explored to ensure that benefits have long-term impacts and do not promote solutions that have negative side effects such as nutritional imbalances or unsustainable practices.

6.4.4 Reducing Inequalities and Promoting Gender-responsive Interventions and Policies

There is evidence around the world that greater involvement of women in forest management improves the condition and sustainability of forests (GEF, 2015). This may be because women are responsible for a majority of household chores that are related to forest products, especially firewood collection, and women are thus more aware of the effects of deteriorating forest conditions. Consequently, many women engage in conservation of forest resources or in environmentally-friendly practices in order to avoid or mitigate future hardship (Agarwal, 1997; Acharya and Gentle, 2006). Studies show, for instance, that in parts of Asia where rural migration to urban centres is widespread, women tend to plant more trees on their lands than men as the intensity of agricultural management declines in response to rising incomes through remittances (Agrawal et al., 2013). In other places, the supposed positive influence of women on the landscape is further encouraged by targeted legislation and programmes, with vast effects at the local level. For example, with support of the Forest Department of India, village forest committees (VFCs) were formed to ensure equal participation of men and women in forest activities. The deliberate inclusion of women in the VFCs provided incentives for women to manage, protect and conserve their forests sustainably (WWF, 2012).

However, women's productive potential in natural resource management is often constrained by socio-cultural factors, particularly in rural areas in developing countries. Often, women's participation in natural resource management is also restricted by land ownership and land tenure rights and agreements, as well as distribution of decision-making powers that favour men (see Chapter 3). As a result of complicated tenure arrangements, women often have to negotiate for rights to land and associated resources. In some areas, women thus engage in collective action to influence such decisions, for example by entering sharecropping arrangements, buying or accessing land collectively, often with the help of NGOs (FAO, 2002; Agarwal, 2009). Many of these NGOs use multilevel approaches by, for instance, including training of leadership with best practice training in technology and innovation that are tailored to women's tasks and needs (USDS, 2011). The potential impact of more efficient land use by women is especially interesting due to the fact that subsistence crops are often "women crops", under the primary responsibility of women (Das and Laub, 2005). Taking climate change into account, the frequent exclusion of women from technology and adaptive innovation is particularly counter-productive (Terry, 2009). Access and better use of land are particularly important in light of study results that show that food security might still be compromised even in food secure households, often to the disadvantage of women and children (Hughes, 2010). Women are often more inclined to reduce the number of meals they take in a day or the quantity and/or quality of food per meal for the benefit of other household members, thereby exposing themselves to enormous health risk (Nelson and Stathers, 2009). Altogether, if women had the same access to productive resources as men, the FAO estimates that women could increase the yields on their farms by 20-30 percent, leading to a total increase of agricultural output of 2.5 percent in developing countries and thus reducing the number of hungry people by 12-17 percent (FAO, 2011).

Woman pounding cassava for fufu, Senegal. Photo © Terry Sunderland

While this outlook is promising, increasing women's contribution to enhanced food security and nutrition at large scale will require a clear commitment for further inclusion of women in decision-making processes concerning land use and land use planning. Although national legislation granting equal access to productive resources is essential for social equity, socio-cultural attitudes and practices that have evolved over decades and centuries are not easily changed and often resist adaptation to new laws and transformation altogether. This is particularly true in rural areas where local norms are often enforced by older and respected, and thus more powerful (frequently male) community members (FAO, 2002). Because saliency, legitimacy and relevance of more equitable laws and resource management rules are critical for community buy-in and effective implementation, some organisations implement participatory awareness campaigns, characterised by local community involvement in design and implementation (Clark et al., 2011). Communication approaches that widely disseminate information and education campaigns can also help. At the same time, an in-depth understanding is needed of the complex relationships between on the one hand, land tenure, use and control and on the other, their influence on food security, forests and tree-based systems, in order to promote gender inclusiveness in decision-making and sustainable benefit sharing. Such an understanding is also a prerequisite to external contribution to more inclusive land tenure policy frameworks (see Section 6.2).

6.4.5 Social Mobilisation for Food Security

Social mobilisation seeks to foster change through awareness creation by engaging a wide range of actors in interrelated and complementary efforts (UNICEF, 2015; FAO, 2003). Engagement processes (historically face-to-face) allow stakeholders to reflect on and understand their situation, organise themselves and initiate action. Traditionally, social mobilisation is an endogenous process through which like-minded persons attempt to exchange ideas, define common purpose and strengthen their voices in order to be heard by their fellow citizens and authorities alike. In addition, social mobilisation has also been used as a tool to increase legitimacy and sustainability of externally encouraged activities in the context of community development.

In the USA and Western Europe popular interest in food and agriculture has skyrocketed in recent years and with it a multiplicity of social "food movements". Some even speak of a "food revolution" (Nestle, 2009). Most of these movements critically assess modern food production technology and the entire heavily subsidised, chemically intensive and cheap labour dependent, industrial, corporate food system. Many movements advocate for and promote more humane and environmentally friendly ways of producing, selling and consuming. The massive disposing of unwanted or wasted food is yet another facet of the same problem (Zerbe, 2010). The FAO's 2013 "Food Wastage Footprint" report for instance specifies that 1.3 billion tonnes of edible food parts from 1.4 billion hectares of land (28 percent of the world's agricultural area) are wasted, leading to a direct economic loss of an estimated USD 750 billion per year (FAO, 2013b). In the same vein, since the inclusion of almost all agricultural products in trade liberalisation in 1994, under the auspices of the World Trade Organization (WTO), developing countries have been further encouraged to reorient their economies towards export to the North and to neglect the production of food for the domestic market. Also, a lack of democracy in basic political institutions has favoured corporate interests of the food industry (Marshall, 2013). This critique has largely informed the Right to Food, food justice and *food sovereignty* movements (Hughes 2010), which have their root in the fair trade movement (Zerbe, 2010). Globally, the perhaps largest social mobilisation concerning food security and environmental sustainability concerns the regulation and restriction of genetically modified foods. While a majority of the main food crops in the US continue to be genetically modified, activists' mobilisation has been very successful in Europe (Alkon, 2014).

Due to its decentralised, often community-centred and sometimes sporadic nature, a characterisation of contemporary food movements is difficult. For the United States, Nestle suggests a separation between movements that address the production side (such as the Slow Food, the farm-animal welfare, the organic foods, or the locally grown food movements) and those that address the consumption side (anti-marketing-foods-to-kids, school food, anti-trans-fat, or the calorie labelling movement), while others unite both purposes (community food security, better farm bill movement) (Nestle, 2009).

Social mobilisation for environmental sustainability and food security is also witnessed in the developing world. For example, the tree planting programmes of the Green Belt Movement (GBM) provide incentives for Kenyans to successfully improve their environments, doubling as a sustainable land management approach with a reliable source of income for women (Shaw, 2011). By early 2015, the GBM published figures indicating that it had planted over 51 million trees in Kenya through its extensive network of over 50,000 female members. Using a multidisciplinary approach, the GBM integrates the promotion of environmental conservation, with women's and girls' empowerment, and a focus on democracy and sustainable livelihoods (GBM, 2015). Starting as a small, local project, promoting gender inclusiveness and conservation, the GBM has become a nationally recognised and internationally acclaimed movement. Its founder, the late Wangari Maathai, received the Nobel Peace Prize in 2004 for her extraordinary contribution to awareness creation, environmental conservation and social transformation.

Another interesting example is the food sovereignty movement "La Via Campesina" (the Peasant Way), an international organisation and platform that assembles peasant farmers, small-scale farmers and activist groups from 73 countries in Africa, Asia, Europe and the Americas. In existence since 1993, La Via Campesina has gained international recognition as a main actor in food and agricultural debates and is heard by institutions such as the FAO and the UN Human Rights Council (Hughes, 2010; La Via Campesina, 2011).

Social mobilisation is also a common tool in rural development and poverty alleviation programmes to strengthen participation of the rural poor in local decision-making. According to UN-HABITAT, communities and stakeholders that take ownership of their own problems, such as conflicts and environmental degradation, take better informed decisions, are able to reach more sustainable solutions and achieve results faster, while fostering their solidarity and capacity to undertake development initiatives (UN-HABITAT, 2014). Collective action has been successful in many regards, for example improved access to social and production services, greater efficiency in the use of locally-available resources, and enhanced asset building by the poorest of the poor (FAO, 2003; NRSP, 2005).

Garlic sellers, Kouana village, Guinea. Photo © Terry Sunderland

There are various examples in the world where social mobilisation has worked in favour of food security and forest conservation. For example, through a UNDP initiative in Tajikistan on agroforestry, communities around the Gissar Mountains were mobilised to plant salt-tolerant trees and other grafted new tree species to alleviate the impacts of overutilisation and degradation of natural resources, civil war and consequent socio-economic hardships. As a result, pressure on forest resources was reduced and household incomes increased, from the trees themselves, as well as from the establishment of tree nurseries. Local farmers also experimented further, using grafting technology to cultivate fruit trees (UNDP, 2015).

Urban dynamics, behavioural change, tackling inequalities and social mobilisation all represent different options to address the drivers that affect forests and tree-based systems, and thus their impacts on food security and nutrition.

6.5 Conclusions

This chapter looked at options to improve food security and nutrition in forests and tree-based systems through governance and policies at national scales, market-based approaches and socio-cultural responses. Based on examples from numerous developing but also developed countries we have shown how changes in policies, corporate strategies, social norms and values, and technical developments can positively influence outcomes in livelihoods and human wellbeing in forest contexts.

The section on governance innovations to better link forests with food security discussed lessons from tenure reform, decentralisation of authority, market regulation and access to knowledge and technology. A central governance issue is how and to what extent policy and regulatory frameworks help ensure equitable access of the poor, women and disadvantaged groups to forests and tree-based systems, and to what extent these regulatory arrangements recognise the rights to direct and indirect benefits for food and nutritional security. On the process side of governance reform it is important to include relevant actors, from local communities to government departments, whereas on the substantive side, tenure security, decentralisation of decision-making and strengthening institutional capacity at local levels have shown to be effective.

The section on private sector-driven reform emphasised initiatives aimed at enhancing the governance of large-scale investors supporting sustainable practices in the commodity value chain, improved benefit sharing and protection of local people's rights. In most cases, these initiatives interact with the public sector and complement policy frameworks. While guidelines and principles to regulate large-scale investments are becoming increasingly important, it is unclear if they will change corporate behaviour toward greater sustainability. Sustainability standards supported by multi-stakeholder processes may therefore foster greater changes on the ground but their adoption is still limited and smallholders not able to comply with more complex terms and conditions may be excluded. The more recent commitments and pledges by corporate actors to zero deforestation and sustainable supply, as well as improving benefits for local people, may therefore have significant influence in shaping future production practices and business models. To achieve more inclusive food systems that not only use appropriate innovations but also value local practices, management systems and knowledge, it may be necessary to promote more structural reform, involving greater intervention from the state to harmonise regulatory regimes. Co-regulatory approaches that involve both public and private sector actors may in the future enhance the governance of food systems.

The section on socio-cultural responses focused on examples from gender research, behavioural change, social mobilisation and urban dynamics to illustrate the importance of education, communication and access to information in achieving better food security and nutrition outcomes while preserving and improving forests and other land-based natural resources. Education plays an important role in empowering rural populations and has the potential to generate tangible and fundamental benefits for households and communities in achieving food security and nutrition, sustainable forest and landscape management, and health. For women and other vulnerable groups appropriate education and training programmes can improve the understanding of healthy and nutritious foods and natural resource management practices, and support traditional rural societies in understanding and incorporating necessary changes that enable gender inclusiveness in decision-making and benefit sharing in forests and tree-based systems. Behavioural change to encourage foods and diets with better environmental footprints, such as low meat consumption diets and an increased use of organically produced foods, can have significant positive impacts on rural populations if the value chains necessary to meet the demand are set up to include smallholders and marginalised groups.

Overall, these public and private sector reforms and social changes achieve greatest impact when they go hand in hand. Whether or not innovation, reform and change at the levels discussed in this chapter are sufficient to transform food systems toward long-term sustainability and food security and nutrition in forests and tree-based landscapes requires continued scrutiny and assessment.

References

Alexandratos, N. and Bruinsma, J., 2012. *World Agriculture Towards 2030/2050: The 2012 Revision.* ESA Working paper No. 12-03. Rome: FAO. http://www.fao.org/docrep/016/ap106e/ap106e.pdf

Acharya, K.P. and Gentle, P., 2006. *Improving the Effectiveness of Collective Action: Sharing Experiences from Community Forestry in Nepal.* CAPRI Working Paper No. 54. Washington DC: IFPRI.

Agarwal, B., 2009. Gender and forest conservation: The impact of women's participation in community forest governance. *Ecological Economics* 68: 2785-2799. http://dx.doi.org/10.1016/j.ecolecon.2009.04.025

Agarwal, B., 2001. Participatory exclusions, community forestry, and gender: An analysis for South Asia and a conceptual framework. *World Development* 29: 1623-1648. http://dx.doi.org/10.1016/s0305-750x(01)00066-3

Agarwal, B., 1997. 'Bargaining' and gender relations within and beyond the household. *Feminist Economics* 3: 1-51. http://dx.doi.org/10.1080/135457097338799

Agrawal, A., 2001. The regulatory community: Decentralization and the environment in the Van Panchayats (Forest Councils) of Kumaon, India. *Mountain Research and Development* 21: 208-211. http://dx.doi.org/10.1659/0276-4741(2001)021[0208:trc]2.0.co;2

Agrawal, A. and Ribot, J., 1999. Accountability in decentralization: A framework with South Asian and West African cases. *Journal of Developing Areas* 33: 473-502. http://www.jstor.org/stable/4192885

Agrawal, A., Cashore, B., Hardin, R., Shepherd, G., Benson, C. and Miller, D., 2013. *Economic Contributions of Forests.* Background Paper 1, United Nations Forum on Forests. http://www.un.org/esa/forests/pdf/session_documents/unff10/EcoContrForests.pdf

Alkon, A.H., 2014. Food justice and the challenge to neoliberalism. *Gastronomica: The Journal of Food and Culture* 14: 27-40. http://dx.doi.org/10.1525/gfc.2014.14.2.27

Antinori, C. and Bray, D.B., 2005. Community forest enterprises as entrepreneurial firms: Economic and institutional perspectives from Mexico. *World Development* 33: 1529-1543. http://dx.doi.org/10.1016/j.worlddev.2004.10.011

Badgley, C. and Perfecto, I., 2007. Can organic agriculture feed the world? *Renewable Agriculture and Food Systems* 22: 80-86. http://dx.doi.org/10.1017/s1742170507001986

Bampton, J.F., Ebregt A. and Banjade, M.R., 2007. Collaborative forest management in Nepal's Terai: Policy, practice and contestation. *Journal of Forest and Livelihood* 6: 30-43. http://www.forestaction.org/app/webroot/js/tinymce/editor/plugins/filemanager/files/4_Collaborative forestry-final.pdf

Bardhan, P., 2002. Decentralization of governance and development. *The Journal of Economic Perspectives* 16: 185-205. http://people.bu.edu/dilipm/ec722/papers/28-s05bardhan.pdf

Barham, B., Callenes, M., Gitter, S., Lewis, J. and Weber, J., 2011. Fair trade/organic coffee, rural livelihoods, and the "agrarian questions": Southern Mexican coffee families in transition. *World Development* 39: 134-145. http://dx.doi.org/10.1016/j.worlddev.2010.08.005

Baron, D., Harjoto, M.A. and Jo, H., 2009. *The Economics and Politics of Corporate Social Performance.* Research Paper No. 1993. Stanford: Stanford University Graduate School of Business. http://dx.doi.org/10.2202/1469-3569.1374

Beardsworth, A. and Keil, T., 1992. The vegetarian option: Varieties, conversions, motives and careers. *The Sociological Review* 40: 253-293. http://dx.doi.org/10.1111/j.1467-954x.1992.tb00889.x

Beddington, J., Asaduzzaman, M., Clark, M., Fernández, A., Guillou, M., Jahn, M., Erda, L., Mamo, T., Van Bo, N., Nobre, C.A., Scholes, R., Sharma, R. and Wakhungu, J., 2012. *Achieving Food Security in the Face of Climate Change: Final Report from the Commission on Sustainable Agriculture and Climate Change.* Copenhagen: CGIAR Research Program on Climate Change, Agriculture and Food Security (CCAFS). https://cgspace.cgiar.org/bitstream/handle/10568/35589/climate_food_commission-final-mar2012.pdf?sequence=1

Belcher, B., Ruíz-Pérez M. and Achdiawan, R., 2005. Global patterns and trends in the use and management of commercial NTFPs: Implications for livelihoods and conservation. *World Development* 33: 1435-1452. http://dx.doi.org/10.1016/j.worlddev.2004.10.007

Bender, A., 1992. *Meat and Meat Products in Human Nutrition in Developing Countries.* Rome: FAO. http://www.fao.org/docrep/t0562e/t0562e00.HTM

Benjamin, C.E., 2008. Legal pluralism and decentralization: Natural resource management in Mali. *World Development* 36: 2255-2276. http://dx.doi.org/10.1016/j.worlddev.2008.03.005

Bettencourt, L.M.A., 2013. The origins of scaling in cities. *Science* 340: 1438-1441. http://dx.doi.org/10.1126/science.1235823

Bettencourt, L.M. A., Lobo, J., Helbing, D., Kuhnert, C. and West, G.B., 2007. Growth, innovation, scaling, and the pace of life in cities. *Proceedings of the National Academy of Sciences* 107: 7301-7306. http://dx.doi.org/10.1073/pnas.0610172104

Beuchelt, T. and Zeller, M., 2011. Profits and poverty: Certification's troubled link for Nicaragua's organic and Fairtrade coffee producers. *Ecological Economics* 70: 1316-1324. http://dx.doi.org/10.1016/j.ecolecon.2011.01.005

Bhattarai, B., Dhungana, S.P. and Kafley, G.P., 2007. Poor-focused common forest management: Lessons from leasehold forestry in Nepal. *Journal of Forest and Livelihood* 6: 20-29.

Biggs, S. and Smith, S., 2003. A paradox of learning in project cycle management and the role of organizational culture. *World Development* 31: 1743-1757. http://dx.doi.org/10.1016/s0305-750x(03)00143-8

Blaikie, P., 2006. Is small really beautiful? Community-based natural resource management in Malawi and Botswana. *World Development* 34: 1942-1957. http://dx.doi.org/10.1016/j.worlddev.2005.11.023

BNDES, 2014. *Social and Environmental Responsibility.* http://www.bndes.gov.br/SiteBNDES/bndes/bndes_en/Institucional/Social_and_Environmental_Responsibility/

Bray, D.B. and Merino-Pérez, L., 2002. *The Rise of Community Forestry in Mexico: History, Concepts, and Lessens Learned from Twenty-Five Years of Community Timber Production.* New York: Ford Foundation.

Brooks, J., 2014. Policy coherence and food security: The effects of OECD countries' agricultural policies. *Food Policy* 44: 88-94. http://dx.doi.org/10.1016/j.foodpol.2013.10.006

Burlingame, B. and Dernini, S. (eds.), 2012. *Sustainable Diets and Biodiversity: Directions and Solutions for Policy, Research and Action.* Rome: FAO. http://www.fao.org/docrep/016/i3004e/i3004e.pdf

Campbell, B., Wamukoya, G., Kinyangi, J., Verchot, L., Wollenberg, L., Vermeulen, S.J., Minang, P.A., Neufeldt, H., Vidal, A., Loboguerrero Rodriguez, A.M. and Hedger, M., 2014. *The Role of Agriculture in the UN Climate Talks.* CCAFS Info Note. Copenhagen: CGIAR Research Program on Climate Change, Agriculture and Food Security (CCAFS). http://hdl.handle.net/10568/51665

Campbell, J.L., 2007. Why would corporations behave in socially responsible ways? An institutional theory of xorporate social responsibility. *Academy of Management Review* 32: 946-967. http://dx.doi.org/10.5465/amr.2007.25275684

Candel, J.L., 2014. Food security governance: A systematic literature review. *Food Security* 6: 585-601. http://dx.doi.org/10.1007/s12571-014-0364-2

Cashore, B., 2002. Legitimacy and the privatization of environmental governance: How non–state market–driven (NSMD) governance systems gain rule–making authority. *Governance* 15: 503-529. http://dx.doi.org/10.1111/1468-0491.00199

CFS, 2014. CFS *Principles for Responsible Investment in Agriculture and Food Systems* First Draft. Committee on World Food Security. http://www.fao.org/fileadmin/templates/cfs/Docs1314/rai/FirstDraft/CFS_RAI_First_Draft_for_Negotiation.pdf

CFS-FAO, 2012. *Voluntary Guidelines on the Responsible Governance of Tenure of Land, Fisheries and Forests in the Context of National Food Security.* http://www.fao.org/docrep/016/i2801e/i2801e.pdf

Clapp, J. and Fuchs, D., 2009. *Corporate Power in Global Agrifood Governance.* Massachusetts: MIT Press. http://dx.doi.org/10.7551/mitpress/9780262012751.001.0001

Clark, W.C., Tomich, T.P., van Noordwijk, M., Guston, D., Catacutan, D., Dickson, N.M. and McNie, E., 2011. Boundary work for sustainable development: Natural resource management at the Consultative Group on International Agricultural Research (CGIAR). *Proceedings of the National Academy of Sciences* (early edition). http://dx.doi.org/10.1073/pnas.0900231108

Colfer, C.J.P., 2013. The Ups and Downs of Institutional Learning: Reflections on the Emergence and Conduct of Adaptive Collaborative Management at the Centre for International Forestry Research. In: *Adaptive Collaborative Approaches in Natural Resource Governance: Rethinking Participation, Learning and Innovation,* edited by H. Ojha, A. Hall and V. Rasheed Sulaiman. London: Routledge. http://dx.doi.org/10.4324/9780203136294

Colfer, C.J.P., 2005. *The Complex Forest: Communities, Uncertainty, and Adaptive Collaborative Management.* Washington DC: Resources for the Future.

Colfer, C.J.P. and Capistrano, D., 2004. *The Politics of Decentralization: Forests, Power, and People.* London: Earthscan. http://dx.doi.org/10.4324/9781849773218

COMTRADE, 2014. Trade Map: Trade Statistics for International Business Development. http://www.trademap.org/Index.aspx

Cotula, L., Vermeulen, S., Mathieu, P. and Toulmin, C., 2011. Agricultural investment and international land deals: Evidence from a multi-country study in Africa. *Food Security* 3: 99-113. http://dx.doi.org/10.1007/s12571-010-0096-x

Crook, R.C., 2003. Decentralisation and poverty reduction in Africa: The politics of local–central relations. *Public Administration and Development* 23: 77-88. http://dx.doi.org/10.1002/pad.261

Dahal, G.R. and Capistrano, D., 2006. Forest governance and institutional structure: An ignored dimension of community based forest management in the Philippines. *International Forestry Review* 8: 377-394. http://dx.doi.org/10.1505/ifor.8.4.377

Das, S. and Laub, R., 2005. Understanding links between gendered local knowledge of agrobiodiversity and food security in Tanzania. *Mountain Research and Development* 25: 218-222. http://dx.doi.org/10.1659/0276-4741(2005)025[0218:ulbglk]2.0.co;2

Deininger, K., 2011. Challenges posed by the new wave of farmland investment. *The Journal of Peasant Studies* 38: 217-247. http://dx.doi.org/10.1080/03066150.2011.559007

Dibsdall, L.A., Lambert, N., Bobbin, R.F. and Frewer, L.J., 2002. Low-income consumers' attitudes and behaviour towards access, availability and motivation to eat fruit and vegetables. *Public Health Nutrition* 6: 159-168. http://dx.doi.org/10.1079/phn2002412

Dimitri, C. and Greene, C., 2002. *Recent Growth Patterns in the U.S. Organic Foods Market.* Agriculture Information Bulletin No. 777. Washington DC: USDA. http://www.ers.usda.gov/media/249063/aib777_1_.pdf

Djama, M., 2011. *Regulating the Globalised Economy: Articulating Private Voluntary Standards and Public Regulations.* Perspective No. 11. Montpellier: CIRAD.

Donovan, J. and Poole, N., 2014. Partnerships in Fairtrade coffee: Close-up look at buyer interactions and NGO interventions. *Food Chain* 4: 34-48. http://dx.doi.org/10.3362/2046-1887.2014.004

Donovan, J., 2011. *Value Chain Development for Addressing Rural Poverty: Asset Building by Smallholder Coffee Producers and Cooperatives in Nicaragua.* PhD Thesis, London: School of Oriental and African Studies, University of London. http://eprints.soas.ac.uk/12762/1/Donovan_3276.pdf

Dryzek, J. S., 2010. *Foundations and Frontiers of Deliberative Governance.* Oxford: Oxford University Press. http://dx.doi.org/10.1093/acprof:oso/9780199562947.001.0001

Durst, P., McKenzie, P., Brown, C. and Appanah, S., 2006. Challenges facing certification and eco-labelling of forest products in developing countries. *International Forestry Review* 8: 193-200. http://dx.doi.org/10.1505/ifor.8.2.193

Equator Principles, 2014. Members and Reporting. http://www.equator-principles.com/index.php/members-reporting

FAO, 2013a. *Climate-smart Agriculture Sourcebook.* Rome: FAO. http://www.fao.org/3/a-i3325e.pdf

FAO, 2013b. *Food Wastage Footprint: Impacts on Natural Resources. Summary Report.* Rome: Food and Agriculture Organization of the United Nations. http://www.fao.org/docrep/018/i3347e/i3347e.pdf

FAO, 2011. *The State of Food and Agriculture 2010-11: Women in Agriculture.* Rome: FAO. http://www.fao.org/docrep/013/i2050e/i2050e.pdf

FAO, 2010. *Global Forest Resources Assessment.* Rome: FAO. http://www.fao.org/forest-resources-assessment/en/

FAO, 2004. *Voluntary Guidelines to Support the Progressive Realization of the Right to Adequate Food in the Context of the National Food Security.* http://www.fao.org/3/a-y7937e.pdf

FAO, 2003. *A Handbook for Trainers on Participatory Local Development.* Regional Office for Asia and the Pacific Publication 2003-07. Bangkok: FAO. ftp://ftp.fao.org/docrep/fao/006/ad346e/ad346e00.pdf

FAO, 2002. *Gender and Law: Women's Rights in Agriculture.* Legislative Study No. 76. Rome: FAO. http://www.fao.org/3/a-y4311e.pdf

FAO, 1997. *Agriculture Food and Nutrition for Africa: A Resource Book for Teachers of Agriculture.* Rome: FAO. http://www.fao.org/docrep/w0078e/w0078e00.HTM

Fischler, C., 1988. Food, self and identity. *Social Science Information* 27: 275-292. http://dx.doi.org/10.1177/053901888027002005

Forster, T. and Escudero, A.G., 2014. *City Regions as Landscapes for People, Food and Nature.* Washington DC: EcoAgriculture Partners. http://peoplefoodandnature.org/publication/city-regions-as-landscapes-for-people-food-and-nature/

Fuller R.J., Norton, L.R., Feber, R.E., Johnson P.J., Chamberlain, D.E., Joys, A.C., Mathews, F., Stuart, R.C., Townsend, M.C., Manley, W.J., Wolfe, M.S., Macdonald, D.W. and Firbank, L., 2005. Benefits of organic farming to biodiversity vary among taxa. *Biology Letter* 1: 431-434. http://dx.doi.org/10.1098/rsbl.2005.0357

Fung, A. and Wright, E.O., 2001. Deepening democracy: Innovations in empowered participatory governance. *Politics and Society* 29: 5-42. http://dx.doi.org/10.1177/0032329201029001002

GACSA, 2014. Framework Document. Global Alliance for Climate-Smart Agriculture. Series Document 1. http://www.fao.org/climate-smart-agriculture/41760-02b7c16db1b86fcb1e55ef e8fa93ffdc5.pdf

GBM, 2015. Green Belt Movement. http://www.greenbeltmovement.org/

GEF, 2015. *Climate Change Calls for Greater Role of Women in Forest Management.* http://www.thegef.org/gef/press_release/woman_and_forest_2011

Gilmour, D., 2011. *Unlocking the Wealth of Forests for Community Development: Commercializing Products from Community Forests.* Rome: FAO.

Glanz, K., Sallis, J.F., Saelens, B.E. and Frank, L.D., 2005. Healthy nutrition environments: Concepts and measures. *American Journal of Health Promotion* 19: 330-333. http://dx.doi.org/10.4278/0890-1171-19.5.330

Grieg-Gran, M., Porras, I. and Wunder, S., 2005. How can market mechanisms for forest environmental services help the poor? Preliminary lessons from Latin America. *World Development* 33: 1511-1527. http://dx.doi.org/10.1016/j.worlddev.2005.05.002

GSCP, 2014. Global Social Compliance Programme. http://www.mygscp.com

Haberli, C. and Smith, F., 2014. Food security and agri-foreign direct investment in weak states: Finding the governance gap to avoid 'land grab'. *The Modern Law Review* 77: 189-222. http://dx.doi.org/10.1111/1468-2230.12062

Halberg, N., Alrøe, H.F., Knudsen, M.T. and Kristensen, E.S. (eds.), 2005. *Global Development of Organic Agriculture: Challenges and Promises.* Wallingford: CAB International.

Hall, A., 2002. Innovation systems and capacity development: An agenda for North-South research collaboration? *International Journal of Technology Management and Sustainable Development* 1: 146-152. http://dx.doi.org/10.1386/ijtm.1.3.146

Hall, A., Sulaiman, V.R., Yoganand, B. and Clark, N., 2003. Post-harvest innovation systems in South Asia: Research as capacity development and its prospects for impact on the poor in developing countries. *Outlook on Agriculture* 32: 97-104. http://dx.doi.org/10.5367/000000003101294334

Hallam, D., 2011. International investment in developing country agriculture: Issues and challenges. *Food Security* 3: 591-598. http://dx.doi.org/10.1007/s12571-010-0104-1

Hawkes, C., 2013. *Promoting Health Diets Through Nutrition Education and Changes in the Food Environment: An International Review of Actions and their Effectiveness.* Rome: FAO. http://www.fao.org/3/a-i3235e.pdf

HCSS, 2014. *Oil Palm Manifesto. High Carbon Stock Study.* http://www.carbonstockstudy.com/Documents/Sustainable-Palm-Oil-Manifesto.aspx

Head, B. W., 2007. Community engagement: Participation on whose terms? *Australian Journal of Political Science* 42: 441-454. http://dx.doi.org/10.1080/10361140701513570

Hemmati, M., 2002. *Multi-Stakeholder Proscesses for Goverance and Sustainability: Beyond Deadlock and Conflict.* London: Earthscan. http://dx.doi.org/10.4324/9781849772037

Hickey, S. and Mohan, G., 2005. Relocating participation within a radical politics of development. *Development and Change* 36, 237-262. http://dx.doi.org/10.1111/j.0012-155x.2005.00410.x

Hughes, L., 2010. Conceptualizing just food in alternative agrifood initiatives. *Humboldt Journal of Social Relations* 33: 30-63. http://www.jstor.org/stable/23263226

IAASTD 2009. *International Assessment of Agricultural Knowledge, Science and Technology for Development.* Washington DC: Island Press. http://apps.unep.org/publications/pmtdocuments/-Agriculture at a crossroads - Synthesis report-2009Agriculture_at_Crossroads_Synthesis_Report.pdf

IFC, 2012. *IFC Performance Standards on Environmental and Social Sustainability.* http://www.ifc.org/wps/wcm/connect/c8f524004a73daeca09afdf998895a12/IFC_Performance_Standards.pdf?MOD=AJPERES

IFOAM, 2005. *The IFOAM Norms Version 2005.* Bonn: International Federation of Organic Agriculture Movements. http://www.ifoam.bio/sites/default/files/page/files/norms_eng_v4_20090113.pdf

Indonesia Ministry of Agriculture, 2012. *Tree Crop Statistics of Indonesia: 2011-2012.* Jakarta: Directorate General of Estate Crops.

IPCC, 2014. Summary for policymakers. In: *Climate Change 2014: Impacts, Adaptation, and Vulnerability. Part A: Global and Sectoral Aspects.* Contribution of Working Group II to the Fifth Assessment Report of the Intergovernmental Panel on Climate Change, edited by C. Field, V.R. Barros, D.J. Dokken, K.J. Mach, M.D. Mastrandrea, T.E. Bilir, M. Chatterjee, K.L. Ebi, Y.O. Estrada, R.C. Genova, B. Girma, E.S. Kissel, A.N. Levy, S. MacCracken, P.R. Mastrandrea and L.L. White. Cambridge: Cambridge University Press. http://dx.doi.org/10.1017/cbo9781107415379

Iversen, V., Chhetry, B., Francis, P., Gurung, M., Kafle, G., Pain, A. and Seeley, J., 2006. High value forests, hidden economies and elite capture: Evidence from forest user groups in Nepal's Terai. *Ecological Economics* 58: 93-107. http://dx.doi.org/10.1016/j.ecolecon.2005.05.021

Jamnadass, R., Place, F., Torquebiau, E., Malézieux, E., Iiyama, M., Sileshi, G.W., Kehlenbeck, K., Masters, E., McMullin, S. and Dawson, I.K., 2013. Agroforestry for food and nutritional security. *Unasylva* 64: 23-29. http://www.fao.org/forestry/37082-04957fe26afbc90d1e9c0356c48185295.pdf

Jamnadass, R.H., Dawson, I.K., Franzel, S., Leakey, R.R.B., Mithöfer, D., Akinnifesi, F.K. and Tchoundjeu, Z., 2011. Improving livelihoods and nutrition in sub-Saharan Africa through the promotion of indigenous and exotic fruit production in smallholders 'agroforestry systems: A review. *International Forest Review* 13: 338-354. http://dx.doi.org/10.1505/146554811798293836

Keller, G.B., Mndiga, H. and Maass, B., 2006. Diversity and genetic erosion of traditional vegetables in Tanzania from the farmer's point of view. *Plant Genetic Resources* 3: 400-413. http://dx.doi.org/10.1079/pgr200594

Kennedy, J., Thomas, J.W. and Glueck, P, 2001. Evolving forestry and rural development beliefs at midpoint and close to the 20th century. *Forest Policy and Economics* 3: 81-95. http://dx.doi.org/10.1016/s1389-9341(01)00034-x

Kissinger, G., Herold, M. and De Sy, V., 2012. *Drivers of Deforestation and Forest Degradation: A Synthesis Report for REDD+ Policymakers.* Vancouver: Lexeme Consulting. https://www.gov.uk/government/uploads/system/uploads/attachment_data/file/66151/Drivers_of_deforestation_and_forest_degradation.pdf

Kooiman, J., Bavinck, M., Chuenpagdee, R., Mahon, R. and Pullin, R., 2008. Interactive governance and governability: An introduction. *The Journal of Transdisciplinary Environmental Studies* 7: 1-11.

Kuhnlein, H.V. and Receveur, O., 1996. Dietary change and traditional food systems of indigenous peoples. *Annual Review of Nutrition* 16: 417-442. http://dx.doi.org/10.1146/annurev.nu.16.070196.002221

La Via Campesina, 2011. La Via Campesina http://viacampesina.org/en/index.php/organisation-mainmenu-44

Larson, A., Barry, D., Dahal, G.R. and Colfer, C.J.P. (eds.), 2010. *Forest for People: Community Rights and Forest Tenure Reform.* London: Earthscan. http://dx.doi.org/10.4324/9781849774765

Leach, M., Rockström, J., Raskin, P., Scoones, I., Stirling, A.C., Smith, A., Thompson, J., Millstone, E., Ely, A., Folke, C. and Olsson, P., 2012. Transforming innovation for sustainability. *Ecology and Society* 17(2):11. http://dx.doi.org/10.5751/es-04933-170211

Lund, C., 2006. Twilight institutions: Public authority and local politics. *Development and Change* 37: 685-705. http://dx.doi.org/10.1111/j.1467-7660.2006.00497.x

Lykke, A.M., Mertz, O. and Ganaba, S., 2002. Food consumption in rural Burkina Faso. *Ecology of Food Nutrition* 41: 119-153. http://dx.doi.org/10.1080/03670240214492

Magdoff, F., Foster J.B. and Buttel, F.H. (eds.), 2000. *Hungry for Profit: The Agribusiness Threat to Farmers, Food, and the Environment.* New York: Monthly Review Press, 7-21.

Mahanty, S., Fox, J., Nurse, M., Stephen P. and McLees, L., 2006. *Hanging in the Balance: Equity in Community-Based Natural Resource Management in Asia.* Honolulu: East-West Centre.

Margulis, M.E., 2013. The regime complex for food security: Implications for the global hunger challenge. *Global Governance* 19: 53-67. http://pubman.mpdl.mpg.de/pubman/item/escidoc:1719957:4/component/escidoc:2075304/GG_19_2013_Margulis.pdf

Marsden, T., Lee, R., Flynn A. and Thankappan, S., 2009. *The New Regulation and Governance of Food Beyond the Food Crisis.* London: Routledge. http://dx.doi.org/10.4324/9780203877722

Marshall, K., 2013. Ten years of food politics: An interview with Marion Nestle. Gastronomica: *The Journal of Food and Culture* 13: 1-3. http://dx.doi.org/10.1525/gfc.2013.13.3.1

McCullough, F. S. W., Yoo, S. and Ainsworth, P., 2004. Food choice, nutrition education and parental influence on British and Korean primary school children. *International Journal of Consumer Studies* 28: 235-244. http://dx.doi.org/10.1111/j.1470-6431.2003.00341.x

Mendez, V.E., Bacon, C.M., Olson, M., Petchers, S., Herrador, D., Carranza, C., Trujillo, L., Guadarrama-Zugasti, C., Cordon A. and Mendoza, A., 2010. Effects of Fair Trade and organic certifications on small-scale coffee farmer households in Central America and Mexico. *Renewable Agriculture and Food Systems* 25: 236-251. http://dx.doi.org/10.1017/s1742170510000268

Meyer, M., 2010. The Rise of the knowledge broker. *Science Communication* 32: 118-127. http://dx.doi.org/10.1177/1075547009359797

Mohan, G. and Stokke, K., 2000. Participatory development and empowerment: The dangers of localism. *Third World Quarterly* 21: 247-268. http://dx.doi.org/10.1080/01436590050004346

Mohan, G., 1996. Adjustment and decentralization in Ghana: A case of diminished sovereignty. *Political Geography* 15: 75-94. http://dx.doi.org/10.1016/0962-6298(95)00009-7

Mwangi, E. and Wardell, A., 2012. Multi-level governance of forest resources. *International Journal of the Commons* 6: 79-103. http://www.thecommonsjournal.org/index.php/ijc/article/view/374/282

NECOFA, 2013. Network for Ecofarming in Africa, Kenya chapter. https://necofakenya.wordpress.com

Nelson, G.C., Rosegrant, M.W., Palazzo A., Gray, I., Ingersoll C., Robertson, R., Tokgoz, S., Zhu, T., Sulser, T.B., Ringler, C., Msangi, S. and You, L., 2010. *Food Security, Farming, and Climate Change to 2050: Scenarios, Results, Policy Options.* Washington DC: IFPRI. http://dx.doi.org/10.2499/9780896291867

Nelson, V. and Stathers, T., 2009. Resilience, power, culture, and climate: A case study from semi-arid Tanzania, and new research directions. *Gender and Development* 17: 81-94. http://dx.doi.org/10.1080/13552070802696946

Nepal DoF, 2009. *Community Forestry Implementation Guideline.* Kathmandu: Department of Forest.

Nestle, M., 2009. Reading the food social movement. *World Literature Today* 83: 36-39. http://www.foodpolitics.com/wp-content/uploads/wlt-jan09-nestle2.pdf

Neufeldt, H., Jahn, M., Campbell, B., Beddington, J. R., DeClerck, F., De Pinto, A., Gulledge, J., Hellin, J., Herrero, M., Jarvis, A., LeZaks, D., Meinke, H., Rosenstock, T., Scholes, M., Scholes, R., Vermeulen, S., Wollenberg, E. and Zougmoré, R., 2013. Beyond climate-smart agriculture: Toward safe operating spaces for global food systems. *Agriculture and Food Security* 2: 12. http://dx.doi.org/10.1186/2048-7010-2-12

NRSP, 2005. *Social Mobilization.* National Rural Support Programme http://nrsp.org.pk/social-mobilization.html

OECD, 2008. *OECD Guidelines for Multinational Enterprises.* Paris: OECD. http://www.oecd.org/investment/mne/1922428.pdf

Ogoye-Ndegwa, C., and Aagaard-Hansen, J., 2003. Traditional gathering of wild vegetables among the Luo of western Kenya: A nutritional anthropology project. *Ecology of Food and Nutrition* 42: 69-89. http://dx.doi.org/10.1080/03670240303114

Ojha, H.R., 2014. Beyond the 'local community': The evolution of multi-scale politics in Nepal's community forestry regimes. *International Forestry Review* 16: 339-353. http://dx.doi.org/10.1505/146554814812572520

Ojha, H.R., 2013. Counteracting hegemonic powers in the policy process: Critical action research on Nepal's forest governance. *Critical Policy Studies* 7: 242-262. http://dx.doi.org/10.1080/19460171.2013.823879

Ojha, H.R., Banjade, M.R., Sunam, R.K., Bhattarai, B., Jana, S., Goutam K.R. and Dhungana, S., 2014. Can authority change through deliberative politics? Lessons from the four decades of participatory forest policy reform in Nepal. *Forest Policy and Economics* 46: 1-9. http://dx.doi.org/10.1016/j.forpol.2014.04.005

Ojha, H.R. and Hall, A., 2012. Confronting Challenges in Applying Adpative Collaborative Appraoches. In: *Adaptive Collaborative Approaches in Natural Resource Governance: Rethinking Participation, Learning and Innovation,* edited by H. Ojha, A. Hall and V. Rasheed Sulaiman. London: Routledge. http://dx.doi.org/10.4324/9780203136294

Ostrom, E., 1990. *Governing the Commons: The Evolution of Institutions for Collective Actions, Political Economy of Institutions and Decisions.* Cambridge: Cambridge University Press. http://dx.doi.org/10.1017/cbo9780511807763

Pacheco, P. and Poccard-Chapuis, R., 2012. The complex evolution of cattle ranching development amid market integration and policy shifts in the Brazilian Amazon. *Annals of the Association of American Geographers* 102: 1366-1390. http://dx.doi.org/10.1080/00045608.2012.678040

Pacheco, P. and Paudel, N.S., 2010. Communities and Forest Markets: Assessing the Benefits from Diverse Forms of Engagement. In: *Forest for People: Community Rights and Forest Tenure Reform,* edited by A. Larson, D. Barry, G.R. Dahal and C.J.P. Colfer. London: Earthscan. http://dx.doi.org/10.4324/9781849774765

Pacheco, P., German, L., van Gelder, J.W., Weinberger, K. and Guariguata, M., 2011. *Avoiding Deforestation in the Context of Biofuel Feedstock Expansion: An Analysis of the Effectiveness of Market-based Instruments.* Working Paper 73. Bogor: CIFOR. http://dx.doi.org/10.17528/cifor/003511

Padoch, C. and Sunderland, T., 2013. Managing landscapes for greater food security and improved livelihoods. *Unasylva* 64: 3-13. http://www.cifor.org/publications/pdf_files/articles/APadoch1401.pdf

Page, H., 2013. *Global Governance and Food Security as Global Public Good.* New York: New York University, Center for International Cooperation. http://cic.nyu.edu/sites/default/files/page_global_governance_public_good.pdf

Pandit, B.H., Albano, A. and Kumar, C., 2009. Community-based forest enterprises in Nepal: An analysis of their role in increasing income benefits to the poor. *Small-scale Forestry* 8: 447-462. http://dx.doi.org/10.1007/s11842-009-9094-2

Patel, R. and McMichael, P., 2009. A political economy of the food riot. *Review* 32: 9-35. http://rajpatel.org/wp-content/uploads/2009/11/patel-mcmichael-2010Review321.pdf

Perez, O., 2007. *The New Universe of Green Finance: From Self-Regulation to Multi-Polar Governance.* Ramat Gan: Bar Ilan University. http://www.biu.ac.il/law/unger/working_papers/1-07.pdf

Pimbert, M., 2012. FPIC and beyond: Safeguards for power-equalising research that protects biodiversity, rights and culture. *Participatory Learning and Action* 65: 43-54.

Pokharel, B., Branney, P., Nurse, M. and Malla, Y., 2008. Community Forestry: Conserving Forests, Sustaining Livelihoods, Strengthening Democracy. In: *Communities, Forests and Governance: Policy and Institutional Innovations from Nepal,* edited by H. Ojha, N. Timsina, C. Kumar, B. Belcher and M. Banjade. New Delhi: Adroit.

Potts, J., Lynch, M., Wilkings, A., Huppé, G., Cunnigham M. and Vora, V., 2014. *The State of Sustainability Initiatives Review 2014: Standards and the Green Economy.* Winnipeg: IISD. https://www.iisd.org/pdf/2014/ssi_2014.pdf

Pretty, J.N., 1995. Participatory learning for sustainable agriculture. *World Development* 23: 1247-1263. http://dx.doi.org/10.1016/0305-750x(95)00046-f

Pulhin, J.M., Inoue, M. and Enters, T., 2007. Three decades of community-based forest management in the Philippines: Emerging lessons for sustainable and equitable forest management. *International Forestry Review* 9: 865-883. http://dx.doi.org/10.1505/ifor.9.4.865

Rayda, N., 2012. *Switch to Sustainable Palm Oil Just a Matter of Time, Industry Figures Says.* The Jakarta Globe. http://www.thejakartaglobe.com

Raynolds, L.T., 2009. Mainstreaming Fair Trade Coffee: From Partnership to Traceability. *World Development* 37: 1083-1093. http://dx.doi.org/10.1016/j.worlddev.2008.10.001

Reardon, T., Timmer, P., Barrett, C. and Berdegué, J., 2003. The rise of supermarkets in Africa, Asia and Latin America. *American Journal of Agricultural Economics* 85: 1140-1146. http://dx.doi.org/10.1111/j.0092-5853.2003.00520.x

Ribot, J.C., 2003. Democratic decentralisation of natural resources: Institutional choice and discretionary power transfers in sub-Saharan Africa. *Public Administration and Development* 23: 53-65. http://dx.doi.org/10.1002/pad.259

Ribot, J.C., 2007. Representation, citizenship and the public domain in democratic decentralization. *Development* 50: 43-49. http://dx.doi.org/10.1057/palgrave.development.1100335

Ribot, J.C., Agrawal, A. and Larson, A.M., 2006. Recentralizing while decentralizing: How national governments reappropriate forest resources. *World Development* 34: 1864-1886. http://dx.doi.org/10.1016/j.worlddev.2005.11.020

Ribot, J.C., 2006. Authority over forests: Empowerment and subordination in Senegal's democratic decentralization. *Development and Change* 40: 105-129. http://dx.doi.org/10.1002/9781444322903.ch5

Rockström, J., Steffen, W., Noone, K., Persson, A., Chapin, F.S. 3rd, Lambin, E.F., Lenton, T. M., Scheffer, M., Folke, C., Schellnhuber, H.J., Nykvist, B., de Wit, C.A., Hughes, T., van der Leeuw, S., Rodhe, H., Sörlin, S., Snyder, P.K., Costanza, R., Svedin, U., Falkenmark, M., Karlberg, L., Corell, R.W., Fabry, V.J., Hansen, J., Walker, B., Liverman, D., Richardson, K., Crutzen, P. and Foley, J.A., 2009. A safe operating space for humanity. *Nature* 461: 472-475. http://dx.doi.org/10.1038/461472a

RRI, 2014. *What Future for Reform? Progress and Slowdown in Forest Tenure Reform Since 2002.* Washington DC: Rights and Resources Initiative. http://www.rightsandresources.org/wp-content/uploads/RRI4011D_FlagshipMAR2014r13B.pdf

RSPO, 2014. *RSPO in numbers 2014.* http://www.rspo.org

Sayer, J., Sunderland, T., Ghazoul, J., Pfund, J.-L., Sheil, D., Meijaard, E., Venter, M., Boedhihartono, A.K., Day, M., Garcia, C., van Oosten, C. and Buck, L.E., 2013. Ten principles for a landscape approach to reconciling agriculture, conservation, and other competing land uses. *Proceedings of the National Academy of Sciences* 110: 8349-8356. http://dx.doi.org/10.1073/pnas.1210595110

Scherr, S.J., Shames, S. and Friedman, R., 2012. From climate-smart agriculture to climate-smart landscapes. *Agriculture and Food Security* 1: 12. http://dx.doi.org/10.1186/2048-7010-1-12

Scherr, S. J., White, A. and Kaimowitz, D., 2004. *A New Agenda for Forest Conservation and Poverty Reduction.* Washington DC: Forest Trends and CIFOR. http://hdl.handle.net/10568/18758

Schot, J., 2001. Towards new forms of participatory technology development. *Technology Analysis and Strategic Management* 13: 39-52. http://dx.doi.org/10.1080/09537320120040437

Schouten, G. and Glasbergen, P., 2011. Creating legitimacy in global private governance: The case of the Roundtable on Sustainable Palm Oil. *Ecological Economics* 70: 1891-1899. http://dx.doi.org/10.1016/j.ecolecon.2011.03.012

Schusler, T.M., Decker, D.J. and Pfeffer, M.J., 2003. Social learning for collaborative natural resource management. *Society and Natural Resources* 16: 309-326. http://dx.doi.org/10.1080/08941920309158

Sen, A., 1999. *Development as Freedom.* Oxford: Oxford University Press.

Sharma, N. and Ojha, H., 2013. *Community Forestry, Ecosystem Services and Poverty Alleviation: Evidence from Nepal.* Digital Library of the Commons. https://dlc.dlib.indiana.edu/dlc/bitstream/handle/10535/8948/PAUDEL_0945.pdf?sequence=1

Shaw, D., 2011. Interview with Wangari Maathai. In: *Forests and Gender*, edited by L. Aguilar, A. Quesada-Aguilar and D.M.P. Shaw. Gland: IUCN. https://portals.iucn.org/library/efiles/documents/2011-070.pdf

Sherman, J., 2003. From nutritional needs to classroom lessons: Can we make a difference? *Food, Nutrition and Agriculture* 33: 45-51. ftp://ftp.fao.org/docrep/fao/006/j0243m/j0243m06.pdf

Smith P., Bustamante, M., Ahammad, H., Clark, H., Dong, H., Elsiddig, E.A., Haberl, H., Harper, R., House, J., Jafari, M., Masera, O., Mbow, C., Ravindranath, N. H., Rice, C. W., Robledo Abad, C., Romanovskaya, A., Sperling, F. and Tubiello, F., 2014. Agriculture, Forestry and Other Land Use (AFOLU). In: *Climate Change 2014: Mitigation of Climate Change. Contribution of Working Group III to the Fifth Assessment Report of the Intergovernmental Panel on Climate Change*, edited by O. Edenhofer, R. Pichs-Madruga, Y. Sokona, E. Farahani, S. Kadner, K. Seyboth, A. Adler, I. Baum, S. Brunner, P. Eickemeier, B. Kriemann, J. Savolainen, S. Schlömer, C. von Stechow, T. Zwickel and J.C. Minx. Cambridge: Cambridge University Press. http://dx.doi.org/10.1017/cbo9781107415416.017

Sobal, J., Bisogni, C.A. and Jastran, M., 2014. Food choice is multifaceted: Contextual, dynamic, multilevel, integrated, and diverse. *Mind, Brain, and Education* 8: 6-12. http://dx.doi.org/10.1111/mbe.12044

Soil Association, 2014. *Organic Market Report 2013.* http://www.soilassociation.org/marketreport

Springate-Baginski, O., Sarin, M. and Reddy, M.G., 2013. Resisting Rights: Forest Bureaucracy and the Tenure Transition in India. *Small-scale Forestry* 12, 107-124. http://dx.doi.org/10.1007/s11842-012-9219-x

Starbucks, 2014. *Starbucks and Sustainable Palm Oil.* http://globalassets.starbucks.com/assets/85d80f17fae84fc9bb4697e9edc38b74.pdf

Stehfest E., Bouwman, L., van Vuuren, D., den Elzen, M., Eickhout, B. and Kabat P., 2009. Climate benefits of changing diet. *Climatic Change* 95: 83-102. http://dx.doi.org/10.1088/1755-1307/6/26/262009

Story, M., Kaphingst, K.M., Robinson-O'Brien, R. and Glanz, K., 2008. Creating healthy food and eating environments: Policy and environmental approaches. *Annual Review of Public Health* 29: 253-272. http://dx.doi.org/10.1146/annurev.publhealth.29.020907.090926

Sunam, R.K., Paudel, N.S. and Paudel, G., 2013. Community forestry and the threat of recentralization in Nepal: Contesting the bureaucratic hegemony in policy process. *Society and Natural Resources* 26: 1407-1421. http://dx.doi.org/10.1080/08941920.2013.799725

Sundar, N., 2000. Unpacking the 'joint' in joint forest management. *Development and Change* 31: 255-279. http://dx.doi.org/10.1111/1467-7660.00154

Sunderland, T., Powell, B., Ickowitz, A., Foli, S., Pinedo-Vasquez, M., Nasi, R. and Padoch, C., 2013. *Food Security and Nutrition: The Role of Forests.* Bogor: CIFOR. http://www.cifor.org/publications/pdf_files/WPapers/DPSunderland1301.pdf

Terry, G. (ed.), 2009. *Climate Change and Gender Justice.* Rugby: Practical Action Publishing. http://www.climateaccess.org/sites/default/files/Terry_Climate Change and Gender Justice.pdf

Thoms, C.A., Karna, B.K. and Karmacharya, M.B., 2006. Limitations of leasehold forestry for poverty alleviation in Nepal. *Society and Natural Resources* 19: 931-938. http://dx.doi.org/10.1080/08941920600902179

Tilman, D., Cassman, K.G., Matson, P.A., Naylor, R. and Polasky, S., 2002. Agricultural sustainability and intensive production practices. *Nature* 418: 671-677. http://dx.doi.org/10.1038/nature01014

Toledo, V.M., Ortiz-Espejel, B., Cortés, L., Moguel, P. and Ordoñez, M., 2003. The multiple use of tropical forests by indigenous peoples in Mexico: A case of adaptive management. *Conservation Ecology* 7: 9. http://www.ecologyandsociety.org/vol7/iss3/art9/

UN-HABITAT, 2014. Social Mobilization. United Nations Human Settlements Programme. http://www.fukuoka.unhabitat.org/docs/publications/pdf/peoples_process/ChapterII-Social_Mobilization.pdf

UNDP, 2015. Our Work. UNDP in Tajikistan. http://www.tj.undp.org/content/tajikistan/en/home/ourwork/overview.html

UNICEF, 2015. *Social Mobilization. Communication for Development (C4D).* http://www.unicef.org/cbsc/index_42347.html

Unilever, 2014. Our targets. http://www.unilever.com/sustainable-living-2014/reducing-environmental-impact/sustainable-sourcing/sustainable-palm-oil/our-targets/

United Nations, 2014. *The Transformative Potential of the Right to Food.* Report of the Special Rapporteur on the Right to Food, Olivier De Schutter. http://www.srfood.org/images/stories/pdf/officialreports/20140310_finalreport_en.pdf

USAID, 2003. *USAID's Response to the Global Coffee Crisis.* Fact Sheet. http://iipdigital.usembassy.gov/st/english/article/2003/03/20030305122512egreen@pd.state.gov0.0325281.html#axzz3OP3gwSRB

USDS, 2011. *Women and Agriculture: Improving Global Food Security.* Washington DC: US Department of State, Bureau of Public Affairs. http://feedthefuture.gov/sites/default/files/resource/files/Clinton Women and Agriculture report.pdf

Valkila, J., 2009. Fair trade organic coffee production in Nicaragua: Sustainable development or a poverty trap? *Ecological Economics* 68: 3018-3025. http://dx.doi.org/10.1016/j.ecolecon.2009.07.002

Vallejo, N. and Hauselmann, P., 2004. *Governance and Multi-Stakeholder Processes.* Winnipeg: International Institute for Sustainable Development. http://portals.wi.wur.nl/files/docs/msp/sci_governance1.pdf

van Dam, J., Junginger, M., Faaij, A., Jürgens, I., Best G. and Fritsche, U., 2008. Overview of recent developments in sustainable biomass certification. *Biomass and Bioenergy* 32: 749-780. http://dx.doi.org/10.1016/j.biombioe.2008.01.018

van Gelder, J.W. and Kouwenhoven, D., 2011. *Enhancing Financiers' Accountability for the Social and Environmental Impacts of Biofuels.* Working Paper 60. Bogor: CIFOR. http://dx.doi.org/10.17528/cifor/003444

van Gelder, J.W., Herder, A., Kouwenhoven, D. and Wolterink, J., 2010. *Close the Gap: Benchmarking Investment Policies of International Banks.* Nijmegen: BankTrack. http://www.banktrack.org/download/close_the_gap/close_the_gap.pdf

Varangis, P., Siegel, P., Giovannucci, D. and Lewin, B., 2003. *Dealing with the Coffee Crisis in Central America: Impacts and Strategies.* World Bank Policy Research Working Paper 2993. Washington DC: World Bank. http://dx.doi.org/10.1596/1813-9450-2993

Weber, J.G., 2011. How much more do growers receive for Fair Trade-organic coffee? *Food Policy* 36: 678-685. http://dx.doi.org/10.1016/j.foodpol.2011.05.007

Westhoek, H., Lesschen, J.P., Rood, T., Wagner, S., De Marco, A., Murphy-Bokern, D., Leip, A., van Grinsven, H., Sutton, M.A. and Oenema, O., 2014. Food choices, health and environment: Effects of cutting Europe's meat and dairy intake. *Global Environmental Change* 26: 196-205. http://dx.doi.org/10.1016/j.gloenvcha.2014.02.004

Whately, M. and Campanili. M., 2014. *Green Municipalities Program: Lessons Learned and Challenges for 2013-2014.* Belém: Government of the State of Pará. http://municipiosverdes.com.br/files/999816d7a617e650c796109566e1337c/c20ad4d76fe97759aa27a0c99bff6710/versao-ingles (1).pdf

Wilmar International, 2014. Sustainability. http://www.wilmar-international.com/sustainability

Wilson, B., 2010. Indebted to Fair Trade? Coffee and crisis in Nicaragua. *Geoforum* 41: 84-92. http://dx.doi.org/10.1016/j.geoforum.2009.06.008

World Bank, 2011. *The World Bank Group Framework and IFC Strategy for Engagement in the Palm Oil Sector.* Washington DC: World Bank. http://www.ifc.org/wps/wcm/connect/159dce004ea3bd0fb359f71dc0e8434d/WBG+Framework+and+IFC+Strategy_FINAL_FOR+WEB.pdf?MOD=AJPERES

World Bank, 2003. *Scaling-Up the Impact of Good Practices in Rural Development: A Working Paper to Support Implementation of the World Bank's Rural Development Strategy.* Washington DC: World Bank. http://www-wds.worldbank.org/external/default/WDSContentServer/WDSP/IB/2004/01/30/000160016_20040130163125/Rendered/PDF/260310White0co1e1up1final1formatted.pdf

Wunder, S., Engel, S. and Pagiola, S., 2008. Taking stock: A comparative analysis of payments for environmental services programs in developed and developing countries. *Ecological Economics* 65: 834-852. http://dx.doi.org/10.1016/j.ecolecon.2008.03.010

WWF, 2012. *Forest Management and Gender. Social Development Briefing.* Surrey: World Wide Fund for Nature. http://assets.wwf.org.uk/downloads/women_conservation_forests_2012.pdf

WWF, 2010. *Certification and Roundtables?* WWF Review of Multi-Stakeholder Sustainability Initiatives. http://assets.wwf.org.uk/downloads/wwf_certification_and_roundtables_briefing.pdf

Zerbe, N., 2010. Moving from bread and water to milk and honey: Framing the emergent alternative food systems. *Humboldt Journal of Social Relations* 33: 4-29. http://www.jstor.org/stable/23263225

7. Conclusions

Coordinating lead author: *Bhaskar Vira*
Lead authors: *Ramni Jamnadass, Daniela Kleinschmit, Stepha McMullin, Stephanie Mansourian, Henry Neufeldt, John A. Parrotta, Terry Sunderland and Christoph Wildburger*

7.1 Forests and Trees Matter for Food Security and Nutrition

Close to one out of every six persons directly depends on forests, with food being one essential aspect of this dependence. An even greater number rely on the ecosystem services of forests – notably soil and water protection and pollination – specifically for their food and nutrition. Forests and tree-based systems are particularly critical for food security and nutrition for the poorest and the most vulnerable, including women.

Forests and tree-based systems have played a major role throughout human history in supporting livelihoods as well as meeting the food security and nutritional needs of the global population. **These systems, ranging from natural forests that are managed to optimise yields of wild foods and fodder, through shifting cultivation and a wide variety of agroforestry systems to single-species tree crop systems and orchards, remain important components of rural landscapes in most parts of the world.**

There is increasing evidence of the importance of forests and other tree-based systems for supporting food production and contributing to dietary diversity and quality, and addressing nutritional shortfalls as underscored in this book. Additional products essential to food production, such as **fuel, fodder or green fertiliser**, are also provided by trees.

Non-timber forest products (NTFPs) and agroforestry tree products (AFTPs), including tree commodity crops within agroforestry systems, are important sources of revenue to local people and governments, which can can contribute to food supply. **Tree-based incomes offer a considerably more diversified livelihood portfolio** given the environmental and economic risks of relying on cash incomes from single commodity crops. More is known about the economic value of tree commodity crops

 http://dx.doi.org/10.11647/OBP.0085.07

than of other products, although recent initiatives have provided a clearer picture of the "environmental income" from NTFPs.

Forests and tree-based systems also provide valuable ecosystem services that are essential for staple crop production and that of a wider range of edible plants. For instance, many globally important crops require pollinators that are supported by forests and diverse tree-based cropping systems within landscape mosaics. **These systems offer a number of advantages over permanent (crop) agriculture given the diversity of food products derived from them and their adaptability to a broader range of environmental conditions (e.g., soils, topography and climate) and changing socio-economic conditions.**

7.2 Governing Multi-functional Landscapes for Food Security and Nutrition

Forests and tree-based systems are embedded within broader economic, political, cultural and ecological landscapes that typically include a mosaic of different, and often competing food production systems and other land uses. How these different land use patches interact with each other in space and time can profoundly influence the productivity and sustainability of forests and tree-based systems as well as their food security and nutrition outcomes. **The integration of biodiversity conservation and agricultural production goals must be a first step, whether through land sharing or land sparing, or more feasibly through a more nuanced, yet complex, multi-functional integrated landscape approach.** Greater attention to the production of and access to nutrient-dense foods is needed in the debate on the respective benefits of land sharing versus land sparing which has focused to date on the impacts of staple crop yields (one important aspect of food security and nutrition) on biodiversity and forest conservation.

A range of diverse drivers – environmental, social, economic and governance – affect forests and tree-based systems for food security and nutrition, usually by influencing land use and management or through changes in consumption, income and livelihood opportunities. These drivers are often interrelated. Thus, designing appropriate and integrated responses to these complex influences that are effective across multiple, nested scales is a major challenge. **Managing resilient and climate-smart landscapes on a multi-functional basis that combines food production, biodiversity conservation, other land uses and the maintenance of ecosystem services should be at the forefront of efforts to achieve global food security.** In order for this to happen, knowledge from biodiversity science and agricultural research and development needs to be integrated through a systems approach at the landscape scale.

Governance shifts from state-focused government to **multi-sectoral and cross-scale governance present better prospects for integration of different interests and goals related to forest and food systems.** The resulting global emphasis on ecosystem services can also bring opportunities for improved synergies between forest and food systems, changing management forms and changes in income and livelihood structures. To maximise future potential, **greater attention from the scientific and development communities** is required, particularly **to develop a supportive policy framework that considers both the forestry and agriculture sectors in tandem.**

Current governance arrangements are imperfect and ambiguous. **Complexity surrounding the forest-food landscape interface dictates the need for different solutions on a case-by-case basis.** Structural reforms involving greater intervention from the state to harmonise regulatory regimes, may be required in some instances to achieve more inclusive food systems that not only foster innovation but also value local practices, systems and knowledge. **Co-regulatory approaches that involve both public and private actors** also **have the potential to enhance the effective governance of forest and tree-based food systems**. Initiatives aimed at enhancing the governance of large-scale investors supporting sustainable practices in the commodity value chain, improved benefit sharing and protection of local people's rights complement state-led regulatory approaches and policy frameworks.

A central **governance issue** is **how and to what extent policy and regulatory frameworks help ensure equitable access of the poor, women and disadvantaged groups to forests and tree-based systems, and to what extent do these regulatory arrangements recognise the rights to direct and indirect benefits for food and nutritional security**. Richer households with more assets (including livestock) are able to claim or make greater use of forest common property resources; yet, poorer households often have a higher dependence, as a proportion of their total income, on forest resources for food security and livelihoods.

The impacts of interventions are also felt differently, depending on social structures and local contexts, and could improve food security and nutrition for some groups while increasing vulnerability for others. Subsidies and incentives (for tree planting and management) are often captured by larger farmers who are, usually, not food insecure in relative terms. **Responses must be sensitive to these differences, and ensure that they meet the needs of the most vulnerable groups.**

7.3 The Importance of Secure Tenure and Local Control

Improving food sovereignty can help to ensure that local people have better access to food, control over their own diets and are engaged in efforts to improve the nutritional quality of their food intake. Community level engagement with local food and agricultural systems will be particularly important for those people facing a nutrition transition and the burden of malnutrition. It creates a setting ideal for more sustainable management of these food and agricultural systems and the broader landscapes in which communities reside and interact.

Tenure regimes in forests and tree-based systems for food security and nutrition are highly complex, and rights to trees or to their produce may be different from rights to the land on which they are grown. **Different bundles of rights are nested and overlap in these different systems, varying according to geographical, social, cultural, economic and political factors, and affecting the access of different population groups to the trees and their products for food, income and other livelihood needs.**

Policies that support communities' access to forests and that encourage the cultivation of tree products are required. While there is a growing trend towards designating *de jure* land and management rights to communities and indigenous peoples who traditionally hold *de facto* rights to forest, some 80 percent of forest land worldwide remains under state ownership. **Improved security of tenure has significant potential to enhance access to nutritious food.**

Since women represent 43 percent of the global agricultural labour force, and there is evidence of feminisation of agriculture in numerous developing countries, **women's weak and often insecure rights of access to land, forests and trees is undermining their engagement in innovation in forests and agroforestry systems with huge costs for their food security and nutrition, and that of their families.**

7.4 Reimagining Forests and Food Security

Applying an integrated landscape approach provides a unique opportunity for forestry and agricultural research organisations to coordinate efforts at the conceptual and implementation levels to achieve more sustainable agricultural systems. As such, **a clear programme of work on managing landscapes and ecosystems for biodiversity conservation, agriculture, food security and nutrition should be central to development aid**. Agroecology (i.e. the application of ecological concepts and principles in the design of sustainable agricultural systems) appears well suited to these geographies, and an approach that combines biodiversity concerns, along with food production demands, provides a more compelling vision of future food production.

Conservation and restoration in human dominated ecosystems requires strengthening connections between agriculture and biodiversity. In such landscapes, characterised by impoverished biodiversity and in particular depopulated of their medium and large-sized vertebrates, **tree-based agriculture in particular may represent an opportunity, and not necessarily a threat, for conservation and ecosystem restoration**.

Most forest and tree-based systems are underpinned by the **accumulated traditional knowledge of local and indigenous communities**. Traditional knowledge has been crucial to the development and modification of these systems over generations under diverse and changing variable environmental conditions and to meet changing socio-economic needs, and **this contribution needs to be acknowledged and incorporated into management practices and policy.**

Agricultural and forest scientists, extension agents and development organisations have only recently begun to understand the importance and relevance of forests and tree-based systems, and the traditional knowledge that underpins many of these systems. **Working with farmers to combine the best of traditional and formal scientific knowledge offers tremendous potential to enhance the productivity and resilience of these systems and the flow of direct (food security and nutrition) and indirect (income) benefits to their practitioners.**

By targeting particular species for improved harvest and/or cultivation, more optimal portfolios of species could be devised that best support communities' nutrition year-round. An overall increase in production through cultivation of a wide range of foods, including tree fruits and vegetables, is required to bridge consumption shortfalls. There is further potential for the domestication of currently little-researched indigenous fruit trees to bring about large production gains, although more information is needed on the nutritional value of many of these species.

The development of "nutrient-sensitive" value chains is also needed, which means improving nutritional knowledge and awareness among value-chain actors and consumers, focusing on promoting the involvement of women, and considering markets for a wider range of tree foods. By promoting tree food processing and other value additions, the non-farm rural economy can also be stimulated.

Dietary choices are complex and depend on more than just what potential foods are available to communities in their environments. Rather than assumptions based on availability, assessments of actual diet through dietary diversity studies and other related estimators are therefore crucial, to allow an exploration of the reasons behind current limitations in usage. **There are multiple targets to improve food choices and nutritional knowledge and awareness, with women and children being key targets, as well as actors across the value chain.**

Education and basic awareness play important roles in empowering rural populations and have the potential to generate tangible and fundamental benefits for their households and communities including food security, sustainable forest management, health, education and general household nutrition. For women and other vulnerable groups appropriate education and training programmes can improve their understanding of healthy and nutritious foods and natural resource management practices. Such programmes can also support traditional rural societies in understanding and incorporating necessary changes that enable gender inclusiveness in decision-making and benefit sharing in forests and tree-based systems for food security and nutrition. Technological innovation, in particular mobile technology can help deliver relevant information to rural populations and is seen as critical in improving existing extension services, education and products to enhance food security and nutrition, dietary choices and health.

7.5 Knowledge Gaps

Through the research of this Global Forest Expert Panel, specific knowledge gaps have been identified concerning the contribution of forests and tree-based systems to food security and nutrition. Although there is a growing body of evidence, much remains to be understood as concerns the role of forests and tree-covered landscapes in food security and nutrition and the provision of nutritious diets.Accurately quantifying the role of forests in food security and nutrition (including dietary diversity) is needed. In particular, better quantification of the relative benefits received by rural communities from different tree production categories is required, supported by an appropriate typology for characterisation. Further research is needed to assess the complementarity and resilience of different crops in agroforestry, particularly in the face of climate change and the need for concomitant adaptation to such change.

Research should support food tree domestication options appropriate for meeting smallholders' needs. To support diverse production systems, genetic selection for commodity crop cultivars that do well under shade may be of particular importance. This may require returning to wild genetic resources still found in shaded, mixed-species forest habitats, reinforcing the importance of their conservation. There are also opportunities to develop valuable new tree commodities that are compatible with other crops and that therefore support more agro-biodiversity.

Specific gaps that have been identified related to the management of forests and tree-based systems to enhance food security and nutrition include the need to refine estimates of land cover in agricultural landscape and the extent of agroforestry practices, including their relationship with factors other than climate and population density. There is a need to assess the actual extent of most management systems, the numbers of people who rely on one or more such systems to meet their household food and/or income needs, and the relative value of different forests and tree-based systems on the diets and health of those who manage them. Further research would also be needed to better understand the food values of forest mosaics from shifting cultivation systems.

There are gaps in our understanding of the inter-relations between drivers affecting the role of forests and tree-based systems in food security and nutrition. In particular, improved understanding is needed on the link between economic valuation of ecosystem services, and their incorporation into global commodity markets, and the ensuing risk of local and indigenous communities being dispossessed of land and related rights and access.

Further research is required at the landscape scale particularly when tackling trade-offs between different stakeholders. There is a need to better understand the economic, environmental and other trade-offs for the different sectors of rural societies when the harvesting of NTFPs is commercialised or they are planted (and perhaps are converted to new commodity crops), as the benefits and costs for different members of society vary. The question of how far research can go in providing useful information about relationships between forest food systems and other land uses at the landscape scale needs to be addressed. In the land sparing/land sharing debate, greater attention is needed on food production and access to nutrient-rich foods.

Gaps remain regarding ways to better link local innovations in resource access and control to a supportive policy and institutional environment. The effectiveness of education-based initiatives for sustainability and food security need to be further explored. More attention is needed on how to effectively link local innovations in relation to management practices, institutions and governance arrangements to an enabling policy environment. Comprehensive information on the complex relationships between land tenure, use, control, ownership and how these relationships impact on food security, forests and tree-based systems is needed to help develop appropriate land tenure policy frameworks (which are also gender sensitive).

7.6 Looking Ahead: The Importance of Forest and Tree-based Systems for Food Security and Nutrition

This book has highlighted the important role that forests and tree-based systems play in complementing agricultural production systems for food security and nutrition. Forests and tree-based systems can contribute to the "Zero Hunger Challenge". To do this, however, requires a much greater understanding of the forest-food nexus, the effective management of landscapes and improved governance. Recognising the role of different configurations of the landscape mosaic, and the ways in which forests and tree-based systems can be managed to effectively deliver ecosystem services for crop production, provide better and more nutritionally-balanced diets, greater control over food inputs – particularly during lean seasons and periods of vulnerability (especially for marginalised groups) – are critical elements of response to global hunger. Through this book, we have identified important opportunities for greater harmonisation and synergy between policies and global commitments to secure more sustainable landscapes for a hunger-free future for all.

Appendix 1: Glossary

Agricultural biodiversity: A broad term that includes all components of biological diversity of relevance to food and agriculture, and all components of biological diversity that constitute the agricultural ecosystems, also named agro-ecosystems: the variety and variability of animals, plants and micro-organisms, at the genetic, species and ecosystem levels, which are necessary to sustain key functions of the agro-ecosystem, its structure and processes (CBD, 2000).

Agrobiodiversity: see *Agricultural biodiversity*

Agroecology: The integrative study of the ecology of the entire food systems, encompassing ecological, economic and social dimensions (Francis et al., 2003).

Agroforestry: A collective name for land use systems and practices in which woody perennials are deliberately integrated with crops and/or animals on the same land management unit. The integration can be either in a spatial mixture or in a temporal sequence. There are normally both ecological and economic interactions between woody and non-woody components in agroforestry (Leakey, 1996; Leakey and Simons, 1998).

Agroforestry tree products (AFTP): refers to timber and non-timber forest products that are sourced from trees cultivated outside of forests. This distinction from the term non-timber forest products (NTFPs) for non-timber extractive resources from natural systems is to distinguish between extractive resources from forests and cultivated trees in farming systems. (Nevertheless, some products will be marketed as both NTFPs and AFTPs (depending on their origin) during the period of transition from wild resources to newly domesticated crops.) (Leakey et al., 2005; Simons and Leakey, 2004).

Biodiversity [Biological Diversity]: The variability among living organisms from all sources including, inter alia, terrestrial, marine and other aquatic ecosystems and the ecological complexes of which they are part; this includes diversity within species, between species and of ecosystems (CBD, 1992).

Climate change: Refers to any change in climate over time, whether due to natural variability or as a result of human activity. This usage differs from that in the United Nations Framework Convention on Climate Change (UNFCCC), which defines *climate change* as: "a change of climate which is attributed directly or indirectly to human activity that alters the composition of the global atmosphere and which is in addition to natural climate variability observed over comparable time periods" (IPCC, 2007).

Deforestation: The conversion of forest to another land use or the long-term reduction of the tree canopy cover below the minimum 10 percent threshold (FAO, 2010). Deforestation implies the long-term or permanent loss of forest cover and implies transformation into another land use. Such a loss can only be caused and maintained by a continued human-induced or natural perturbation. Deforestation includes areas of forest converted to agriculture, pasture, water reservoirs and urban areas. The term specifically *excludes* areas where the trees have been removed as a result of harvesting or logging, and where the forest is expected to regenerate naturally or with the aid of silvicultural measures. Deforestation also includes areas where, for example, the impact of disturbance, over-utilisation or changing environmental conditions affects the forest to an extent that it cannot sustain a tree cover above the 10 percent threshold (FAO, 2001).

Degradation: see *Forest degradation and Land degradation.*

Dietary diversity: Dietary diversity defined as the number of different foods or food groups consumed over a given reference period (Ruel, 2003) is increasingly recognized as a key element of high quality diets and a sustainable way to resolve health problems such as micronutrient deficiencies.

Ecosystem: A dynamic complex of plant, animal and micro-organism communities and their non-living environment interacting as a functional unit (CBD, 1992).

Ecosystem services: Ecosystem services are the benefits people obtain from ecosystems. These include i) provisioning services such as food, water, timber, and fibre; (ii) regulating services that affect climate, floods, disease, wastes, and water quality;(iii) cultural services that provide recreational, aesthetic, and spiritual benefits; and (iv) supporting services such as soil formation, photosynthesis, and nutrient cycling (MA, 2005).

Food insecurity: A situation that exists when people lack secure access to sufficient amounts of safe and nutritious food for normal growth and development and an active and healthy life. It may be caused by the unavailability of food, insufficient purchasing power, inappropriate distribution or inadequate use of food at the household level. Food insecurity, poor conditions of health and sanitation and inappropriate care and feeding practices are the major causes of poor nutritional status. Food insecurity may be chronic, seasonal or transitory (FAO, IFAD and WFP, 2014).

Food security: A situation that exists when all people, at all times, have physical, social and economic access to sufficient, safe and nutritious food that meets their dietary needs and food preferences for an active and healthy life. Based on this definition, four food security dimensions can be identified: food availability, economic and physical access to food, food utilization and stability over time (FAO, IFAD and WFP, 2014).

Food sovereignty is the right of peoples to define their own food and agriculture; to protect and regulate domestic agricultural production and trade in order to achieve sustainable development objectives; to determine the extent to which they want to be self-reliant; to restrict the dumping of products in their markets; and to provide local fisheries-based communities the priority in managing the use of and the rights to aquatic resources (Via Campesina website: www.viacampesina.org).

Food systems: Food systems encompass the entire range of activities involved in the production, processing, marketing, consumption and disposal of goods that originate from agriculture, forestry or fisheries, including the inputs needed and the outputs generated at each of these steps. Food systems also involve the people and institutions that initiate or inhibit change in the system as well as the socio-political, economic and technological environment in which these activities take place (adapted from FAO, 2012b).

Forest: Land spanning more than 0.5 hectares with trees higher than 5 metres and a canopy cover of more than 10 percent, or trees able to reach these thresholds *in situ*. It does not include land that is predominantly under agricultural or urban land use (FAO, 2010). Includes areas with young trees that have not yet reached but which are expected to reach a canopy cover of 10 percent and tree height of 5 meters. It also includes areas that are temporarily unstocked due to clear-cutting as part of a forest management practice or natural disasters, and which are expected to be regenerated within 5 years. Local conditions may, in exceptional cases, justify that a longer time frame is used (FAO, 2010).

Forest degradation: The reduction of the capacity of a forest to provide goods and services (FAO 2010b).

Forest fragmentation: Any process that results in the conversion of formerly continuous forest into patches of forest separated by non-forested lands (CBD website: http://www.cbd.int/forest/definitions.shtml).

Forest management: The processes of planning and implementing practices for the stewardship and use of forests and other wooded land aimed at achieving specific environmental, economic, social and/or cultural objectives. Includes management at all scales such as normative, strategic, tactical and operational level management (FAO, 2004).

Forests and tree-based systems: for the **purposes of this book,** this includes the spectrum from management of forests to optimise yields of wild foods and fodder, to shifting cultivation, to the broad spectrum of agroforestry practices and to single-species tree crop management.

Fragmentation: see *Forest fragmentation*

Governance: refers to the formation and stewardship of the formal and informal rules that regulate the public realm, the arena in which state as well as economic and societal actors interact to make decisions (Hydén and Mease, 2004).

Greenhouse gas: Gaseous constituents of the atmosphere, both natural and anthropogenic, that absorb and emit radiation at specific wavelengths within the spectrum of infrared radiation emitted by the Earth's surface, the atmosphere, and clouds. This property causes the greenhouse effect. Water vapour (H2O), carbon dioxide (CO2), nitrous oxide (N2O), methane (CH4) and ozone (O3) are the primary greenhouse gases in the Earth's atmosphere. As well as CO2, N2O, and CH4, the Kyoto Protocol deals with the greenhouse gases sulphur hexafluoride (SF6), hydrofluorocarbons (HFCs) and perfluorocarbons (PFCs) (IPCC, 2007).

Hidden hunger: refers to vitamin and mineral deficiencies, or micronutrient deficiencies. Micronutrient deficiencies can compromise growth, immune function, cognitive development, and reproductive and work capacity (FAO, 2012c).

Invasive species: Any species that are non-native to a particular ecosystem and whose introduction and spread causes, or are likely to cause socio-cultural, economic or environmental harm or harm to human health (FAO website: http://www.fao.org/forestry/aliens/en).

Land degradation: Reduction or loss in arid, semiarid and dry sub-humid areas of the biological or economic productivity and complexity of rainfed cropland, irrigated cropland, or range, pasture, forest and woodlands resulting from land uses or from a process or combination of processes, including processes arising from human activities and habitation patterns, such as: (i) soil erosion caused by wind and/or water; (ii) deterioration of the physical, chemical and biological or economic properties of soil; and (iii) long-term loss of natural vegetation (UNCCD, 1994).

Landscape: Drawing on ecosystem definitions, we define a landscape as an area delineated by an actor for a specific set of objectives (Gignoux et al., 2011). It constitutes an arena in which entities, including humans, interact according to rules (physical, biological, and social) that determine their relationships (Sayer et al., 2013).

Landscape approach: Aims to reconcile competing land uses and to achieve both conservation and production outcomes, while recognizing and negotiating for inherent trade-offs (Milder et al., 2012; Sayer et al., 2013).

Land-sparing: For the **purposes of this book**, defined as "The promotion of agricultural techniques that encourage the highest possible yields in a given area (even if it involves reduced in-farm biodiversity) with the goal of meeting agricultural needs in the minimum possible area, so as to reduce the pressure over wild areas."

Land-sharing: For the **purposes of this book**, defined as "The promotion of agricultural techniques, mainly agroforestry, that are 'friendly' to wild species, aimed at fostering the co-existence of managed (crops or livestock) and wild species in the same area."

Livelihoods: The capabilities, assets – both material and social resources – and activities required for a means of living. A livelihood is sustainable when it can cope with and recover from stresses and shocks, maintain or enhance its capabilities and assets, and provide net benefits to other livelihoods locally and more widely, both now and in the future, while not undermining the natural resource base (Chambers and Conway, 1991).

Malnutrition. An abnormal physiological condition caused by inadequate, unbalanced or excessive consumption of macronutrients and/or micronutrients. Malnutrition includes undernutrition and overnutrition as well as micronutrient deficiencies (FAO, IFAD and WFP, 2014).

Managed forests: For the **purposes of this book**, managed forests are those whose structure, and the diversity and density of edible plant and animal species, have been modified by various management practices to improve their nutritional, economic and biodiversity values for people.

***Non-timber forest products* (NTFP)**: All biological materials other than timber, which are extracted from forests for human use. Forest refers to a natural ecosystem in which trees are a significant component. In addition to trees, forest products are derived from all plants, fungi and animals (including fish) for which the forest ecosystem provides habitat (IUFRO, 2005).

Nutrition: the consequence of the intake of food and the utilization of nutrients by the body (CFS, 2012).

Nutrition security: A situation that exists when secure access to an appropriately nutritious diet is coupled with a sanitary environment, adequate health services and care, in order to ensure a healthy and active life for all household members. Nutrition security differs from food security in that it also considers the aspects of adequate caring practices, health and hygiene in addition to dietary adequacy (FAO, IFAD and WFP, 2014).

Primary forest: Naturally regenerated forest of native species, where there are no clearly visible indications of human activities [including commercial logging] and the ecological processes are not significantly disturbed (FAO, 2010b).

Resilience: Capacity of the system to cope with all kind of shocks and disturbances, and so be able to avoid crossing all thresholds, known or unknown, to alternate regimes sometimes referred to as "coping capacity" and synonymous with "adaptive capacity" (O'Connell et al, 2015).

Secondary forest: forests regenerating largely through natural processes after significant removal or disturbance of the original forest vegetation by human or natural causes at a single point in time or over an extended period, and displaying a major difference in forest structure and/or canopy species composition with respect to pristine primary forests (FAO, 2003).

Shifting cultivation: Also referred to as *slash-and-burn cultivation* or *swidden agriculture*. A land use system that employs a natural or improved fallow phase, which is longer than the cultivation phase of annual crops, sufficiently long to be dominated by woody vegetation, and cleared by means of fire (Mertz et al., 2009)

Slash-and-burn cultivation: see *Shifting cultivation*

Sustainable intensification: where the yields of global agriculture are increased without adverse environmental impact and without the cultivation of more land (The Royal Society, 2009).

Swidden agriculture: see *Shifting cultivation*

Tenure: Systems of tenure define and regulate how people, communities and others gain access to land, fisheries and forests. These tenure systems determine who can use which resources, for how long, and under what conditions. The systems may be based on written policies and laws, as well as on unwritten customs and practices (FAO, 2012a).

Traditional (ecological) knowledge: A cumulative body of knowledge, practice and belief, handed down through generations by cultural transmission and evolving by adaptive processes, about the relationship between living beings (including humans) with one another and with their forest environment (Berkes, 1999).

Tree crops (also Tree commodity crops): Generally defined as food products from trees that are exported and traded widely in international commodity markets. These crops may be produced by smallholder- and/or in plantation-production systems. Examples include coffee, cocoa, tea and oil palm (Jain and Priyadarshan, 2004).

References

Berkes, F., 1999. *Sacred Ecology. Traditional Ecological Knowledge and Resource Management.* Philadelphia and London: Taylor and Francis.

CBD, 1992. Convention on Biological Diversity, Art. 2. Montreal: UNEP.

CBD, 2000. Convention on Biological Diversity, Decision V/5, Annex. Fifth Meeting of the Conference of the Parties to the Convention on Biological Diversity, Nairobi, Kenya, 15-26 May 2000.

CBD website: http://www.cbd.int/forest/definitions.shtml

CFS (Committe on World Food Security), 2012. *Coming to Terms With Terminology.* Rome: FAO.

Chambers, R., and Conway, G., 1992. *Sustainable Rural Livelihoods: Practical Concepts for the 21st Century.* London: Institute of Development Studies.

FAO, 2012a. *Voluntary Guidelines on the Responsible Governance of Tenure of Land, Fisheries and Forests in the Context of National Food Security.* Rome: FAO.

FAO, 2012b. *Sustainability Assessment of Food and Agriculture Systems (SAFA)* 2012. Rome: Food and Agriculture Organization of the United Nations.

FAO, 2012c. *State of Food Insecurity in the World.* Rome: FAO.

FAO, 2010. *Global Forest Resources Assessment.* Forestry Paper 163. Rome: Food and Agriculture Organization of the United Nations.

FAO, 2010b. *Global Forest Resources Assessment, Terms and Definitions.* Working paper 144. Rome: Food and Agriculture Organization of the United Nations.

FAO, 2004. *Global Forest Resources Assessment Update 2005 Terms and Definitions.* Final version. Rome: Food and Agriculture Organization of the United Nations.

FAO, 2003. Proceedings of the Workshop on Tropical Secondary Forest Management in Africa: Reality and Perspectives, Nairobi, Kenya, 9-13 December 2002. Forestry Department. Rome: Food and Agriculture Organization of the United Nations.

FAO, 2001. *Global Forest Resources Assessment FRA 2000: Main Report.* Rome: Food and Agriculture Organization of the United Nations.

FAO website: Invasive species. http://www.fao.org/forestry/aliens/en/

FAO, IFAD and WFP. 2014. *The State of Food Insecurity in the World 2014. Strengthening the Enabling Environment for Food Security and Nutrition.* Rome: FAO.

Francis, C., Lieblein, G., Gliessman, S., Breland, T.A., Creamer, N., Harwood, R., Salomonsson, L., Helenius, J., Rickerl, D., Salvador, R., Wiedenhoeft, M. Simmons, S., Allen, P., Altieri, M., Flora, C. and Poincelot, R., 2003. Agroecology: The ecology of food systems. *Journal of Sustainable Agriculture* 22(3): 99-118.

Gignoux, J., Davies, I., Flint, S., Zucker, J.-D., 2011. The ecosystem in practice: Interest and problems of an old definition for constructing ecological models. *Ecosystems* 14(7): 1039-1054.

Hydén, G., and Mease, K., 2004. *Making Sense of Governance: Empirical Evidence from Sixteen Developing Countries.* Boulder and London: Lynne Rienner Publishers.

IPCC, 2007. *Climate Change 2007: Impacts, Adaptation and Vulnerability.* Contribution of Working Group II to the Fourth Assessment Report of the Intergovernmental Panel on Climate Change. Appendix 1: Glossary. Parry, M.L., Canziani, O.F., Palutikof, J.P., van der Linden, P.J. and Hanson, C.E. (eds.). Cambridge: Cambridge University Press.

IUFRO 2005. *Multilingual Pocket Glossary of Forest Terms and Definitions, Compiled on the Occasion of the XXII IUFRO World Congress.* August 2005, Brisbane, Australia. Vienna: IUFRO.

Jain, M.S., and Priyadarshan, P.M. (eds.), 2009. *Breeding Plantation Tree Crops. Tropical Species.* New York: Springer Science+Business Media.

Leakey, R., 1996. Definition of agroforestry revisited. *Agroforestry Today* 8(1): 5-7.

Leakey, R.R.B., Tchoundjeu, Z., Schreckenberg K., Shackleton, S.E. and Shackleton, C.M., 2005. Agroforestry tree products (AFTPs): Targeting poverty reduction and enhanced livelihoods. *International Journal of Agricultural Sustainability* 3(1): 1-23.

Leakey, R.R.B. and Simons, A.J., 1998. The domestication and commercialization of indigenous trees in agroforestry for the alleviation of poverty. *Agroforestry Systems* 38: 165-176.

MA (Millennium Ecosystem Assessment), 2005. *Ecosystems and Human Well-being: Synthesis.* Washington DC: Island Press.

Mertz, O., Padoch, C., Fox, J., Cramb, R.A., Leisz, S.J., Lam, N.T. and Vien, T.D., 2009. Swidden change in Southeast Asia: Understanding causes and consequences. *Human Ecology* 37(3): 259-264.

Milder, J. C., Buck, L. E., DeClerck, F. and Scherr, S. J., 2012. Landscape approaches to achieving food production, natural resource conservation and the millennium development goals. In: *Integrating Ecology and Poverty Reduction*, edited by J.C. Ingram, F. DeClerck and C. Rumbaitis del Rio. New York: Springer.

O'Connell, D., Walker, B., Abel, N., Grigg, N., Cowie, A., Durón, G., 2015. *An Introduction to the Resilience Adaptation Transformation Assessment (RATA) Framework.* Working Paper for the Scientific and Technical Advisory Panel of the Global Environment Facility. February 2015.

Ruel, M.T., 2003. Operationalizing dietary diversity: A review of measurement issues and research priorities. *Journal of Nutrition* 133: 3911S-3926S.

Sayer, J., Sunderland, T., Ghazoul, J., Pfund, J.-L., Sheil, D., Meijaard, E., Venter, M., Boedhihartono, A.K., Day, M., Garcia, C., van Oosten, C. and L. Buck. 2013. The landscape approach: Ten principles to apply at the nexus of agriculture, conservation and other competing land-uses. *Proceedings of the National Academy of Sciences* 110(21): 8345-8348.

Simons, A.J., and Leakey, R.R.B, 2004. Tree domestication in tropical agroforestry. New vistas in agroforestry. *Advances in Agroforestry*, Vol. 1.

The Royal Society, 2009. *Reaping the Benefits Science and the Sustainable Intensification of Global Agriculture.* London: Royal Society of London.

UNCCD, 1994. *United Nations Convention to Combat Desertification.* Bonn: UNCCD.

Via Campesina website: http://viacampesina.org/en/index.php/main-issues-mainmenu-27/food-sovereignty-and-trade-mainmenu-38/262-declaration-of-nyi

Appendix 2: List of Panel Members, Authors and Reviewers

GFEP Panel meeting in Cambridge, UK. Photo © Eva-Maria Schimpf

Bina Agarwal
University of Manchester, UK
and Institute of Economic Growth,
Delhi, India
E-mail: bina_india@yahoo.com

Sarah Ayeri Ogalleh
Centre for Training & Integrated Research in ASAL Development (CETRAD)
Nanyuki, Kenya
E-mail: sarahayeri@yahoo.com

Frédéric Baudron
International Maize and Wheat Improvement Centre (CIMMYT)
Hawassa, Ethiopia
E-mail: f.baudron@cgiar.org

Sammy Carsan
World Agroforestry Centre (ICRAF)
Nairobi, Kenya
E-mail: s.carsan@cgiar.org

Paolo Cerutti
Center for International Forestry Research (CIFOR)
Nairobi, Kenya
E-mail: p.cerutti@cgiar.org

Josephine Chambers
University of Cambridge
Cambridge, UK
E-mail: jc706@cam.ac.uk

Beatrice Darko Obiri
Forestry Research Institute of Ghana
Kumasi, Ghana
E-mail: bdobiri@csir-forig.org.gh
or bdobiri@yahoo.com

Ian K. Dawson
World Agroforestry Centre (ICRAF)
Nairobi, Kenya
E-mail: iankdawson@aol.com

Neil M. Dawson
University of East Anglia
Bunachton, Inverness, Scotland
E-mail: Neilm_dawson@yahoo.co.uk

Elizabeth Deakin
Center for International Forestry Research (CIFOR)
Bogor, Indonesia
E-mail: L.deakin@cgiar.org

Ann Degrande
World Agroforestry Centre (ICRAF)
Yaoundé, Cameroon
E-mail: a.degrande@cgiar.org

Jennie Dey de Pryck
Senior Gender Adviser and Consultant
Brussels, Belgium
E-mail: jenniedeydepryck@yahoo.com

Jason Donovan
World Agroforestry Centre (ICRAF)
Lima, Peru
E-mail: j.donovan@cgiar.org

Samson Foli
Center for International Forestry Research (CIFOR)
Bogor, Indonesia
E-mail: s.foli@cgiar.org

Lisa Fuchs
World Agroforestry Centre (ICRAF)
Nairobi, Kenya
E-mail: L.Fuchs@cgiar.org

Amos Gyau
World Agroforestry Centre (ICRAF)
Nairobi, Kenya
E-mail: a.gyau@cgiar.org

Gordon Hickey
McGill University
Quebec, Canada
E-mail: gordon.hickey@mcgill.ca

Amy Ickowitz
Center for International Forestry Research (CIFOR)
Bogor, Indonesia
E-mail: a.ickowitz@cgiar.org

Miyuki Iiyama
World Agroforestry Centre (ICRAF)
Nairobi, Kenya
E-mail: m.iiyama@cgiar.org

Ramni Jamnadass
World Agroforestry Centre (ICRAF)
Nairobi, Kenya
E-mail: r.jamnadass@cgiar.org

Katy Jeary
University of Cambridge
Cambridge, UK
E-mail: kew60@cam.ac.uk

Gudrun Keding
Bioversity International, Rome, Italy and GeorgAugust-University, Goettingen, Germany
E-mail: g.keding@cgiar.org

Katja Kehlenbeck
World Agroforestry Centre (ICRAF)
Nairobi, Kenya
E-mail: k.kehlenbeck@cgiar.org

Daniela Kleinschmit
University of Freiburg,
Freiburg, Germany
E-mail:
daniela.kleinschmit@ifp.uni-freiburg.de

Christophe Kouame
World Agroforestry Centre (ICRAF)
Abidjan, Côte d'Ivoire
E-mail: c.kouame@cgiar.org

Godwin Kowero
African Forest Forum (AFF)
Nairobi, Kenya
E-mail: g.kowero@cgiar.org

Patti Kristjanson
World Agroforestry Centre (ICRAF)
Washington, DC, USA
E-mail: p.kristjanson@cgiar.org

Stephanie Mansourian
Consultant, Environment and Development
Gingins, Switzerland
E-mail: smansourian@infomaniak.ch

Adrian Martin
University of East Anglia
Norwich, UK
E-mail: adrian.martin@uea.ac.uk

Stepha McMullin
World Agroforestry Centre (ICRAF)
Nairobi, Kenya
E-mail: s.mcmullin@cgiar.org

Henry Neufeldt
World Agroforestry Centre (ICRAF)
Nairobi, Kenya
E-mail: H.Neufeldt@cgiar.org

Mary Njenga
World Agroforesty Centre (ICRAF)
Nairobi, Kenya
E-mail: m.njenga@cgiar.org

Vincent O. Oeba
Kenya Forestry Research Institute (KEFRI)
Nairobi, Kenya
E-mail: V.Oeba@cgiar.org

Daniel Ofori
World Agroforestry Centre (ICRAF)
Nairobi, Kenya
E-mail: d.ofori@cgiar.org

Hemant R. Ojha
School of Social Sciences
University of New South Wales,
Syndey, Australia
E-mail: h.ojha@unsw.edu.au

Pablo Pacheco
Center for International Forestry Research (CIFOR)
Bogor, Indonesia
E-mail: p.pacheco@cgiar.org

Christine Padoch
Center for International Forestry Research (CIFOR)
Bogor, Indonesia
E-mail: c.padoch@cgiar.org

John A. Parrotta
U.S. Forest Service, Research & Development
Washington DC, USA
E-mail: jparrotta@fs.fed.us

Bronwen Powell
Research Consultant, Forests and Food Security
CIFOR
Marrakesh, Medina, Morocco
E-mail: b.powell@cgiar.org

Nitin D. Rai
Ashoka Trust for Research in Ecology and the Environment
Bangalore, India
E-mail: nitinrai@atree.org

Patrick Ranjatson
University of Antananarivo
Antananarivo, Madagascar
E-mail: pranjatson@yahoo.fr

James Reed
Center for International Forestry Research (CIFOR)
Bogor, Indonesia
E-mail: j.reed@cgiar.org

Mirjam Ros-Tonen
University of Amsterdam
Amsterdam, Netherlands
E-mail: M.A.F.Ros-Tonen@uva.nl

Chris Sandbrook
University of Cambridge
Cambridge, UK
E-mail: cgsandbrook@gmail.com

Jolien Schure
Forest and Nature Conservation Policy Group
Wageningen University, Netherlands
E-mail: jolien@schure-research.com

Anca Serban
University of Cambridge
Cambridge, UK
E-mail: as2344@cam.ac.uk

Bimbika Sijapati Basnett
Center for International Forestry Research (CIFOR)
Bogor, Indonesia
E-mail: b.basnett@cgiar.org

Carsten Smith-Hall
University of Copenhagen
Copenhagen, Denmark
E-mail: cso@ifro.ku.dk

Barbara Stadlmayr
World Agroforestry Centre (ICRAF)
Nairobi, Kenya
E-mail: B.Stadlmayr@gmx.at

Terry Sunderland
Center for International Forestry Research (CIFOR)
Bogor, Indonesia
E-mail: t.sunderland@cgiar.org

Céline Termote
Bioversity International
Nairobi, Kenya
E-mail: c.termote@cgiar.org

Tran Nam Tu
Hue University of Agriculture and Forestry
Hue City, Vietnam
E-mail: trannamtu@gmail.com

Patrick Van Damme
Plant Production Department
Ghent University
Ghent, Belgium
E-mail: Patrick.VanDamme@UGent.be

Nathalie van Vliet
Wildlife and livelihoods expert, CIFOR
Bogota, Colombia
E-mail: vanvlietnathalie@yahoo.com

Barbara Vinceti
Bioversity International
Rome, Italy
E-mail: b.vinceti@cgiar.org

Bhaskar Vira
Department of Geography
University of Cambridge
Cambridge, UK
E-mail: bv101@cam.ac.uk

Solomon Zena Walelign
University of Copenhagen
Section for Global Development
Copenhagen, Denmark
E-mail: szw@ifro.ku.dk

Christoph Wildburger
Consultant, Environmental Policy and Natural Resource Management
Vienna, Austria
E-mail: office@wildburger.at

List of Reviewers

Eduardo Brondizio
Indiana University
Bloomington, IN, USA
E-mail: ebrondiz@indiana.edu

Carol Colfer
Visiting scholar, Cornell University
Ithaca, USA
E-mail: cjc59@cornell.edu

Martina Kress
Food and Agriculture Organization of the UN (FAO)
Rome, Italy
E-mail: Martina.Kress@fao.org

Eric Lambin
Stanford University
University of Louvain
Stanford, USA
E-mail: elambin@stanford.edu

Kae Mihara
Food and Agriculture Organization of the UN (FAO)
Rome, Italy
E-mail: Kae.Mihara@fao.org

Sarah Milne
Australian National University
Canberra, Australia
E-mail: sarah.milne@anu.edu.au

Ellen Muehlhoff
Food and Agriculture Organization of the UN (FAO)
Rome, Italy
E-mail: Ellen.Muehlhoff@fao.org

Ben Phalan
University of Cambridge
Cambridge, UK
E-mail: btp22@cam.ac.uk

Dominique Reeb
Food and Agriculture Organization of the UN (FAO)
Rome, Italy
E-mail: Dominique.Reeb@fao.org

Patricia Shanley
People and Plants International
Bristol, VT, USA
E-mail: p.shanley@cgiar.org

Ingrid Visseren-Hamakers
Wageningen University & Research Centre (WUR)
Wageningen, the Netherlands
E-mail: ingrid.visseren@wur.nl

www.ingramcontent.com/pod-product-compliance
Lightning Source LLC
LaVergne TN
LVHW060056220926
848484LV00008BA/1986